POOR MAN'S FORTUNE

POOR MAN'S

JAROD ROLL

FORTUNE

WHITE WORKING-CLASS CONSERVATISM IN AMERICAN METAL MINING, 1850-1950

THE UNIVERSITY OF
NORTH CAROLINA PRESS
Chapel Hill

This book was published with the assistance of the
Anniversary Fund of the University of North Carolina Press.

Manufactured in the United States of America
Set in Miller and Champion
by Tseng Information Systems, Inc.

The University of North Carolina Press has been a
member of the Green Press Initiative since 2003.

Cover illustrations: black powder image © iStockPhoto/Pattadis Walarput;
shovelers image courtesy Historical Mining Photographs Collection,
Joplin History and Mineral Museum

Library of Congress Cataloging-in-Publication Data
Names: Roll, Jarod, author.
Title: Poor man's fortune : white working-class conservatism in
American metal mining, 1850–1950 / Jarod Roll.
Description: Chapel Hill : University of North Carolina Press, [2020] |
Includes bibliographical references and index.
Identifiers: LCCN 2019052079 | ISBN 9781469656281 (cloth : alk. paper) |
ISBN 9781469656298 (paperback : alk. paper) | ISBN 9781469656304 (ebook)
Subjects: LCSH: Miners—Tri-State Mining District—History—19th century. |
Miners—Tri-State Mining District—History—20th century. | Working-class whites—
Attitudes. | Working-class men—Attitudes. | Conservatism—Tri-State Mining
District—History. | Masculinity—Economic aspects. | White nationalism.
Classification: LCC HD8039.M72 U687 2020 | DDC 305.9/622344097309034—dc23
LC record available at https://lccn.loc.gov/2019052079

For R. S. and M. S. R.

The fault, dear Brutus, is not in our stars,
but in ourselves, that we are underlings.
—Cassius, *Julius Caesar*, Act 1, Scene 2

Dirty deeds, done dirt cheap.
—AC/DC

CONTENTS

ILLUSTRATIONS

POOR MAN'S FORTUNE

INTRODUCTION

This book is about white working-class American men who opposed social democratic labor unions and politics in the century that culminated in the New Deal. It follows five generations of miners who, beginning in the 1850s, discovered and developed a rich swath of zinc and lead that straddled the boundaries between Missouri, Kansas, and Oklahoma. By the 1920s, the Tri-State district led the nation in the production of these unheralded but essential metals. From the beginning, the miners pursued class interests that differed, to varying degrees, from those of the men who controlled the land, bought the ore, and smelted the metal. Yet for sixty years, from 1880 to 1940, national labor unions could not organize the Tri-State miners. This outcome mattered. The miners developed a powerful animus against the idea of class-based solidarity, particularly as practiced by the Western Federation of Miners (WFM), a pioneer of radical unionism, and later by the Congress of Industrial Organizations (CIO). Tri-State miners worked, willingly and re-peatedly, as strikebreakers against the WFM in a series of clashes across the western United States between 1896 and 1910. Their actions helped to de-feat and nearly destroy the WFM. These outcomes also mattered in Tri-State mining communities. Miners resisted government efforts, often backed by unions, to impose health and safety regulations despite the obvious dangers, the worst of which was silicosis, a fatal lung disease. Even during the Great Depression, when the federal government encouraged workers like them to organize for higher pay and greater security, Tri-State miners remained ob-stinate. The district's majority crushed a promising drive by some of their peers to realize the full benefits of New Deal collective bargaining rights. Rarely, it seemed, had so many American workers fought so long to remain at the raw edge of industrial capitalism.

Tri-State miners baffled, frustrated, and enraged those who tried to get them to change. WFM leaders called them "a dangerous class" with a "de-plorable lack of intelligence." Twenty years later, an American Federation of Labor (AFL) organizer blamed the absence of unions in the district on "the stupid miner himself." Reformers likewise struggled to make sense of them. A social worker concluded that a "feverish unsteadiness" warped their "social instincts and ideals." Government health and safety investigators, meanwhile, found that the miners "seem indifferent, even fatalistic, and will take precautions only if compelled to do so." These commentators concluded, as we might also conclude, that something was wrong with Tri-State miners and that it made them act against their own interests.[1]

The story of the Tri-State miners runs counter to what we know about American labor and working-class history in the decades between the Civil War and World War II. The new labor historians focused on the organizing story of how different kinds of workers banded together in common cause through unions and social movements to improve their working conditions, to articulate, defend, and exercise their rights, and to challenge employers, the state, and capitalism more generally. These stories were often about how workers and activists overcame obstacles and divisions to build solidarity through collective action. Their focus tended to be on the industrial unions that welcomed most workers, generally regardless of skill, race, nativity, or gender, such as the Knights of Labor, United Mine Workers of America (UMW), WFM, Industrial Workers of the World (IWW), and the wider CIO. The impediments that American workers struggled with, sometimes successfully and sometimes not, were usually seen as coming from external sources, often through elite instruments of power.[2]

Of course, we know that fear and vulnerability hindered the labor movement in this period. We know, too, that many unions were limited by animosities attuned to racial, ethnic, gender, and religious differences that were often manipulated by employers.[3] Yet we also know that many American workers overcame these encumbrances, even if only slowly and partially, to perceive common class interests and to form groups to defend them against economic and social exploitation, particularly in the New Deal era, when organized labor's influence was strongest.[4] Those expressions of class interest often included demands for safer, healthier workplaces and communities.[5] Whether or not these histories explore union successes or failures, all take as their central subjects those workers who sought some form of collective organization as a means to blunt the experience of industrial capitalism and emphasize those who pursued political and economic changes ranging from reform to revolution. Even craft unions, once considered "a conservative social force" because of exclusive policies and an overriding focus on individual material gains, were shown to be allies, however inconsistent and flawed, in the broader working-class struggle for security.[6] More than any other group, miners—in coal and metal—have served as the field's lodestar because their unions, especially the UMW and the WFM, led the social democratic vanguard.[7]

Our understanding of American workers has been guided by an assumption that they would join unions and welcome government regulation if only they had the knowledge and freedom to do so. Less conspicuous but nonetheless enduring is a related assumption that working-class democracy would prevail, sooner or later, over divisions of race, ethnicity, and gender. These assumptions rest on a scholarly faith in working-class mutualism. David Montgomery articulated that faith best when he argued that workers in industrial America developed "an ideology of mutualism" that taught them that their "only hope of securing what they wanted in life was through concerted action." Despite differences of race, ethnicity, gender, skill, and politics, he argued, their "working-class bondings and struggles" informed "the shared presumption that individualism was appropriate only for the prosperous and wellborn." Because workers were mutualists, Montgomery concluded, they rejected "the ideology of acquisitive individualism, which explained and justified a society regulated by market mechanisms and propelled the accumulation of capital." "A whole generation of research and writing on working-class history," he wrote elsewhere, rests on the finding that mutualism, as idea and practice, prevailed in the "workplace, community life, and local politics" of most working-class Americans. The concept is so powerful that even our understanding of working-class conservatism has been framed, in most cases, by studies of craft unions, such as those in the AFL, the nation's largest and most enduring labor organization.[8]

Until recently we have given little attention to the workers who did not join unions, even though they always outnumbered those who did. Roughly 20 percent of nonagricultural American workers belonged to unions in the early 1920s—the labor movement's strongest years before the New Deal. The miners' unions usually fared best but still struggled to organize a majority of workers in coal or metal. The WFM, at its height in 1910, claimed only 20 percent of metal miners. Unions gained more members after federal legislation in the 1930s made it easier to organize but never more than 35 percent of nonfarm workers, the 1950s pinnacle. At best, most scholars have treated those not in the labor movement as prospective unionists—needing only the right political leadership, union appeal, or social conditions to act on their true mutualist interests. Otherwise, we explain them in terms that privilege the determining power and strategies of elites, whether corporate bosses, right-wing politicians, or conservative cultural leaders. Negative prefixes define these workers as what they were not—nonunion, antiunion, unorganized—revealing a big blind spot that obscures what they thought and why, especially when those thoughts led to a persistent pattern of action, as with the Tri-State miners. Most of the few studies to take workers like these seriously adopted short-term views, examining the events of a single year or decade, that yielded situational explanations about how momentary exceptions or contingencies produced counterpoints to the dominant labor movement narrative.[9] Our assumptions about mutualism have left little room to interpret and understand those who opposed labor unions and social democratic politics over time, particularly in the decades before the New Deal, on their own terms, as historical actors with the same choices and choice-making ability as the unionized minority. When it comes to stories like these, we are little wiser than the contemporaries who reached for easy, dismissive explanations for why Tri-State miners acted as they did for so long.

Poor Man's Fortune reconstructs the century-long story of the Tri-State miners, treating its subjects as creative agents whose decisions and actions across generations reflected a logic of self-interest, both material and ideal, that they themselves crafted.[10] It reveals a tradition of working-class conservatism, from the age of Jackson through the New Deal era, made by white men who identified their interests with the acquisitive market functions of capitalism and the social and political privileges of their race, nativity, and gender. As an ideal, that tradition offered poor white men a good chance to share the national prosperity through hard work and in turn uphold manly paternal responsibilities. In practice, amid many obstacles, it encouraged working-class white men, particularly the native born, to pursue narrowing economic opportunities through reckless physical action and often violent assertions of racial and nativist advantage. As much as larger structural forces influenced, and elites took advantage of, that tradition, multiple generations of Tri-State miners sustained and shaped it in dynamic ways with their own choices against often compelling alternatives offered by unions of all stripes, social reformers, and political allies of the labor movement. Far from ignorant pawns, they acted consciously and consistently for decades according to their own interests—as they understood them, past, present, and future. The great irony, indeed tragedy, is that their cumulative decisions yielded a future of early death, widespread poverty, and diminished freedom.[11]

To a great extent, Tri-State miners authored their own fates. Across five generations, Tri-State miners saw their interests served best by capitalist markets and a culture of individual acquisitiveness that scholars have come to see as anathema to the working-class experience. For a long time, until around 1895, the social and political fraternities of white manliness gave them remarkable opportunities as owner-operators of small mining ventures. The next generation continued to expect the future to be like the past, even after 1900, when most men faced a system of permanent wage labor in real and imagined competition with new European immigrant groups they considered nonwhite—and acted on their expectation in ways that closed them off to alternative visions of the future. The narrower the terms of advancement, the more these men asserted racial and gendered claims to the promises of capitalism. They embraced wage labor with the entrepreneurial zeal of an earlier era, first as mercenary strikebreakers, then as mine workers who insisted on personal wage incentives tied to market prices. These men transformed hard, dirty jobs into potentially lucrative opportunities that demanded reckless physical strength. In doing so, they created a more disruptive but still transactional working-class culture that abandoned older ideas of manly responsibility for a new logic of aggressive, heedless masculinity. As white American men, they expected their dangerous work in pursuit of individual gain to deliver special freedom from the new controls and restraints of corporate capitalism—whether by employers, the government, or other workers, especially those considered enemies of the competitive system. Such was the durable logic of white working-class conservatism that led most Tri-State miners to reject wider solidarities, attack organizations with the boldest ideas of collective security, and embrace the most restrictive forms of American nationalism.[12]

Workers like these who remained outside the labor movement are the "dark matter" of American working-class history: We have witnessed the consequences of their actions but have not mapped their motivations. We have seen their effects in failed strikes, weak and divided unions, and the political vulnerability of the regulatory and welfare state. We have registered their impact in popular support for immigration restriction, racial segregation, and policies that favor capital and business. We have even detected their shadow in the New Deal era, a period otherwise portrayed as a "working-class interregnum" when American workers compelled the federal government to deliver unprecedented "collective economic rights" with social and political campaigns led by a surging union movement, at the forefront of which was the social democratic CIO. Yet even in the most optimistic retellings, scholars caution us that labor's New Deal triumph was short and tenuous. The CIO was a "fragile juggernaut," its power "truncated and brittle." The labor movement's gains depended on a federal labor regime that was slowed by conservatives in Congress, in business, and in organized labor itself, particularly the skilled workers in the AFL.[13] These opponents, usually portrayed as elites, drew upon a lineage of conservative nationalism that combined the "ethos of 'rugged individualism' and the closely associated ideology of liberal capitalism," white supremacy, male sexism, and "suspicion of foreigners."[14] *Poor Man's Fortune* shows that these conservative

ideas were widespread among Tri-State miners in the 1930s and before—
the result of their own decades-long grassroots practice of white working-
class faith in capitalism. Their story reflected the experiences of other white
working-class men in rural areas, towns, and small cities across the United
States where it remained possible to imagine individual opportunities for
economic advancement, whether these men ran farms, owned and operated
small businesses, or contracted their labor by the piece or job. The Tri-State
miners were an extreme case, perhaps, but not an exceptional one.

The first two generations of Tri-State miners sought individual economic
opportunity as they pursued promises born during the Jacksonian market
revolution. From the 1850s to the 1890s, the district, which would cut across
five counties in three states—Jasper, Lawrence, and Newton in Missouri;
Cherokee in Kansas; and Ottawa in Oklahoma—was known as a "poor man's
camp," where individual prospectors with very little capital could secure
speculative mining leaseholds on land they hoped would yield ore deposits
and make them socially and economically independent. The possibility of
becoming an owner-operator miner on the basis of one's muscle power, mining skill, and diligence attracted thousands of ambitious men in the years
surrounding the Civil War. White men, particularly the native born, had the
freedom of movement and the access to legal and financial resources required to take up these opportunities. Their racial advantages were both
psychological and material, as tangible as a mine shaft or a chunk of lead
mineral. Many succeeded, a few got rich, and many more did not. Together
they built prosperous communities that championed a democratic spirit of
fairness and opportunity between risk-taking white men. That poor man's
culture explained achievement and made no excuse for failure. Proof of its
efficacy abounded in this forty-year period, when hundreds of mining companies, most of them owner operated, discovered, mined, and sold lead mineral and zinc ore worth more than $36 million.

The entrepreneurial ambitions of Tri-State miners had deep roots in the
broader region. The district's stories of men who developed prospects into
profitable small mining companies inspired thousands more newcomers
from other mining districts and farms across the Midwest, Upper South,
and Great Plains. They came because those stories were familiar and made
sense; the Tri-State district was an organic part of a wider society and economy, not an outlier. Many, especially in the beginning, came from old lead-
mining districts in the Mississippi River valley. They were first to develop the
district's deposits in a serious way and to insist on the poor man's terms for
doing so. More ambitious but poor white men came from the surrounding

Arkansas and Missouri Ozarks, rural places similarly shaped since the 1850s by lively commercial markets that favored small-scale producers. They had worked in lumbering, milling, tanning, or iron mining or were farmers who grew crops for sale and took seasonal work in these industries to make extra money. Others came from farms across the wider region, from Tennessee to Iowa. They all shared an economic experience and culture as white men who produced for the market with the goal of attaining independence for themselves and their immediate families. Their transition to the Tri-State mining district, where the barriers to entry were low for men like them, was smooth and logical.[15]

We know that many other rural white workers across the country navigated the market economy with a similar entrepreneurial outlook. In Appalachia, rural workers ran diversified household economies that combined farm production with wage work in mines and factories. Similarly, many rural workers in the Midwest combined farming with seasonal work in other industries, particularly shallow coal mining. By the 1930s and after, independent truck drivers would think and act according to a similar logic. Whether they were West Virginia farmers, Illinois coal miners, or truckers on the open road, we know that they often understood their interests as separate from the solidarity and collective action of the emerging labor movement. Company domination of local economies and communities ultimately pushed many of these workers toward class-based confrontation, particularly in the UMW. Tri-State miners, however, enjoyed opportunities for small producers longer than most in a district defined by competition between hundreds of separate companies, none of them in control.[16]

Small producers registered their claims on capitalism at the ballot box in the late nineteenth century. Despite punishing depressions in the 1870s and 1890s and increased corporate consolidation of power, white men, particularly the native born, continued to rally to the Republican Party's free labor ideology that heralded economic opportunity through hard work in a competitive marketplace and asserted the closeness of worker and employer interests. Those engaged in domestic manufacturing and resource extraction especially valued Republican pledges to protect American labor from foreign competition with tariffs and, in time, immigration restriction. At the same time, many Democrats also believed that white workers could claim a share of capitalist prosperity, that the divide between them and the rich was not total. Presidential candidate William Jennings Bryan said as much in his 1896 Cross of Gold speech, an occasion said to mark the arrival of a popular challenge to industrial capitalism. Responding to charges that his campaign would damage American business, Bryan declared that his critics were "too limited" in their "definition of a business man." "The man who is employed for wages is as much a business man as his employer," he claimed. "The miners who go down a thousand feet into the earth ... are as much business men as the few financial magnates who, in a back room, corner the money of the world." Tri-State miners, along with many other workers, considered themselves among Bryan's "broader class of business men."[17]

Even those most attuned to the threats corporate consolidation posed to small producers—after all, that was Bryan's point—imagined restorative remedies that reaffirmed democratic access to market prosperity. The main popular insurgencies that culminated in Bryan's 1896 candidacy—organizations such as the Grange, the Greenback-Labor Party, and ultimately, the People's Party—all revolved around foundational commitments to antimonopoly reform and inflationary monetary policies designed to give poor men fair economic opportunity. Significant constituencies in the era's labor movement, including within the Knights of Labor and the AFL, shared this vision and goal.[18]

By the 1890s, Tri-State miners saw themselves as entrepreneurs who combined hard labor with business acumen and thus shared little common interest with the growing ranks of wage laborers, many of whom took a more critical stance against capitalism. That did not make them passive. Like many others, Tri-State miners were vigilant against monopolies, particularly land and smelting companies that bought interests in the district in the 1870s and after. That vigilance led some to resist the concentration of power. A minority joined with area farmers to support the Greenback-Labor and People's Parties but only on the basis of antimonopoly proposals that honored their market-oriented aims. They were also familiar with the Knights of Labor. The Knights were active in the Kansas coalfields beginning in the late 1870s and tried to organize the Tri-State district for a decade with no lasting success. Tri-State miners could not square their economic ambitions with the Knights' vision of collective security and solidarity, especially after the mid-1880s when the Knights adopted a more combative stance with a series of prominent strikes. The AFL, generally more conservative and accommodating of capitalism, might have fared better, but the federation had no presence in metal mining until 1899 when it challenged the WFM, which emerged, along with the UMW, from the early 1890s dissolution of the Knights. By then, the vast majority of poor man's camp miners were nonunion, not antiunion, although some were growing wary of union tactics and aims as the upheaval against capitalism intensified with violent clashes across the country. Like many other workers who aspired to ownership, they rejected strikes because such direct action assumed a fundamental division between workers and owners when they saw none for men like them. They also ignored or resisted the mine safety laws that the Knights helped to pass because the cost of compliance threatened their small-scale operations, whether present or future. The legacies of the poor man's camp—mental as well as material—encouraged Tri-State miners even in the economic crisis of the 1890s when opportunities for becoming an owner-operator slowly gave way to a permanent regime of wage labor.

Tri-State miners understood the loss of self-determination at work as a threat to their ideal of white manliness, which had valorized responsibility, perseverance, and autonomy. After the depression of the 1890s, outside investors intensified district mining practices that foreclosed future owner-operator possibilities for small-scale miners. Like other white men facing subordination in this era, Tri-State miners embraced rough masculinity, a way of understanding manhood that "emphasized toughness, physical strength, aggressiveness, and risk taking."[19] Informed by a long-standing emphasis on independence and freedom of action, they began acting forcefully for their own benefit, with little care for the detriment to others—first as strikebreakers against the WFM, which launched a wave of strikes against wealthy mining corporations across the West in the 1890s. With skillful negotiation, these men exploited their nonunion, native white status to take jobs, often at high wages, from mainly foreign-born union miners. At first, most did it temporarily, eager to return home to invest their earnings. While these divisions benefited mining companies foremost, the option of strikebreaking became an important means for many men to weather bad economic conditions and see new entrepreneurial possibilities in wage labor.

Tri-State strikebreakers devastated the WFM, setting it on a radical course that would roil the labor movement for decades. They helped break nearly every major WFM strike: at Leadville, Colorado, in 1896; at Coeur d'Alene, Idaho, in 1899; at Cripple Creek, Colorado, in 1903; and at Lead and Deadwood, South Dakota, in 1910. Tri-State strikebreakers sparked simmering tensions that led the WFM to challenge the AFL by forming the Western Labor Union, a direct forerunner to the anticapitalist IWW, in 1898. While subsequent strike defeats pushed moderate WFM leaders to seek peace with the AFL in 1910, attempts by radicals to regain control would torment and divide the union to the point of collapse by the end of World War I. Tri-State miners haunted the WFM: in the minds of radicals, as mortal foes to be crushed; in the minds of moderates, as a potent force that should be unionized with whatever accommodation might be required. Both the WFM and the AFL launched repeated, unsuccessful efforts to organize them, first in competition with one another from 1899 to 1910 and then in concert until the 1930s when the New Deal presented the best chance yet.

Over a fifteen-year cascade of conflict, Tri-State miners learned to regard the WFM, especially its members born in eastern and southern Europe, as a threat to their economic opportunities and social advantages as white American men. Strikebreaking against foreign union miners—in an increasingly racist and xenophobic national culture, often defended by military force—drew from patriotism the confidence to disregard the concerns of immigrants and the nonwhite, no matter the cruelty. This feeling sharpened as WFM leaders denounced strikebreaking as an immoral act that deviated from the expectations of traditional manliness. Like other white men in this age of Jim Crow and imperial aggression, Tri-State miners asserted their claim of racial authority through violence against perceived enemies. They attacked African Americans, fought unionized immigrants in strike zones, and forcibly ejected foreign-born miners from the Tri-State. They saw their fears realized in WFM and IWW radicalism; fighting these groups became a main way of proving one's worth as a white man, akin to soldiering. Tri-State miners not only deepened their opposition to radical unions but began to understand themselves in a new way—as free, patriotic workers whose respect for capitalism earned them special privileges, a view that alienated many from the labor movement and social democratic politics more broadly.[20]

Tri-State miners doubled down on the performance of rough masculinity to keep alive the risk-and-reward ethic of the poor man's camp, albeit on narrower, tougher terms. While small-scale prospecting and mine leasing ended after 1900, men found that their physical labor was in high demand in the district's still hypercompetitive, undercapitalized operations. Mining companies relied especially on unskilled shovelers, workers who loaded ore into cans for hoisting, and began paying them a piece rate to boost productivity. Shovelers embraced piecework, which was plentiful as American industrial and military expansion consumed ever-greater amounts of zinc and lead from the Spanish-American War to World War I. Shovelers made themselves indispensable by treating their bodies as capital. Between 1900 and 1930, men wielding standard-size shovels that held twenty-one pounds a scoop moved more than 600 billion pounds of ore in the Tri-State—the equivalent of more than 820 Empire State Buildings. They made money and gained status. Now the largest occupational group in the mines, shovelers were heralded for embodying the rugged white masculinity that elite nationalists like President Theodore Roosevelt championed. They soon dominated the district's working-class culture and defined what it meant to be a strong white man. Their example emphasized youth, reckless power, and short-term incentives; it belittled ideas of safety, sustainability, and public aid. Shovelers took risks with their bodies, endured pain and suffering, and showed no weakness—all requisites for a good payday under the logic of the piece rate. As they came to see it, to give in to the threat of injury or death not only reduced pay but revealed masculine failure, a signal of physical and mental inadequacies that risked association with the foreign-born and nonwhite people ruled inferior in the era's racist politics. This performance of white working-class masculinity further distanced Tri-State miners from union workers who sought to make work safer and uphold older ideas of manly responsibility. It also created a problem for how they advocated for themselves: they were committed to doing work that broke their bodies but prejudiced the weak and damaged.[21]

As risk-taking wageworkers, Tri-State miners also strained against their employers as they demanded a share of the district's expanding profits. They still expected capitalism to work for white men like them. After 1905, the miners, not mining companies, insisted on tying wages to production incentives, including bonuses and a sliding scale that tracked the market price of metal. They defended this raw claim on prosperity with physical defiance that reflected both their social privileges and the harsh realities of their labor. When companies tried to cut wages or replace them with machines or cheaper foreign workers, these men disrupted operations without fully rebelling. They turned again to temporary migration as strikebreakers, switched employers without notice, sued over injuries, forcibly excluded foreign workers, and after the Panic of 1907, withheld their labor in small, isolated wildcat strikes. Their tactics were often successful; mining companies, divided and usually small, could not control them. Some miners even flirted with the idea of union organization, at first on an independent basis and then in bids to align with the AFL, which was increasingly regarded as a trustworthy ally of white, native-born workingmen. And yet most Tri-State miners could not accommodate union demands for safety and security. They remained committed to an ethic of individual physical risk for market-based incentives—an ethic given new life by American entry into World War I and the rise of a new boomtown in Oklahoma.

Despite a pattern of resistance to government regulation, Tri-State miners expected the nation to reward their rugged, patriotic fidelity to capitalism. For them, American nationalism delivered crucial racial and nativist advantages that promised ongoing access to market prosperity for white men who no longer enjoyed full economic independence. Their expectation was based on evidence. The federal government had helped create the conditions that gave rise to the poor man's camp by forcing Native Americans off the land, supporting railroad construction, erecting tariff barriers that protected domestic metal markets from foreign competition, allowing rampant discrimination against African Americans and others considered nonwhite, restricting immigration, and suppressing radical, anticapitalist unions and groups, such as the IWW. Above all, the federal government encouraged nationalist economic expectations with military excursions and wars against demonized enemies that also created rising demand for zinc and lead, no more so than during World War I. Tri-State miners benefited directly from the political economy of belligerence and came to see their interests entwined with the white nationalist ideas and policies that fueled lucrative American militarism.

By the late 1920s, however, they were no longer sure that the government was on their side. Federal agencies encouraged a new, more rational regime of managerial capitalism, aiming to empower corporations to bring order and efficiency to the larger economy. In the Tri-State, that meant helping mining companies address their risky workplaces. Federal modernizers like Thomas Parran, who would become U.S. surgeon general a decade later, were horrified by the physical toll of life in the district, counted in rates of infectious disease, particularly syphilis and tuberculosis, as well as the injuries, fatalities, and incidence of silicosis among miners. Mining companies were worried about the rising cost of workmen's compensation insurance, now a legal requirement, as the mining economy slowed after 1927. The government and the companies joined forces, together with leading insurance companies, to impose health and safety reforms and, most critical, prevent the most damaged miners from working. Led by the U.S. Bureau of Mines and the Public Health Service, this paternalistic strategy scrutinized the bodies and behaviors of men in ways that threatened what it meant to be a Tri-State miner as a worker and a white man. As the downturn became a depression, companies pursued this strategy of control with ruthless energy.[22]

While opposed to reforms that seemed to restrict their livelihoods, Tri-State miners looked to the New Deal—especially with its nationalist allusions to wartime precedent—to restore the economic and social standing of men like them. They wanted President Franklin D. Roosevelt to revive the nation's economy and roll back the health and safety regime so that they could share in prosperity once again. But the New Deal presented a dilemma: in order to regain what they had lost, Tri-State miners would have to deal with the labor unions they had long opposed. In the national wave of organizing that followed the passage of the National Industrial Recovery

Act in 1933, a substantial minority of district miners joined the International Union of Mine, Mill and Smelter Workers (Mine Mill), the WFM's successor and at the time an AFL affiliate. When companies refused to bargain, Mine Mill launched a strike in 1935 that closed the whole district. The clash exposed a stark divide between past and future: on one side were men who remained committed to the deep-seated verities of race, masculine risk, and the market, and on the other was a new, smaller group that embraced the possibilities of social democratic unionism as Mine Mill left the AFL for the new CIO.

The conservative tradition prevailed. The majority of the Tri-State miners turned against their allies in Mine Mill by joining a company union that broke the strike, encouraging and empowering employers keen to further exploit divisions of race, gender, and nationalism. While the federal National Labor Relations Board ultimately sided with Mine Mill by outlawing the company union, it could not change the minds of the district's miners. If the New Deal was a "decades-long experiment in the economic enfranchisement of the American working class," men like the majority of Tri-State miners viewed its benefits in conservative terms, as privileges that should flow once again to white Americans, especially men, who kept their faith in capitalism.[23]

What made Tri-State miners conservative was a sustained desire to return to an earlier, greater era when ordinary white men could attain some material benefit and personal autonomy from their mental and physical skills in competition with other white men. Their conservatism was primarily economic and social with deep roots in the Jacksonian market revolution. It championed hard work, democratic markets, the prerogatives of manliness, and the privileges of white people in the United States. Some might call it individualism; without qualification, however, that term erases the ways hierarchies of race, gender, and nativity structure individual opportunities. While Tri-State miners were self-interested, they claimed those interests as white men at the expense of others with increasing vehemence; the power to exclude became integral to the claim.[24] Over time, Tri-State miners, and other Americans, synthesized these commitments into a belligerent white nationalism. They judged new circumstances against these old certainties, viewing any attempt to challenge or limit their prerogatives—whether through safety regulations or radical labor unions—as a threat.

Tri-State miners knew exactly what they were doing and why. They thought and acted within a political structure and culture that generally encouraged their ideas about economics, race, gender, and the nation. Powerful groups certainly tried to exploit their racism and nativism for their own ends—mining companies, strikebreaker recruiters, and labor organizers alike. But Tri-State miners were not simply duped by elite propaganda; they made these views their own over time through dynamic practice. They neither were victims of businesses and politicians nor fooled into acting against their own interests. They did what they did for a reason. Tri-State miners vexed employers and commanded a grudging respect.

Their story was in the mainline white American tradition, a product of its self-styled sunlit heartland, where promises of family independence, economic opportunity, and cultural cohesion seemed to bear real fruit. The Tri-State district and its surrounding counties were prosperous. Residents credited that prosperity to the dominance of native-born whites like them; they violently expelled many African Americans and foreigners to make it so. Some called the region a "white man's heaven." By 1910, Jasper County, the district's heart, not only led Missouri in mining output but was first among its 114 counties in cattle production, twenty-first in total crop value, and sixth in value-added manufacturing. It was surrounded by counties with similar agricultural profiles. The state's fifth-largest county by population, Jasper boasted a literacy rate of 96.6 percent and youth school attendance rate of 87.9 percent, both more than three points above the state average. Joplin, the largest city, was modern; a streetcar and light-rail network provided transportation between its neighborhoods and to district towns in all three states. Four railroads provided passenger and freight connections in every direction. In addition to high-end hotels and theaters, the district also offered popular leisure activities through cinemas, amusement parks, and sporting events, including professional baseball games. While the people who lived and worked in the Tri-State inhabited a world different from the places labor historians usually study, whether big cities, single-industry areas, or plantation zones, their experience would have been very familiar to many, if not most, contemporary white Americans.[25]

Some readers might expect religion to explain the miners' conservatism. But the evidence does not show that religion, organized or not, had a causal influence. Like other miners elsewhere, they were decidedly not pious. Those who went to church could choose from a dozen or more different denominations in the Tri-State, most of them Protestant. Miners encountered little in the churches that challenged their social and economic views, and they did not air their objections to labor unions or their political allies in religious language.[26] Theirs is certainly a story about belief, however—about white working-class men believing, both as thought and expectant action, in capitalism, the nation, whiteness, and their own physical power.[27]

Tri-State conservatism was surely political but did not produce a neat pattern of partisan politics. Democrats and Republicans alike were well organized and ran strong in the district, often neck and neck, even in precincts where miners predominated. Miners also had ready access to political ideas beyond the mainstream parties. The Grange, Greenback-Labor Party, Union Labor Party, People's Party, and Socialist Party all canvassed the district. The Socialist Party's main newspaper, the *Appeal to Reason*, was published in nearby Girard, Kansas. At the same time, the nation's leading anti-Catholic, nativist newspaper, the *Menace*, was published in Aurora, Missouri, a mining town on the district's eastern edge. While a few miners were won over by the radicalism of the *Appeal to Reason*, many more were drawn to the reactionary views of the *Menace*. Tri-State miners seemed more likely to favor Republicans before the depression of the 1890s, when they could reasonably aspire to be owner-operators. As a general trend, they turned toward favoring Democrats afterward, once they faced permanent wage labor. But it was only a trend. In the 1920s, many miners gave their support to Republicans promising protective tariffs and immigration restriction in three successive elections. Third parties drew some support, usually when the economy was bad. For example, the Greenback-Labor Party and Union Labor Party each got around 15 percent of the vote in the 1880s, a period of crisis in the lead industry. Socialist Eugene Debs won a plurality in two precincts in 1912, a rare moment when union organizers seemed to make headway, but more miners voted for William H. Taft, Theodore Roosevelt, or Woodrow Wilson, who won a plurality in the district as a whole. Overall, Tri-State miners tended to choose candidates who promised government measures that would preserve or restore their prerogatives as white men.

The better we understand workers like these, the better we will understand the durable power of a politics that in the present defies explanation. The small-business owner, the specter of burdensome government regulation, the injustices of welfare, and the allure of "right to work" count among the enduring tropes of right-wing populism. Those who claim these ideas often act to deny chances or choices to other workers, particularly along racial, nativist, and gendered lines. Scholars of working-class conservatism since the 1950s have interpreted this as "backlash," a term that suggests a negative reaction produced in response to change. The story of the Tri-State miners reveals a longer lineage.[28]

Some Tri-State miners did heed the appeals of union organizers and progressive reformers. For decades, labor unions doggedly tried to win them over. To their immense credit, these activists did not give up. Unions that persisted against Tri-State strikebreakers in spite of defeat may have had little choice but to fight on. But they also believed that a politics of solidarity, collective action, and economic and physical security would ultimately prevail. Even the most social democratic unions, the WFM and Mine Mill, won converts in the district. But these remained minority voices, neither able to convince the majority nor stay mobilized for long. Most miners only came close to joining the labor movement when doing so entailed no risk of associating with foreigners, African Americans, or political radicals. Even then, the lure of economic self-interest, particularly in times of prosperity, beggared and eroded union pledges of collective advancement. Some miners did support health and safety regulations in the mines. Most did not.

The defeat of progressive unions and health and safety regimes testified, time and again, to the difficulty of convincing native-born white men to subordinate their desire for personal opportunity and independence, real or imagined, to the public good. Most would not abandon the advantages of race and the dominant mode of aggressive masculinity, even in the 1930s when they enjoyed few prospects in a political economy of physical suffering and low pay. Those who did were genuine revolutionaries in context. They were also rare. The struggle of reformers to change the Tri-State miners revealed a broader truth, that it was always difficult to convince Americans, particularly those who benefited most from racial and gender inequality, to embrace a new vision of the future. That should make any progressive successes in the Tri-State, rare as they were, seem all the more remarkable and important.

More commonly, the Tri-State miners emboldened the most conservative elements within organized labor—both in the leadership and at the grass roots. Their attachments to capitalism, pugnacious masculinity, and white nationalism encouraged many white union members to conclude that more democratic options were not viable and to opt instead for defensive choices. This process, driven by the actions of workers outside of organized labor, moved the center of gravity in the labor movement to the right, where it was already rewarded by capital and state. In the radical WFM, for example, native-born members followed strikebreakers back to work. In the AFL, national leaders hoped to organize Tri-State miners into a conservative metal-mining union to thwart radical groups in the WFM and ultimately in the CIO. Even moderate leaders of the WFM and Mine Mill appeased the racism and nativism of Tri-State miners in the hopes of building stronger unions. By the late 1930s, the AFL stood in robust defense of capitalism and white Americanism, revealing how ideas that had motivated generations of Tri-State miners were now reflected in the central assumptions and positions of the nation's largest federation of workers. By the 1950s, the same was also true of large parts of the CIO, as "new" European immigrants and their children learned the dark power of white American masculinity. That says less about the Tri-State miners than it does about the enduring influence of their conservative ideals in the minds of other white working-class men then and since.[29]

Sylvan Bruner, a local lawyer and former miner, never imagined that a metal miner's union would hold its annual convention in Joplin, Missouri, the center of the Tri-State district, as Mine Mill did in August 1941. In the preceding year, the union, with the help of the National Labor Relations Board, had made progress among the district's workers. Mine Mill held its convention in the Tri-State as a show of collective determination to finally organize them. "We are very glad to have all of you here," Bruner told the gathered delegates. They might have hoped for a sunnier welcome than what came next. "I think I can say that the miners in the Tri-State District have paid a most tragic price for their lack of organization," he explained, and "that their families have paid a tragic price." Few people wanted to see the union's latest attempt succeed more than Bruner, who had witnessed all the union failures since 1907. "I think I can say, without fear of contradiction," Bruner repeated, "that every worker in the Tri-State District has paid a tragic price through the lack of organization and through open shop conditions in what is known as the Tri-State District." But he was not yet satisfied that his statement conveyed the role those workers had played in that history, so he said it again, clearer. "What I mean to say is that the hard rock miners in the Tri-State District have paid with their lives and broken bodies for 30 years because they have not realized that they should organize."[30]

Tri-State communities indeed bore a terrible cost for the decisions of generations of miners. Their story is full of economic failures, life-shattering injuries, and premature deaths from sudden accidents and from the slow, bloody suffocation of silicosis. Women shared that cost as wives, mothers, and sisters. We know that some of them challenged the outlook of the district's men, disagreed and pleaded with them, called for a different life. We also know that many women went along, often with enthusiasm. They sought their own fortunes in the Tri-State and often profited from their labor in the mining camps. Many who married miners shared the ambitions of their husbands and sent their sons into the mines. Whatever their perspectives, the voices of women in the Tri-State mining communities are rare in the documentary record. What follows amplifies those voices as much as possible.

In local memory, collected and shared mainly by the daughters of these people, the story of the Tri-State district was told with defiant pride, despite the costs, after mining stopped in the 1950s. Their tellings emphasized the hard work of poor men in discovering and developing the mines. These were tales of ingenuity and achievement amid hardship. They talked about making something from nothing: profitable mining camps, thriving towns, and a city, Joplin, whose streets still bear the names of the most successful miners. They invoked the district's contributions to the nation, supplying vital metals for two world wars. They insisted on positive legacies, despite devastating environmental consequences that have plagued many Tri-State communities and destroyed some. The district's history of opposition to unions, government regulation, and the promise of social democratic change went unspoken.

When people spoke of the miners, they talked about the poor men seeking fortune: the prospector, the owner-operator made good, the shoveler, even the district's favorite son of a miner, Mickey Mantle. Today, the mining memorials at Joplin, Webb City, and Baxter Springs each reflect that image with a statue of a miner, always alone.

CHAPTER 1
FINDING'S KEEPING

In the two decades before the Civil War, in the course of a market revolution that quickened national commerce and territorial expansion, Americans used metal in more ways and in greater quantities than they ever had before. They used iron for the engines, machines, and tools that shaped the new steam-powered economy. They used copper to make household items, roofing material, and sheathing for ships but also as wire to conduct electrical signals over the new telegraph system. By the late 1850s, they mixed copper with zinc to make brass for buttons, precision instruments, and the first metallic bullet cartridges. They used lead in paint, sheet glass, and water pipes to improve growing towns and cities and for ammunition in armed conflict against Native Americans, foreign powers, and ultimately, other Americans. Of course, they used gold and silver to pay for these things. Yet as this new age began, the American market craved more metal than domestic sources could deliver. The development of the nation's industrial economy needed metal miners keen to take advantage of its insatiable demand.

Beginning in the 1840s, miners responded with conquests and discoveries that exploited ore deposits across the continent with unprecedented fervor. Most famously, the Argonauts, who went to California, mined more gold in the 1850s than the whole world had in the previous 150 years combined. In 1859, miners prospecting for gold in Utah Territory (present-day Nevada) struck the Comstock Lode, the nation's first major silver mine. These prodigious supplies of gold and silver increased the world's currency by 600 percent, which further stimulated demand for base metals. In 1860, the United States produced twenty times more pig iron than it had in 1840. On Michigan's Upper Peninsula, where in 1845 European Americans took copper mines long worked by the dispossessed Ojibwe, miners produced over 7,300 tons of the reddish metal in 1860. In the 1850s in New Jersey, miners opened the continent's first zinc mines. Even though Americans had not yet mastered the methods to smelt zinc, manufacturers used 726 tons of it in 1860. In most cases, white miners backed their claims with violence against Native peoples, Spanish speakers, and others deemed inferior. The expansion of metal mining during the market revolution exemplified the powerful enticements of American capitalism and white assertions of dominance—and the consequential relationship between the two.[1]

The early development of metal mining in southwest Missouri followed this pattern. In the 1840s, the first non-Natives to settle in Jasper and Newton Counties found deposits of galena, the most common mineral form of lead, on land taken from the Osage and Delaware peoples. In some cases, these settlers simply identified old Native mines; others were new discoveries. They developed small but profitable mines and smelters that soon attracted the attention of several hundred experienced lead miners, who in the mid-1850s rushed to this isolated region on the border of Indian and Kansas Territories. By 1860, they were producing one-fourth of all the lead used in the United States. However, compared to the miraculous possibilities of the gold rush, which by 1860 drew over 100,000 miners to the West with hopes of sudden riches, the mining of lead required a more patient and elaborate economic imagination. While gold sold for twenty dollars per ounce in the 1850s, 1,000 pounds of lead sold for just sixteen dollars. In order to make money from galena, then, miners needed to extract it by the ton. In an era without much machine help or dynamite, this required not only immense physical effort but also a substantial investment of time. Although we might speak of rushes, lead mining privileged the sober investment of hard work and the careful consideration of prices and costs, including the risk of physical injury or death, over a relatively long period. Why, we might ask, in the midst of multiplying opportunities to mine other, more immediately valuable metals, did some miners find the promise of lead so attractive?[2]

In fact, despite promises of riches for the many, most American mining districts offered opportunities that proved more limited than initially claimed. Although small groups of working miners made most of the initial discoveries of gold, silver, and copper in this period, wealthy investors soon dominated these industries through companies that increasingly used large-scale, mechanized methods and a system of wage labor. The mining of iron ore had always been hierarchical in the United States, whether on Pennsylvania's "iron plantations" or in the slave labor camps of Virginia. Adopting this general form, well-financed, large companies bought out or financed the most successful small miners. These new firms organized deeper mining in search of the lode, or main source of surface deposits. These shafts soon extended several hundred and sometimes over 1,000 feet down. Deep mining relied increasingly on machine power to pump water and hoist material to the surface. As big companies mechanized operations, they employed miners on a wage basis, usually hiring Americans or European immigrants but sometimes, especially in gold mining, Chinese workers. Individual wage miners, if they had experience and skill, could make good money, about three dollars a day in western goldfields. In large base metal mines, such as those in Michigan, however, miners earned much less, only about $1.25 a day in 1860. Although both wage levels exceeded the average earnings of general laborers in the 1850s, employment with the large companies dominating California's goldfields, Michigan's copper belt, and Nevada's Comstock Lode meant that miners answered to other men who pocketed all of the real wealth they produced. Despite the heady possibilities of the western rushes, the interests of most metal miners quickly narrowed to questions of wages and working conditions as the market revolution gave rise to a new form of capitalism defined by industrial consolidation, mechanization, and regimented hierarchies.[3]

The lead industry stood apart, continuing to offer opportunities for white men with little capital to work on an independent basis and to make money producing a valuable commodity with their own skilled labor. By the time of the Civil War, American lead mining was old. While Native Americans in the Upper Mississippi River valley had mined galena on a small scale for generations, French traders developed the first European-controlled lead mines in North America in 1720, at a site in the eastern Ozarks of Upper Louisiana, sixty miles southwest of St. Louis in present-day Missouri. The eastern Missouri field, redeveloped and expanded by Spanish and later American miners, produced most of the lead in North America for the next century. As demand for lead increased, French and American miners seized control of mines from the Sauk and Fox peoples near present-day Dubuque, Iowa, in the 1790s and along the Fever River (now the Galena River) in present-day Illinois and Wisconsin in the 1820s. Although these mines soon surpassed the output of the eastern Missouri field, all lead miners continued to use preindustrial methods little changed from the previous century. That was possible because geological formations had left shallow deposits of galena at depths that could be reached with hand and animal power. Meanwhile, these deposits were far enough west of major European settlements—often beyond the effective reach of any government—to privilege those miners willing to take the risks of claiming and developing them. The lack of close governance allowed lead miners considerable sway with successive regimes in negotiating and maintaining a generous legal code of access that, from 1720 into the 1850s, allowed them to lease or claim mineral rights in exchange for royalty payments equal to a portion, usually from 10 to 20 percent, of the total mineral they extracted. Some aspirants, French, Spanish, and American among them, experimented with the use of enslaved labor in all of these fields. Over time, however, free mining proved less costly and less legally cumbersome and came to predominate, especially once Cornish miners began arriving in 1830. Free miners generally worked for themselves in small-scale operations, sometimes alone but often in small basic partnerships according to the leasehold terms, from which they could make far more money—from a few hundred to a few thousand dollars more a year—than most other white men who worked with their hands. They were confident in their claims, especially after American forces defeated Native efforts to retake land in the 1832 Black Hawk War. By the 1840s, when farmers made the first discoveries of galena in southwest Missouri, miners working these older fields produced all of the lead that Americans needed and more, and made St. Louis, by virtue of its proximity on the Mississippi River, the western center of the lead industry.[4]

Surging domestic demand for lead in the 1850s destabilized the lead industry. Americans used more than double the amount they had in the decade before, creating powerful new incentives to exploit the galena in Jasper and Newton Counties. Miners went there looking to re-create the terms that governed lead production elsewhere. In return for the risk of investing their time, money, and labor in physically demanding and dangerous work, they gained claims with the potential for high rewards. That investment reflected confidence in their own skill and prerogatives as free miners, as well as a careful appraisal of market conditions. While some urban workers in the 1850s resisted wage labor and industrial discipline with republican critiques of base materialism, lead miners took advantage of the market revolution and the opportunities it created for them. They sought the economic independence and social status expected of white men. Neither soft-palmed speculators nor dependent wage hands, these miners staked a position that straddled the social and economic chasm being wrenched open by industrial capitalism.[5]

We do not know who first discovered galena in the southwest corner of Missouri. In 1819 the geologist and geographer Henry Rowe Schoolcraft found abandoned surface mines and a small smelting furnace along the banks of the James River, about sixty miles southeast of present-day Joplin. He attributed the rudimentary but apparently successful workings to the Osage people, who dominated the northern Ozarks into the eighteenth century. "The Indians have been in the habit of procuring lead for bullets at that place," Schoolcraft wrote, "by smelting the ore in a kind of furnace, made by digging a pit in the ground." The Osage took their knowledge of lead production with them, however, after losing their remaining claim to land within Missouri's borders in 1825. The first white settlers who moved into the area in the 1830s came looking for fertile farmland and good water, not lead.

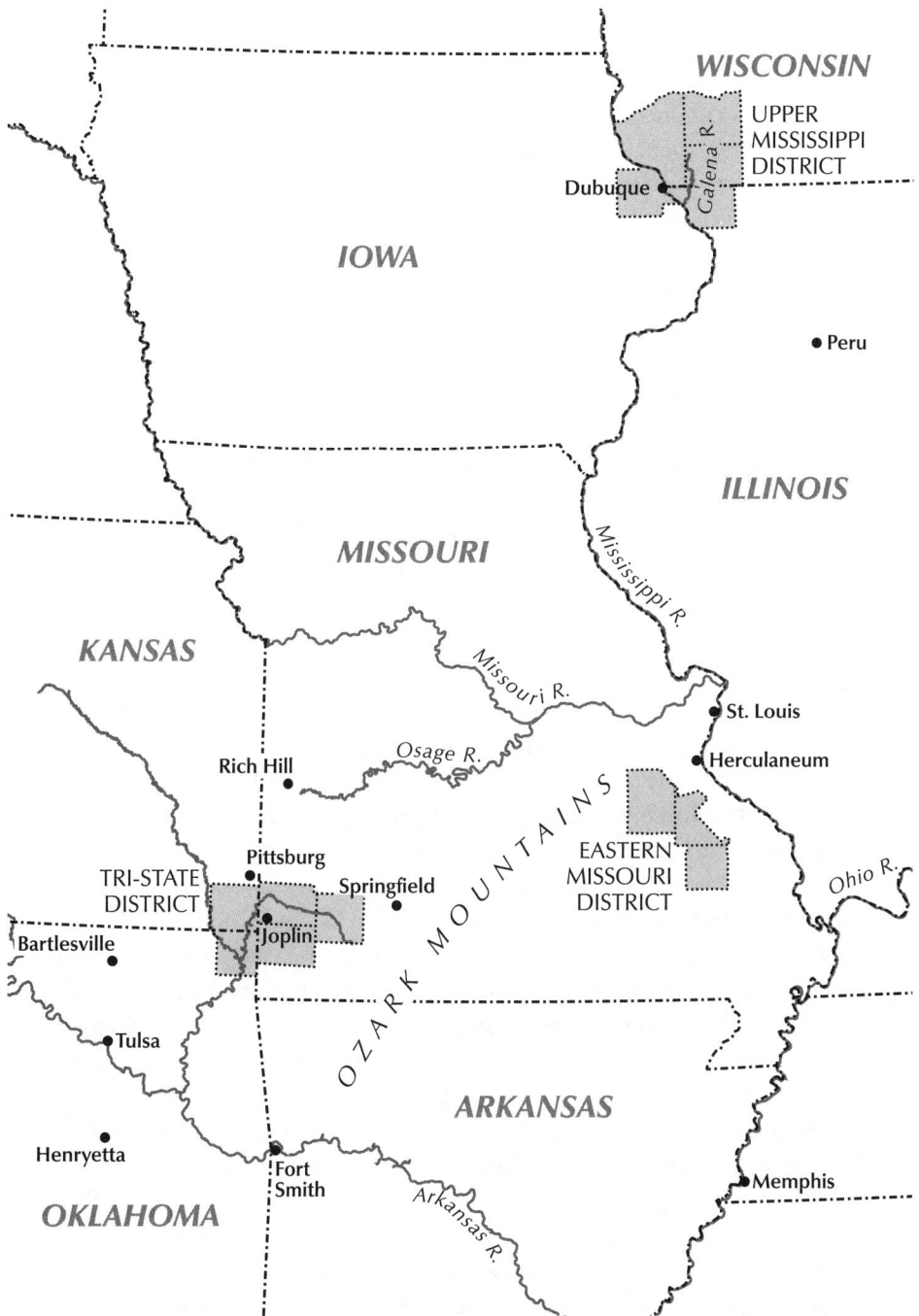

Map 1. Midwestern lead and zinc fields. Courtesy INCase, LLC.

Local lore reported the occasional uncovering of shafts along area streams, but few farmers investigated the holes until the late 1840s, when a quick succession of discoveries of relatively shallow but rich deposits of galena sparked new commercial considerations.[6]

Settlers in Jasper and Newton Counties lived in isolated communities on rich land at the western edge of the contiguous United States. The first few arrived in 1831 when the federal government forced the last remaining Osage and Delaware people into Indian Territory. Around 2,000 families, over 8,400 people, followed during the next two decades, mainly from Kentucky, North Carolina, and Tennessee. Most established small farms along rivers and creeks. Some owned slaves but in small numbers. Only 123 slave owners and 454 enslaved people lived in the two counties in 1850. Nine slave owners held ten or more people; most owned three or fewer. Large-scale slavery did not pay so long as the new settlements lacked adequate means to transport agricultural products. No roads of any kind ran to St. Louis, the closest major market, nearly 300 miles away. Farmers could send goods overland to the Osage River, 100 miles north, and then by boat to the Missouri and Mississippi Rivers. Wagon haulage was expensive, however, and painfully slow. Or they could ship downriver through Indian Territory to Memphis and New Orleans, but the route was long and dangerous. Instead, farmers produced mainly for subsistence but also for sale in the small regional market: wheat, corn, oat, sheep for wool and meat, and some tobacco.[7]

The first Americans who tried to make money from lead mining in the area were local farmers and merchants. According to most accounts, Thomas Shepherd and Simpson Oldham began mining galena from an outcropping along Shoal Creek in present-day Newton County in 1847. We do not know if they went looking for mineral, as many called galena, or found it by accident. In 1848 Amos Spurgeon discovered mineral while plowing on his farm three miles west of the Shepherd and Oldham mine. In the spring of 1849, David Campbell found a deposit on his uncle William Tingle's land along Turkey Creek in Jasper County, sixteen miles to the northwest. John Cox, Tingle's new neighbor, recognized the mineral as the same material his enslaved man had unearthed on the banks of Joplin Creek, a tributary of Turkey Creek. In 1850, Andrew McKee found galena a few miles north of Tingle at his homestead; in 1851 he discovered more along Center Creek, ten miles north of Cox's land, apparently at the site of an old Osage surface mine. Of these settlers, it seems only Campbell had prior mining experience; local accounts stated that he soon gave up on lead, however, to go search for California gold. We know that Cox, Tingle, and Spurgeon were among the very first white farmers to settle in the area. Cox also ran a store and served as a

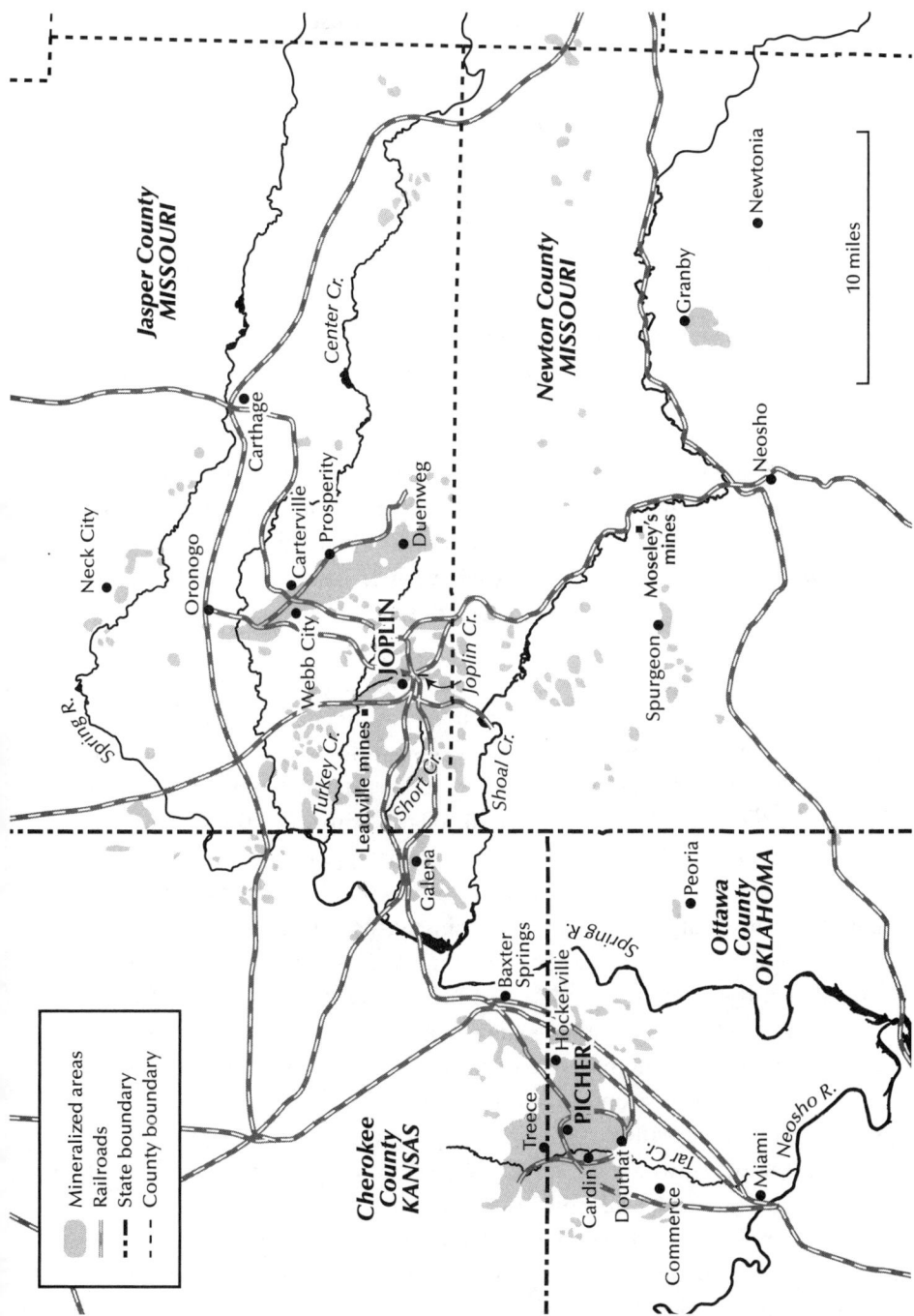

Map 2. The Tri-State district. Courtesy INCase, LLC.

Legend:
- Mineralized areas
- Railroads
- State boundary
- County boundary

10 miles

Jasper County **MISSOURI**

Newton County **MISSOURI**

Ottawa County **OKLAHOMA**

Cherokee County **KANSAS**

Neck City

Carthage

Oronogo

Carterville

Prosperity

Duenweg

Webb City

JOPLIN

Leadville mines

Galena

Baxter Springs

Hockerville

Treece

PICHER

Cardin

Douthat

Commerce

Miami

Peoria

Spurgeon

Moseley's mines

Granby

Newtonia

Neosho

Spring R.

Center Cr.

Turkey Cr.

Short Cr.

Joplin Cr.

Shoal Cr.

Spring R.

Tar Cr.

Neosho R.

local postmaster. In 1850, the census enumerator listed Tingle's profession as "merchandizing" but reported no real estate of any value. McKee ran a store that supplied goods into Indian Territory, worked a small farm, and served as a county judge. No records indicate what work Shepherd and Oldham pursued, although they continued to mine. All of them knew the value of lead, at least as ammunition. With their commercial ties, Cox and Tingle would have known more about its marketability.[8]

Their discoveries revealed a rare geological environment in Jasper and Newton Counties. Shoal, Turkey, and Center Creeks cut valleys across a rich field of surface ore fifteen miles wide and over twenty miles long. Here, where the Ozark uplands descend gradually onto the western prairie, ancient cataclysmic events in the earth left behind shallow, irregular pocket deposits of different mineral forms of lead and zinc, particularly galena (lead sulfide) and sphalerite (zinc sulfide), tossed in a disturbed formation of clay, fragmentary limestone, and conglomerate chert, a fine-grained, silica-rich sedimentary rock. Miners would come to call this shattered formation the "soft ground" because it rested, fifty to sixty feet in depth, on a substratum of solid limestone. Farmers first discovered the soft ground minerals near area streams because the fast-running waters had gradually eroded surface soils to uncover the harder, heavier rocks, among which galena predominated.[9]

The production of lead from surface mines was relatively straightforward. After his discovery, Amos Spurgeon and his brother, John, worked a set of shallow pits, probably no more than twenty feet deep. As one man worked in the ground, the other hoisted the dislodged mineral to the surface with a rope and bucket. They could stop or start as opportunity allowed. Since lead melts at a relatively low temperature, $621.5°F$, easily attainable by burning wood, the Spurgeons smelted the mineral themselves on a log-roasting furnace built into a hillside, like the Osage had. After washing the galena in a sluice, they placed it with a mix of wood chips onto a bed of burning logs in the furnace, a process known as charging. They then stoked the charge through an opening, called an eye, at the bottom of the slope. Once in the furnace, the carbon in the burned fuel combined with the sulfide in the galena to make sulfur dioxide gas, which dissipated in the open air, and molten lead oxide, which the miners drew from the eye into a clay pot. Although not pure lead, lead oxide was a common industrial form used in the production of glass and pigments. A local report estimated that by 1851 the Spurgeons had mined 113 tons of galena, from which their inefficient smelting methods still yielded 40 tons of pig lead worth more than $4,000.[10]

A few settlers had their enslaved laborers do all or parts of the work. The young man John Cox owned did all of the mining on his land, a decision that was notable enough that Cox's neighbors called the mine "Nigger Diggings." In five years this enslaved miner produced five tons of mineral. Cox's neighbors Tingle and McKee likewise used slaves to produce pig lead, although records do not reveal who did the actual mining. Their mines, on either side of Turkey Creek, yielded a total of twenty-five tons of mineral in 1850. Tingle and McKee were more interested in marketing lead, however, and invested in a wood-fired smelter that became the hub of lead production in their neck of Jasper County, soon dubbed "Leadville" by locals. They sold the smelted lead "in the small towns and in the Indian country on our western border."[11]

Most of these settlers aspired to be farmers, however, not miners or lead traders. Some used their lead earnings to buy and expand homesteads. Until 1848, few settlers in southwest Missouri possessed legal land titles. The federal government's 1825 treaty with the Osage brought present-day Jasper and Newton Counties into Missouri as public land. Under the Land Act of 1820 and the Preemption Act of 1841, the white farmers who settled there established a right to purchase up to 160 acres for $1.25 per acre. By the mid-1840s only a few settlers in southwest Missouri had paid for their title deeds, mainly because they lacked cash but also because isolation and sparse settlement made official papers a low priority. In August 1848, not long after they began mining, Amos and John Spurgeon made the seventy-mile trip to the U.S. Land Office in Springfield to buy title to eighty acres each. Further mining allowed the brothers to add to their holdings, which they turned into successful farms. When the census enumerator listed them as "farmers" in 1860, they each claimed property worth more than $3,000. John Cox did likewise. He bought his first title, for forty acres, in January 1849 and continued to add to his holdings, but not his enslaved workforce, over the following decade until he was one of the richest farmers in Jasper County. Cox worked the land himself with his son and one or both of his slaves. In 1860 he claimed $12,000 in property, up substantially from $750 ten years before. Andrew McKee bought title to eighty-four acres of farm and mineral land on Turkey Creek in July 1852. McKee clearly envisioned a future as a farmer because he traded his preemption right to known mineral land along Center Creek in exchange for a wood frame for his new farmhouse. The real or potential profitability of the early mineral discoveries made other settlers keen to convert their preemption rights into title deeds in the years after 1848, whether or not their land contained mineral. Most, however, still bought land for its agricultural promise.[12]

Unprecedented rates of economic growth in the 1850s sparked new demand for lead that would almost immediately challenge these agrarian priorities.

American use of lead increased from an average of 20,000 tons a year in the 1840s to over 40,000 tons a year in the 1850s. Much of this demand came from manufacturers in booming western cities, such as St. Louis, Chicago, Cincinnati, and Milwaukee, where new industries produced pipes, glass, and paint for residential and business construction. Miners in the Upper Mississippi and eastern Missouri fields could not meet their needs. From producing an all-time high of 31,000 tons in 1847, these mines yielded diminishing returns. In 1852 American lead miners produced only 18,800 tons, the lowest total since 1840. In 1855 American manufacturers used 48,000 tons of lead, more than double what domestic mines produced. Prices for smelted lead soared, up from 3.30 cents per pound in 1845 to 4.60 cents per pound in 1850 and 5.75 cents per pound in 1855. Lead miners had never had such a good opportunity to make money or such strong incentives to look for new sources of mineral.[13]

The rudimentary mines of southwest Missouri soon attracted the attention of a motley group of frontier entrepreneurs with bigger ambitions. In 1850 Shepherd and Oldham partnered with George Moseley and his nephew William Moseley to develop their mine. Born in Kentucky, the Moseleys had only recently arrived in Newton County, George to run a store and William to start a law practice. William Moseley entertained grand ideas and liked to write about them. In September 1850 he wrote to the *Western Journal*, a St. Louis–based newspaper devoted to regional economic issues, describing his efforts to exploit "a recent discovery of very rich and valuable lead ore." This was the first published account of successful mining in the area. Over the summer, Moseley announced, he and his uncle, who both lacked mining experience, had expanded two shafts on the Shepherd and Oldham claim. He explained that the "six hands" who worked the mines, each sixty feet deep, had produced "about 100,000 pounds of ore." Moseley did not say who did the work or how they were paid. They might have been slaves hired out from local farmers. In the autumn, the partners, operating as Moseley, Oldham, and Company, built their own smelter, a Drummond-style furnace that used a horse-powered fan to circulate heat above and below a raised platform that held the wood and mineral charge. First developed in 1835 and used widely in eastern Missouri, the Drummond furnace captured more lead than did log furnaces with less fuel. The Moseleys apparently financed the operation themselves; an 1851 credit report did not mention any outstanding debts. According to the 1850 census, the pair owned $11,000 in real estate, as well as three enslaved women and one enslaved man, between them. In exchange for their investment, Shepherd and Oldham gave the Moseleys a share of the land.[14]

Moseley, Oldham, and Company seemed to enjoy bright prospects, as William Moseley explained in 1854 in a second letter to the *Western Journal*. Rising prices in the early 1850s had made it profitable to sell pig lead to buyers beyond southwest Missouri, although it was still too expensive and difficult to ship to St. Louis. In 1851, the company sold lead to New York and Boston buyers in New Orleans at the St. Louis price, 4.3 cents per pound, which took into account onward shipping costs. The company shipped this lead, worth forty-three dollars per 1,000 pounds, by water to New Orleans, first on flatboats through Indian Territory to Fort Smith, and then on steamboats along the Arkansas and Mississippi Rivers. This method was profitable enough by 1852 to finance the construction of a water-powered blast furnace with a 70 percent recovery rate. The company smelted mineral from its own mines and mineral it bought from smaller operations nearby. In four years, Moseley figured that his firm smelted 825,000 pounds of galena that would have yielded around 500,000 pounds of pig lead worth more than $24,000 at prevailing market prices.[15]

Inspired by the Moseleys and rising lead prices, others began investing in larger, more efficient smelters with hopes that they could buy enough galena to make it pay. Between 1850 and 1854, the price of lead in St. Louis rose from 4.00 cents per pound to 6.19 cents per pound, a 55 percent increase. Tingle replaced his log furnace in Leadville with a Drummond. In 1853, William Harklerode built a similar blast furnace near Center Creek in Jasper County. But they all soon ran into trouble. Both Tingle and Harklerode struggled to stay in business. The Moseleys, meanwhile, were found with "heavy debts" in late 1853. With lead prices high, their collective misfortunes likely stemmed from short supplies of mineral. In 1854, G. C. Swallow, Missouri's first state geologist, reported that smelters were raising the prices paid for galena to as high as twenty dollars per 1,000 pounds. For entrepreneurs such as the Moseleys, the decision to pay more for mineral suggests that smelting capacity had outpaced mine production and that there were not enough miners, enslaved or free, with the know-how to find more.[16]

Word of a new lead district in southwest Missouri, however, soon attracted experienced miners to the area. The first professional miners had arrived in late 1850 when Francis Reando and David Sunday, both of whom had mined in Washington County, Missouri, bought titles to public land along Turkey Creek. More followed from the eastern Missouri and Upper Mississippi districts. Although declining production in the older fields set them searching, these miners went to southwest Missouri to take advantage of the skyrocketing price of pig lead and competition between area smelters for mineral. According to Swallow, experienced miners "say they can make more money in these mines, raising mineral at $20 per thousand, than they could in any other mines they have seen." The first wave of skilled miners soon expanded the existing diggings. They did so without the benefit of a scientific geological survey. In fact, when Swallow arrived to conduct his first survey of Jasper and Newton Counties, he relied on "several experienced miners who have worked in the mines of Iowa and Wisconsin, and of the eastern counties of Missouri" to guide him. With area smelters weak and divided, the newcomers soon gained authority in the area's nascent mining industry.[17]

News of these developments attracted the attention of outside investors. Between November 1852 and April 1853, Ferdinand Kennett and John Casey bought 355 acres near known mineral deposits along Turkey and Center Creeks in Jasper County. Born in Kentucky in 1813, Kennett was wealthy and powerful. In 1846, along with his brother, Luther Kennett, and with James White, he formed the St. Louis Shot Tower Company, which controlled lead-smelting and processing facilities in St. Louis and nearby Herculaneum that turned galena into ammunition for federal soldiers stationed at Jefferson Barracks and for settlers embarking for the West. In 1850 Kennett owned fifteen slaves and real estate worth $84,500. Although not as wealthy as Kennett, the Irish-born Casey had developed and operated several lead mines in Washington County, Missouri, since the 1830s. The 1850 census listed him as a farmer with $50,000 in real estate and twelve slaves. He had served as the local postmaster and was also Kennett's neighbor. Their investment in Jasper County eyed future needs. They did not immediately open this land to mining.[18]

Kennett knew that a railroad might one day bring these mineral deposits within easier reach of St. Louis markets. After the state of Missouri chartered the proposed transcontinental Pacific Railroad Company in 1849, surveyors outlined two potential routes from St. Louis to Missouri's western boundary: one going west to Kansas City along the thirty-ninth parallel and another going southwest toward Indian Territory to follow the thirty-fifth parallel. The company chose the northern route and began construction in St. Louis in July 1851. When the U.S. Congress gave Missouri a public land grant to facilitate internal improvements in 1852, the state used a share of the land to help the Pacific Railroad build a second line to follow the southwestern route, which became known as the Southwest Branch Railroad. The state granted the railroad alternating sections of land along the surveyed path from the new town of Pacific to Springfield and on to the state line with a right to preempt any settlers who occupied these sections. Ferdinand Kennett was well placed to take advantage of the plan. His brother, Luther, was not only the mayor of St. Louis at the time but also a director of the Pacific Railroad. No one knew when the railroad would actually be built, but Ferdinand Kennett was convinced that once it was, the mines of southwest Missouri would become more valuable than ever. He bought land there in the weeks before the state legislature officially approved the deal.[19]

While investors like Kennett waited, miners went to work. They read topographical signs and surface rocks for clues as to what lay in the ground. With this knowledge, miners explored the area in the hopes of intersecting previously known deposits or making new discoveries. In 1853, William Foster, a Cornish miner who had recently arrived from eastern Missouri, struck a vein of galena on unclaimed land along Shoal Creek in Newton County, ten miles east of the Moseley mines. It was the richest discovery yet and more would follow. "There are a great number of shafts sunk in many places in this neighborhood," Swallow reported. On Turkey Creek, he found five mines, including those operated by Sunday and Reando, which had produced 260,000 pounds of mineral. On Center Creek, miners worked six new mines that yielded over 400,000 pounds. "From what I could see of the veins and learn of the amount of mineral raised, and from the general satisfaction of the miners," Swallow concluded, "I would judge that mining at the Center-Creek Diggings has been very profitable."[20]

These miners looked to replicate the generous terms that had governed their efforts in the older lead fields of the Mississippi Valley. On unclaimed public land, they assumed unfettered access. Where someone owned the land, they had to negotiate a mining lease, a straightforward proposition between white men. Under these leaseholds, Swallow reported, miners paid one-eighth of the total mineral they raised to the landowner. They were then free to sell the remainder to local smelters at the market rate, which in 1854 was twenty dollars per 1,000 pounds. Far riskier but also potentially far more remunerative than wage labor, this leasing system had governed lead mining in North America since the eighteenth century. The administrators of New Spain first granted leases in what became eastern Missouri as a means to encourage the renewed development of derelict mines originally opened by the French. Subsequent American miners, led by Moses Austin, continued the generous Spanish leasing system because it provided the cheapest, most effective means to encourage mine exploration in such a remote area. In 1807, the U.S. Congress adopted the same practice on public lands, which would cover most of the new lead discoveries in eastern Missouri and, beginning in the 1820s, in the Upper Mississippi field. Under the U.S. regime, miners could secure plots 300 yards square provided that they worked the claim at least once every seven days and paid a 10 percent royalty. Isolation made government enforcement difficult, however; many miners avoided royalty payments.[21]

Although working on public land, lead miners in the old fields claimed entitlement to their diggings by virtue of the risks and effort required to extract the mineral. Their claims rested on the informal rule of "finding's keeping." The leasing system "favored individuals, families, and small companies," according to a local historian, and created a culture of mining in which "miners were intensely interested in small claims," often fiercely so. The system remained in place until 1847 when the federal government began selling public mineral lands in forty-acre lots. The government might have facilitated the consolidation of the lead-mining industry if the yields on those mineral lands had not unexpectedly and almost immediately collapsed. The miners who left the eastern Missouri and Upper Mississippi fields went looking not only for new places to mine lead but also for places to defend and sustain the proprietary claims that the leasing system made possible.[22]

Experienced miners shaped the regime in southwest Missouri to meet their expectations with relative ease. On private land, leaseholds offered benefits to all parties. For landowners, leasing was a means to profit from the growing mineral industry without disrupting their agricultural efforts or risking their own bodies and money. With overextended smelters desperate to buy mineral, meanwhile, skilled miners could earn handsome sums based on their productivity. While leasing required miners to risk investing their labor in ground that might not yield much, their geological knowledge and the apparent richness of the soft ground deposits tilted the odds of making money in their favor. At one Newton County mine, for example, three miners raised 70,000 pounds of mineral in eighteen months. After paying 8,750 pounds to the landowners as a royalty, the miners sold the remainder and shared the proceeds, which totaled more than $1,000 after deducting the cost of tools, fuel, and food. Not all made such high sums. Two miners working another claim shared fifty dollars in profits after mining 3,500 pounds of mineral in three months in 1854.[23]

Whether the newcomers experienced wild success or simply broke even, they had confidence that the prospects, based on the trend of profitable discoveries, would only yield more galena. "Mineral is found over this whole region," Swallow concluded, where "scarcely a shaft has been sunk ... without obtaining mineral sufficient to render the labor profitable." "The day is not far distant," he predicted, "when this will prove to be one of the richest mineral districts in the country." In search of marketable metal, miners focused on extracting only galena despite its natural occurrence among large quantities of sphalerite, the most common mineral form of zinc. Swallow noted that sphalerite, "called Black-Jack by the miners, is almost as abundant as the galena in many of the mines" but that "many thousand pounds have been thrown out with the rubbish." Although techniques for smelting zinc had been developed in Europe and were in use in New Jersey and Pennsylvania, no one west of the Appalachians had mastered the process. While Swallow and others anticipated the future profitability of zinc, it was not yet so.[24]

More and more miners arrived to seek their share of the wealth. By early 1855 prospectors swarmed the hillsides along Shoal, Turkey, and Center Creeks, some working on proper leasehold agreements with landowners, others digging where they liked on land that no one seemed to claim. Among the latter was Robert Brock, an experienced miner who had recently arrived from Wisconsin. Digging along a seasonal creek bed near William Foster's discovery, Brock struck an uncommonly rich vein of galena on uninhabited, seemingly public land. When he inquired about the land, Brock learned that the tract had been included in the 1852 grant to the Southwest Branch Railroad. Since construction had not even commenced, and many doubted whether it ever would, Brock continued to mine, rent-free. But he could not keep his discovery a secret. Word of his find "was electric," one commentator reported. The excitement, according to the *Daily Missouri Republican*, "caused the miners to come pouring in from the 'diggings' which had previously been discovered in this county and Jasper; and, as usual in such cases, began prospecting as near the famous discovery already made as possible." By the end of the year, the stampede, as it became known, had brought several hundred miners onto the land surrounding Brock's discovery. Merchants, saloonkeepers, lawyers, and other backcountry schemers on the make soon followed. Most of the miners joined Brock as squatting prospectors on the railroad's section, none of them with any clear legal right to do so. They called their settlement Granby, perhaps after the town in Connecticut near where, in the 1730s, Samuel Higley had minted Granby coppers, the first American coins, from metal he mined himself.[25]

White men, most of them native born, predominated at Granby. When the census enumerator surveyed Newton County in July 1860, the vast majority, over 80 percent, of the men who gave their occupation as "miner" had been born in the United States. The enumerator considered all but one to be white. Jacob Blackwell, a twenty-six-year-old Missourian, was the only person of color; he was free. Many miners, such as Brock and Foster, had come from the older lead-mining fields. The successes of the skilled, however, also attracted those with little or no experience. These men came from Tennessee, Kentucky, Arkansas, and even as far afield as Alabama and North Carolina.

None owned much property, certainly not slaves. Most of the foreign-born miners came from the British Isles, particularly from Cornwall and Wales, where metal mining had a long history.[26]

In contrast to the early western precious metal camps, where transient single men predominated, most of the Granby miners, especially the more experienced ones, traveled with their families. These men and women organized households with traditional gender roles while pursuing entrepreneurial mining ventures. George and Sarah Benge, who were twenty-six and twenty-eight years old, respectively, came from Iowa, where their daughter had been born. The details of their lives plausibly suggest a pattern of movement through the lead fields of the Mississippi Valley. George was born in Ohio, but Sarah was born in Missouri; the two could have met in the mining camps of eastern Missouri or near Dubuque. Although over thirty years older than the Benges, William and Sarah Linton, who were both born in South Carolina, moved to southwest Missouri with their five children, who had all been born in Illinois. Their neighbors, A. B. and Catherine Fowler, who were from Ohio and Georgia, respectively, had two children under the age of five, who had been born in Illinois. Jonathan and Agnes Tisdall, meanwhile, came from England, but their five children, all under the age of ten, had been born in Wisconsin and Iowa. Although Granby soon featured several boardinghouses, family groups anchored the mining camp by providing homes for single miners. For example, David and Mary Holland, who had two children, boarded four single miners, two born in England and two born in Wisconsin. Other families did likewise. While all shared interests in lead mining, the centrality of family groups revealed longer-term commitments and provided much needed stability and cohesion in a new mining camp. These social groups also reflected the era's ideal of respectable white manhood with experienced miners as paternal figures who were expected to be responsible, industrious, and independent.[27]

The first image we have of the Granby camp, sketched during Swallow's second visit in the summer of 1857, shows two hillsides covered with dwellings and small-scale mines. Although the lithograph does not show us the men as they worked in the ground, it depicts surface methods sufficient for us to imagine what their work was like. In the foreground we see men gathered near what miners called an "armstrong" windlass, a large crank with handles on either end that wound a rope around a central axle. The windlass straddled the mouth of the mine. Everyone and everything entered and exited by means of the windlass and the strong arms that turned it. The shallow depth of the "soft ground" mineral here—from twenty to seventy feet deep—allowed for mining methods powered entirely by human labor. In

Granby, Missouri, 1857. From Swallow, Geological Report, *36–37.*

this pre-dynamite era, miners wielding picks and shovels dug out the galena from irregular deposits that pitched, narrowed, and finally pinched out. They hoisted the mineral to the surface in buckets. The work was difficult, hot, and often dangerous and required cooperation between miners, in small groups or even in pairs. This foregrounded scene is replicated in the image almost fifty times, with each shaft marked by a windlass, of which there are almost as many as there are residences. Whether they lived in log cabins or simple tents, the Granby miners worked their own holdings, close by one another. These crowded conditions encouraged an ethic of rough equality and fairness between mining groups so that they did not interfere with each other's chances to make money. Everyone understood that those chances were good. According to Swallow, miners regularly located chunks of galena "so large that it is found somewhat difficult to raise them to the surface."[28]

Although trespassing on railroad land, they claimed mining rights by virtue of their discovery and labor. Once word of the Granby boom reached St. Louis, representatives of the railroad asserted control. According to one account, the miners, who by early 1857 numbered nearly 1,000, argued that their investment of "labor and industry" to locate and remove "the immense mineral wealth" gave them rights to continued access that preempted any claim the railroad made on the land or on them; in other words, they invoked the older "finding's keeping" principle. When railroad agents tried to collect rent, the miners refused to pay. With enough problems elsewhere, and a long, rough wagon ride between St. Louis and Granby, the railroad did not pursue the issue with much energy. And so, for a time, the miners enjoyed complete control over the mines. Market conditions also favored them. In 1856–57, lead sold for over six cents per pound in St. Louis, the highest price in living memory. An observer estimated in January 1857 that miners had sold 5 million pounds of galena from the Granby land in less than two years.[29]

The sudden productivity of the Granby camp, however, soon exposed the limits of its isolation. Several smelters had opened in the vicinity after Brock's discovery, but they lacked the currency to buy all of the galena being produced. The Panic of 1857 compounded problems by further reducing the amount of money in circulation as well as demand. Soon most of the local smelting outfits were broke; Tingle and Harklerode were gone and the Moseleys "hopelessly insolvent." "Capital is wanted to pay for mineral as it is brought to the furnace," a local physician reported. Those able to overcome the financial obstacle of paying for mineral usually lacked the means to pay the expensive transportation costs required to reach distant markets. "The smelters are generally responsible men, but owing to the great difficulty of getting lead to the river, their means have become exhausted." According to the *Daily Missouri Republican*, "the greatest obstacle to the progress of Granby has been a want of market for their mineral." These structural and financial limitations created a bottleneck that placed sharp limits on profitability. The weakness of the smelters meant that "little money was thrown into circulation," the paper reported in early 1858, and "general hard times with the miners was the natural result."[30]

Southwest Missouri miners and St. Louis–based manufacturers needed one another. As the military and commercial entrepôt to the western frontier, St. Louis provided a growing, lucrative market for manufacturers of all kinds but especially for those of lead-based goods. At first, smelters in Carondelet and Herculaneum processed mineral from the eastern Missouri and Upper Mississippi fields into pig lead for sale to manufacturers in the East. With the growth of Jefferson Barracks in the 1830s and 1840s, then the nation's largest military post, the U.S. Army created local demand for lead, primarily for ammunition to supply the soldiers securing the nation's expansion. Soon the steamboats pushing off the riverfront wharves headed west instead of east. As merchants packed them with goods such as white lead paint, red lead cosmetic rouge, and lead type for printing presses, they widened the market for local manufacturing. The demand for lead goods made in St.

Louis also increased as the population of the city itself grew, from 16,000 in 1840 to 160,000 in 1860. As production from the Mississippi Valley mines collapsed in the 1850s, St. Louis–based manufacturers struggled to access an adequate supply of mineral to meet these demands. The completion in 1853 of the Galena & Chicago Union Railroad linking the mines of the Upper Mississippi field to an eastern shipping route only made matters worse. New, more intense competition over a diminishing resource forced St. Louis firms to seek alternative sources of mineral.[31]

Ferdinand Kennett had planned for this. In June 1857, he used his brother's influence to lease the entire Granby section of land, 640 acres, from the Southwest Branch Railroad. He did so in partnership with brothers Peter E. Blow and Henry T. Blow, both prominent St. Louisans. Politically, the three made an odd combination. Elected to the state senate in 1854, Henry Blow opposed slavery, particularly in his own family. He supported the freedom suit of Dred Scott, whom his parents had owned and Blow himself manumitted in May 1857. His partners, by contrast, were slaveholders; in 1860, Kennett owned fifty slaves and Peter Blow owned six. The pursuit of profit, however, united them. The formation of their new company, Blow & Kennett, consolidated major interests in Missouri lead. Kennett owned the St. Louis Shot Tower Company as well as mineral lands in eastern Missouri and, since 1852, in Jasper County. Henry Blow had built the Collier White Lead and Oil Company into the city's largest manufacturer of paint and pigment and thus its largest commercial buyer of lead. The company bought much of its lead from Peter Blow, Henry's brother, who owned mining interests in Washington County, Missouri, not far from Kennett's mines. The lease Blow & Kennett signed with the railroad gave the company the right to act as the owner for the purposes of mining on the entire tract for ten years. In exchange, Blow & Kennett agreed to pay the railroad a rent of either two dollars for every 1,000 pounds of mineral mined or 10 percent of its value, depending on prevailing prices.[32]

From the outset, Blow & Kennett aimed to buy and smelt lead, not mine it directly, and so needed the skilled miners already at Granby to remain. Soon after signing the lease in 1857, the company offered clear terms to the squatters: they "should continue upon their claims, and work them as heretofore," on a sublease basis from the company. The terms required, however, that the miners sell their ore to Blow & Kennett at the company price, sixteen dollars per 1,000 pounds, and pay rent of two dollars for every 1,000 pounds they produced. The company stipulated that its future price would fluctuate on a sliding scale in proportion to market trends in St. Louis. In cash terms, Blow & Kennett's offer paled in comparison to the price that miners had received in the area only two years earlier, when smelters paid twenty dollars per 1,000 pounds. On the other hand, the offer was now the best they could get after local smelters had collapsed. It also allowed squatters to continue working their claims, thus recognizing their investment of time and labor and creating preemptive rights that defended them against other miners who were continuing to arrive in the area. Furthermore, the deal made the labor of these miners less economically risky because it provided access to a reliable buyer. From the company's perspective, these terms promised a supply of galena at below-market prices, which would compensate for the high transportation costs to St. Louis and rent payments to the railroad. Blow & Kennett sealed its investment in late 1857 by constructing what was then the largest lead-smelting furnace in the United States at Granby. With eight eyes, the steam-powered blast furnace boasted a recovery rate of 80 to 90 percent and cost $20,000.[33]

Despite a moribund market, many miners rejected the offer as a threat to their prerogatives as free men who had invested time and labor. Squatters feared that they could "be driven from their claims and lose the rights which they regarded as belonging to them, and deprived of all profits arising from the working of their lots or claims." They cooperated in defense of their position, as white men were expected to do. "Factions of disaffected miners were consequently formed," one observer noted, and began "calling meetings both private and public in reference to the rights of Blow & Kennett, the legality of which they questioned." They made a skilled case that showed considerable knowledge. The 1852 state law that granted land for the construction of a southwestern railroad stipulated that the railroad company could not dispose of any parcel of land until the actual construction of the line had progressed to within twenty miles of that parcel. Miners argued that the lease agreement violated this condition, since the nearest construction was over 200 miles away. Until the line neared Granby, they insisted, miners had the right to work as if the land was still in the public domain. While apparently happy to lease from those they considered legitimate landowners, the miners echoed older customary claims of "finding's keeping" when they refused to acknowledge the rights of the railroad and Blow & Kennett, even though the company was a buyer with money. The holdouts continued to mine illegally into 1858 while trying to sell their mineral to local smelters.[34]

A mutual desire for Granby's mineral wealth soon brought the company and the miners together. In early 1858, Blow & Kennett sued two miners, John Plummer and Eli Powers, in Newton County Circuit Court for illegally mining 50,000 pounds of galena from the company's leasehold. The company argued that the terms of the lease gave it sole legal control over the land and any mineral found there. The county court, however, sided with the miners, who had argued that the lease was illegitimate and that their labor gave them preemption rights. While miners continued to work their claims at Granby, Blow & Kennett appealed the ruling to the Missouri Supreme Court, which overturned the lower court's ruling in March 1859. The high court declared that the company's lease gave it the rights and privileges of full ownership of the land in question for as long as it lasted. "The miners raved," one local recalled, "but were powerless to overthrow the monopoly." Their resistance, however, did not go unrewarded. Keen to establish good working relations with the miners, Blow & Kennett reiterated its original offer while raising the price for galena to twenty dollars per 1,000 pounds. The deal was too lucrative to pass up. Former squatters who had rich claims but struggled to find buyers now seized the chance to sell galena for a good price. According to one source, the offer from Blow & Kennett "soon brought about a mutual good feeling between the proprietors and miners." Even Plummer and Powers took advantage of the deal. In 1860, the federal census surveyor reported that both men enjoyed substantial wealth: Plummer owned property worth over $16,000, including one enslaved man, while Powers claimed over $7,000.[35]

The accord fueled Granby's economic and social prosperity. When Albert Richardson, a *New-York Daily Tribune* reporter, visited in 1859, he found 2,000 people living and working in Granby, a "rude village ... dotted with log buildings, and like a prairie-dog town, with mounds of red loam gravel and stone thrown up from hundreds of shafts." Where two years before little social life existed beyond the mines and the smelter, a thriving community of families and single men now enjoyed the services of several grocers, merchants, physicians, attorneys, shoemakers, tailors, carpenters, blacksmiths, barkeepers, and hotelkeepers. The place had become respectable. Residents could send their children to one of at least five schools and could even attend a church. G. C. Swallow could hardly believe how much had changed when he returned to southwest Missouri in 1859. "In the fall of 1854," he recalled, "there was not a cabin on the site where Granby now stands with several thousand inhabitants; and only one shaft had been sunk beneath the soil into the rich mineral veins, which are now penetrated by thousands." In the eighteen months to November 1859, the miners working in these shafts, now numbering almost 1,000, produced over 7.5 million pounds of galena, more than five times as much as the entire region had yielded between 1850 and 1854. All "seemed to be agreed," Swallow noted, "that the Granby Mines are the best they have ever seen."[36]

Miners made money and maintained autonomy in the new relationship with Blow & Kennett. Although the smelting company legally held the land, the miners, working individually or in small companies, retained complete authority over the work on their leaseholds, from the windlass down. They financed their operations with backing from local merchants. Take, for example, the progress of Joseph Hopkins, a forty-two-year-old miner in 1860 with extensive experience in the lead fields of eastern Missouri. He first began work on his claim as a squatter. In 1858, however, he signed a sublease with Blow & Kennett. Hopkins then entered a partnership with B. K. Hersey, a local merchant, who provided what was known as a "grubstake," a cash investment to buy supplies and equipment in exchange for a share of the mine's profits, often 10 percent. Hopkins used the money to hire laborers to open new tunnels from the original shaft and to construct wooden rails to speed transportation between the different parts of the mine. Laborers, such as those Hopkins hired, earned $1.25 per day with hopes of soon investing in their own sublease. From April 1858 to November 1859, Hopkins, Hersey, and Company mined over 1.2 million pounds of mineral. Meanwhile, William Frazer took a succession of grubstake partners, including Hersey and Hopkins, as he developed his claim. He mined more than 1 million pounds of galena in 1859, which amounted to an average monthly profit of $1,400 after paying expenses and his partners.[37]

Miners performed hard, dangerous labor in pursuit of such achievements. After descending seventy feet down a shaft while standing in a bucket and clinging to the lone rope, Richardson toured a "labyrinth of passages, at times not more than two feet high," where he saw miners "lying flat upon their backs, digging [galena] with picks" and "perched high in a gallery, breaking off the blocks and rolling them down." "Sometimes there are huge masses nearly pure," he explained, "again it is mingled with flint rock; and again the vein seems to run out, but re-appears in unexpected directions." The miners he met embraced the risks involved, both physical and mental. They "sometimes obtained no reward for many days, and again cleared a hundred and fifty dollars a week." One miner admitted that the pursuit of mineral wealth was "a slave's life," but he "was unable to content himself in any other pursuit."[38]

Enslaved to what? These miners bound themselves not to Blow & Kennett but to the vicissitudes of the market price of lead, which determined the value of mineral. With prices rising to unprecedented levels in the 1850s, they invested hard, physical effort in their own ability, often honed by experience, to locate and exploit the area's rich but often fragmented deposits of galena. Although Blow & Kennett diminished some of the freedom that miners had enjoyed in southwest Missouri, these white men could still pros-

Miners at work, Granby, Missouri, ca. 1859. From Richardson,
Beyond the Mississippi, *211.*

per from their skill as workers—slavery's opposite. A life committed to mining metal was indeed hard, but the potential rewards—independence, respect, and profit—equaled any that free men with little property could hope
to attain in 1860 and were difficult to resist.

Enough miners experienced sufficient success after 1855 to sustain the
field's lucrative and democratic promise. Frazer "came to Granby poor," an
1859 credit report noted, "but has made it is said some money." He invested
in a store while he continued "makg money mining." George and Sarah
Benge, meanwhile, claimed personal assets worth more than $4,600 in 1860.
More common were mining families like A. B. and Catherine Fowler, who
possessed property worth $150, or David and Mary Holland, who claimed
assets worth $200, or Jacob Blackwell, who reported $125 in personal property. A census survey of ninety-nine Granby miners revealed forty-two with
property worth an average of $279. Of those forty-two miners, six owned as-
sets worth $1,000 or more, five owned $500 or more, eleven owned $100 or
more, and twenty-two owned $25 or more. Most of these people were decid-
edly not rich, but their accumulations of wealth were not insignificant when
in 1860 the average gross annual income was only $297. Fifty-seven miners
in the sample, however, owned no property at all.[39]

The future looked bright. The success of Granby inspired more mining in Jasper County. In 1859 Swallow revisited the cluster of mines along Center Creek. He found only thirty miners at the small camp, which they called Minersville, but their pace was accelerating. The three most profitable mines were all new since his previous visit. William Orchard's success there demonstrated that good mineral land could be found outside Granby. By 1860 Orchard claimed over $2,000 in personal wealth. Meanwhile, at Leadville, miners continued to work the land leased from Cox and McKee. Swallow reported that even though their mines extended down only about fifteen feet, the deposits "have been worked with success and profit" and were "much esteemed by the miners." Although no production figures survive, geologist Arthur Winslow estimated that the Jasper County mines produced around 100 tons of galena a year in the 1850s. Still, the surging productivity at Granby led the way. In 1860, according to one report, Blow & Kennett bought over 7 million pounds of mineral, 28 percent more than the year before, making the southwestern lead field the richest in the state of Missouri.[40] Those riches would bring unforeseen dangers as the secession crisis gave way to national civil war.

The border conflict that bled Kansas and Missouri farming communities in the late 1850s had barely registered in Granby and Leadville. In the election of 1860, voters there chose sectional conciliation; Northern Democrat Stephen Douglas carried Newton County, while Constitutional Unionist John Bell won Jasper County. Southwest Missourians soon felt the effects of the Civil War, however, due to Governor Claiborne Jackson's efforts to join the Confederacy in 1861. After failing to seize Federal installations in St. Louis, Jackson fled with his rebellious Missouri State Guard toward Arkansas, which had seceded in May. A U.S. force met Jackson's army in July near Carthage, in Jasper County, in what many considered the first major land battle of the war. Jackson's tactical victory led to further attempts by Federal forces to drive his army out of the state, most notably at the Battle of Wilson's Creek in August 1861, where the Missouri State Guard and the Confederate Army of the West repulsed the U.S. effort. The Confederate victory bolstered popular support for Jackson among Missouri secessionists and gave him tenuous control over the southwestern counties. In October 1861, Jackson and secessionist legislators convened in Neosho, the seat of Newton County, to reestablish what they considered to be the legitimate government of the state. Soon after, this group passed a secession ordinance, which the Confederacy recognized in November, and declared Neosho the capital of Confederate Missouri. While the main Army of the West set up its winter camp forty-five miles to the south at Camp Jackson, Arkansas, the Missouri

State Guard remained in Newton County. The combined force of more than 16,000 soldiers gave the Confederacy control of the surrounding countryside, including Granby.[41]

Rebel leaders soon realized the strategic fortune of their makeshift Newton County headquarters. While Granby occupied "a range of bare, desolate, bleak-looking hills," an amazed Louisiana soldier observed, "the mines are the richest in the known world." The Confederacy needed lead for ammunition but lacked mines of any significance west of Virginia. An agent for the Missouri State Guard offered to buy lead from Blow & Kennett, but Peter Blow, who was in sole charge of the Granby operations following Ferdinand Kennett's death earlier that year, refused "to sell a solitary pig of lead or anything else belonging to the firm to the *so-called* Southern Confederacy." Undeterred, the Missouri State Guard took forcible control of the mines and the smelter. News of the seizure worried metal industry leaders. "We are very sorry to learn that the richest lead mine in Missouri, and, indeed, probably on the globe, is now in the hands of the insurgents," a mining journal noted in late 1861. "With the mines and furnaces at Granby in their possession, the rebels can supply themselves with lead to any required extent."[42]

For that to happen, however, the Confederates would have to master the same problems that had bedeviled southwest Missouri miners since the late 1840s. The first involved transportation. The Rebels captured 32,000 pounds of pig lead and well over 100,000 pounds of unsmelted mineral when they took Granby. They chose to ship it to Fort Smith on wagon supply trains through the Ozarks, a rugged journey of over 120 miles. Secretary of War Judah Benjamin believed that the quartermaster could haul 200,000 pounds of lead a month this way. The Confederates never reached Benjamin's target.[43]

Although Jackson's army had the mines, it did not have many miners. Thousands of people fled southwest Missouri during the violent events of 1861, including Peter Blow and most of the miners, who scattered, many to fight, some for the Confederacy, others for the United States. By the time Benjamin began planning to receive Missouri lead, few experienced miners or smelters remained to produce it. Aware of the problem, the Confederate government contracted a Memphis company in December 1861 to work the mines with enslaved labor, but the new labor force never arrived. In February 1862, the U.S. Army of the Southwest defeated the Army of the West at the Battle of Pea Ridge, Arkansas, which pushed regular Confederate forces out of the area.[44]

Despite the Federal victory, Blow & Kennett struggled to overcome war-related disruptions and restart operations at Granby. While Kennett's death in May 1861 gave the Blows sole control of the company, President Abraham

Lincoln appointed Henry Blow ambassador to Venezuela that summer. To complicate matters, U.S. forces maintained a loose hold on the area that allowed guerillas sympathetic to the Confederacy to make periodic raids on Granby. What these raiders did not seize, Federal soldiers did. Between guerilla attacks in the summer of 1862, the Army of the Southwest requisitioned 1,182 pigs of lead, enough to fill forty-six wagons. Its officers promised to pay later. The threat of violence and the risk of working for little or no payment demoralized the remaining miners. "There are a few miners still holding on waiting in daily expectation of receiving some encouragement from you," a company agent reported to the Blows that July, "but nothing is being done in the mines." Henry Blow made direct requests to President Lincoln for a formal contract to supply lead for the military effort if only more soldiers could be deployed to protect the operations but to little effect. Rather, weak Federal authority in southwest Missouri encouraged a small Confederate force to attempt to retake control of Granby, which culminated in a major battle in September 1862 at Newtonia, seven miles east of the mines. The Confederates won that battle but could not hold the position.[45]

The muddled outcome of the Battle of Newtonia marked the beginning of a new period of intense guerilla raiding and casual violence that brought life and work to a standstill for the duration of the war. In the absence of any firm military control, secessionist guerillas established zones of control from which they frequently harassed Federal soldiers and area residents. The largest, most notorious guerilla band, led by Thomas Livingston, operated from a base near Minersville on Center Creek, where Livingston and his business partner William Parkinson had developed small, paying mines and built two blast furnaces just before the war. Able to provide their own ammunition, these guerillas controlled the western half of Jasper County until Federal soldiers killed Livingston in the summer of 1863. By then most of the prewar residents had fled. Mining stopped and did not begin again until 1865.[46]

When people returned in the final months of the war, they discovered towns and homesteads overgrown and in disrepair, if not completely destroyed. Granby was in ruins, the Blow & Kennett smelter a wreck. During four years of more or less constant violence, one correspondent reported, "the mines were unworked, the miners had sought other employment, and the fields were permitted to go to waste."[47]

The war exacerbated the problems of lead mining in southwest Missouri, but it accentuated the opportunities as well. After the Confederate surrender, Blow & Kennett not only rebuilt but expanded its operations at Granby, incentivized by the price of lead soaring to ten cents per pound in 1865, almost double the value of the fat years of the 1850s. Before the guns had even gone silent in April 1865, the Blows reorganized Blow & Kennett as the Granby Mining and Smelting Company. Investors included some of the leading figures in the triumphant Republican Party, including warship builder James B. Eads; Barton Bates, who was Attorney General Edward Bates's son; and St. Louis banker Thomas Dickson. Granby Mining used the fresh capital to begin repairing the smelter, despite legal uncertainty over the status of its lease with the Southwest Branch Railroad, which had gone bankrupt during the war and was in foreclosure with the state. To solve that problem, the firm used its pooled political clout to convince Governor Thomas Fletcher, a Republican who came from the lead-smelting town of Herculaneum and had worked as a land agent for the Southwest Branch in the 1850s, to extend its lease another ten years. When John C. Fremont, the Republican Party's 1856 presidential candidate, bought the railroad in 1866, he named Eads an incorporator. Although still 140 miles short of Newton County, the reorganized and renamed Atlantic & Pacific Railroad promised the long-sought link not only to St. Louis but also to eastern cities via Eads's proposed bridge across the Mississippi.[48]

With profitability nearly assured, Granby Mining now more than ever needed the kind of skilled, experienced miners who had transformed the district before the war. To attract them, the company offered the same liberal enticements of autonomy and profit-making potential. "This company expect," the *Daily Missouri Republican* reported, "by large outlays, patience and enterprise, to make Granby the most attractive spot in the United States for industrious and skillful miners." In October 1865, a correspondent who wrote under the name E. Pluribus Unum reported that the company was buying mineral for thirty dollars per 1,000 pounds, 50 percent more than it had paid five years earlier.[49]

The opportunity to make money reunited white miners. "You can now see number [*sic*] of windlasses at work all over Granby," he said, with "'Rebs' and 'Feds' all mixed together ... delving in the bowels of the earth in search of greenbacks instead of fighting for it as they have done for the past four years." "Miners," another observer urged, "this is the place for you."[50]

CHAPTER 2
THE FAVORITE OF FORTUNE

As Gilded Age Americans fashioned a new industrial society out of iron and steel along transcontinental railroads, suspension bridges, and unspooling lines of barbed wire, they met an old menace: rust. "The decay of iron and steel by corrosion," metallurgical engineer Alfred Sang warned, "is far more rapid than that of wood and other materials of construction." Manufacturers and builders fought it with lead paint and, after 1870, a newer technique: zinc galvanization. Unlike a coat of paint, which needed regular reapplication, a galvanizing coat of zinc on iron or steel offered decades of protection. More and more manufacturers offered galvanized products as a premium alternative to bare metal goods. At the 1876 Centennial Exposition in Philadelphia, for example, exhibitors displayed galvanized iron in sheets and pipes, fences, railings, gates, and less obvious items such as birdcages and poultry nets. Visitors browsed most of these displays in the Sheet Metal Pavilion, itself made with galvanized iron. Elsewhere, the Brooklyn Bridge, the rising icon of modern America in the 1870s, featured almost 2,000 steel suspension wires protected by "the newly established art of galvanizing." Even paint manufacturers took advantage of zinc's anticorrosive properties by using zinc oxide to make a new type of rust-resistant paint to compete with lead-based products.[1]

The Granby Mining and Smelting Company and those who returned to work on its lands after 1865 reaped the benefits as demand for lead and zinc continued to grow, even after other commodities lost value following the Panic of 1873. Manufacturers used twice as much lead in 1878 as in 1868, which held the price of pig lead above the highest prewar levels for the first three years of the ensuing national economic crisis. In 1878 manufacturers used 17,781 tons of zinc, a fivefold increase since 1868 that gave zinc ore new value that continued to rise despite the depression. Initially, however, miners and smelters in southwest Missouri looked only to lead. Miners found zinc in its three main mineral forms—sphalerite, also known as blende and "black jack" (zinc sulfide); calamine (zinc silicate); and "dry bone" (zinc carbonate)—in considerable quantities around, and often mixed with, almost every run of galena. Although everyone knew what zinc ore was, no one knew what to do with it until the late 1860s, when a new smelter opened near St. Louis. Working miners, not Granby Mining or the leasing companies that followed it, would be the first entrepreneurs to take advantage of the growing market for zinc ore. Their success not only sparked the most frenzied period of prospecting yet but also reopened old questions about who controlled the mines.[2]

Like their antebellum counterparts, the white miners who came to the district after 1865 wanted to work for themselves on an independent basis, preferably as owner-operators, and not as permanent employees of one company. Although Granby Mining offered generous leasehold terms to attract miners, it also ultimately considered them its employees in what resembled the inside contractor systems used in contemporary factories and most coal mines.[3] Some miners found the relationship with Granby Mining too lucrative to leave, while hundreds of prospectors used it to accumulate capital to finance their own independent mining ventures. The emergence of outside buyers of zinc ore after 1870 provided a new, seemingly depression-proof market incentive for doing so. So did the gradual weight of the depression on the lead industry, which led Granby Mining and new leasing companies to tighten contract terms. While a few miners responded to company control with violence, others explored antimonopoly politics. Most, however, looked to escape the company grip, not fight it. As prospectors fanned out from developed holdings, they discovered the richest ore and mineral deposits yet across a wide swath of Jasper and Newton Counties and into Cherokee County, Kansas. Their finds launched dynamic mining camps—Joplin, Webb City, Carterville, Oronogo, and Galena—that made the district into a national leader.

Successful prospectors in turn created dozens of small mining companies that employed a new generation of miners. They built thriving communities that fostered miners' ambitions and their sense of white social equality, particularly between men—in contrast to paternalistic company towns that often defined extractive industries. These achievements made the mining district famous as a poor man's camp, a place where men with little means could still make money on their own account. At a time when some workers across the country organized in resistance to the logic of industrial capitalism, often in spectacular, violent strikes, miners in southwest Missouri deepened their commitment to the risks and rewards of the metal market and the ideals of the poor man's camp.

The Granby Mining and Smelting Company led the reconstruction of the district in the late 1860s to take advantage of the dynamic market for pig lead and the lead-based products of its affiliated companies in St. Louis. The galena found at Granby yielded lead of uncommon purity and softness that makers of white lead paint valued above all other grades. By the early 1870s, traders in New York considered the "soft Missouri" grade "superior to any refined lead produced anywhere in this country or in Europe." Granby Mining, with its powerful connections in the lead industry, the state government, and the region's burgeoning railroad network, stood to make handsome profits. Miners also looked to capitalize on these favorable conditions, first by working on Granby Mining land but increasingly with an eye to mining for themselves.[4]

In order to attract miners to work its holdings in the late 1860s, the company offered advantageous contract terms very similar to those that had gained the loyalty of prospectors in the 1850s. In notices that appeared weekly in area newspapers, the company invited miners to return to their claims on the same basis as before the war: they would have freedom to work as they saw fit so long as they sold their mineral to the company smelter. When Granby Mining purchased hundreds of acres near Minersville in Jasper County after 1865, however, it needed prospectors to explore these new holdings, where shallow deposits ran as irregularly and unpredictably as in Granby. To re-create the kind of entrepreneurial prospecting that had worked so well in the 1850s, the company advertised "the greatest inducements to all working men, seeking remunerative employment, and healthy incomes." Granby Mining would pay prospectors one dollar per foot to dig exploratory shafts; miners in the 1850s had either assumed this cost themselves or sought third-party financing. If the prospector discovered mineral, he could then work the claim on a contract basis by returning the digging wages. In exchange for unfettered access to the claim, the miner agreed to sell all of the galena he raised to the Granby Mining smelter at the contract price, thirty dollars per 1,000 pounds in the late 1860s, which followed the market price of pig lead, and to pay a rent of two dollars per 1,000 pounds of mineral. Prospectors who dug shafts but did not hit mineral kept their wages. "This offer," recruiting literature explained in 1868, "guarantees to each miner a support—whether he discovers mineral or not, and in every case where mineral is found, secures him the actual benefit of the profits of mining."[5]

The recruitment efforts worked. "There are a goodly number of experienced and energetic miners returning to Granby to seek their fortunes," an observer reported in late 1865. William Frazer and Joseph Hopkins, both of whom had mined over a million pounds of mineral in the late 1850s, were among them. So were miners who had not yet achieved such success. Martin

Jarrett, who had lived next door to Frazer in 1860, did not make much money before the war, but he returned to Granby with his wife, Nancy. Now thirty years old, Jarrett resumed mining alongside a new partner, Jasper Moon, with a grubstake from Hopkins. Moon grew up around lead miners in the eastern Missouri field in the 1840s. In the fall of 1866, Moon and Jarrett produced 96,000 pounds of galena. Others achieved similar results. "The amount of lead ore being taken from the ground at the present time far exceeds that of any time previous," a correspondent declared, "and the prospect looks flattering for the discovery of larger deposits than was ever thought of in the first opening of the mines in 1855–6." Although this report exaggerated the comparison, the results suggested that contract miners found the company's terms agreeable. In 1867, Granby Mining bought 2.5 million pounds of mineral from its miners. "Granby is flourishing," Henry Blow informed his partner, James B. Eads, the following year, because "the prospecting was really like that under Blow + Kennett." The place looked the same. A visitor in the fall of 1867 noted the "queer combination of people and pits, houses and hills, business and brush, called Granby." "All around," he said, "the face of the earth is scarred by pit-holes, and streets and roads and buildings have to give way to the requirement of the original idea of Granby—lead."[6]

White men with experience and property predominated at Granby, where they were well positioned to take advantage of the terms of sublease mining. Of 126 miners listed in the 1870 census, eighty reported owning personal property or real estate, with most claiming holdings worth between $50 and $500. Six miners owned property worth more than $500. Although these miners owned less property than their farming neighbors, their holdings compared favorably to those of other men their age, whether farmers or skilled workers. Records do not indicate who had worked in Granby before the war, but most probably mined lead somewhere in the Mississippi Valley. Bazil Meek, for example, mined in Granby in 1870 with his sons, Robert and John, who had been born in Illinois in the late 1830s; they claimed property worth $100. Meanwhile, John Trevaskis, who valued his personal estate at $300, was born in Cornwall in 1840 but had likely worked in more than one American metal camp. Many of these miners, like their prewar counterparts, were married with children. The majority, perhaps two-thirds, were native-born whites, although many of these had at least one foreign-born parent; most of those born abroad came from England, Ireland, Cornwall, or Wales. The 1870 census returns for Jasper and Newton Counties listed only one African American miner, Thomas Walker. Born in Missouri in 1841, Walker mined in Granby and claimed fifty dollars in property. Records do not reveal whether Granby Mining actively avoided subleasing land to African Americans after 1865, but we do know that almost half of the 827 African Americans who lived in both counties in 1860 had since left. Many fled during the violent and chaotic war years. Those who stayed had to deal with former owners and many former Confederates. As mining restarted in the district, then, white men with common social backgrounds and national origins enjoyed almost exclusive access to its opportunities.[7]

The sense of democratic openness was greatest at Minersville, where prospectors were younger, possessing far less working capital than their counterparts at Granby. A visitor in early 1870 found 500 residents in the camp, many with prior mining experience. For example, John Sergeant and Elliott Moffett left Wisconsin, where they had mined lead together, in 1867 to seek better opportunities in Jasper County. Michael Brady, who was born in Ireland in 1836 and had also mined in the Upper Mississippi field, did likewise with his Iowa-born wife, Amanda, and their two young children. Re-creating a prewar pattern of generational support, the Bradys boarded Wisconsin-born James Cummins, a seventeen-year-old miner, in their new home. Unlike the miners at Granby, however, those seeking claims at Minersville had little property or savings. Of thirty-seven miners listed in the census, thirty-one, including Moffett, Sergeant, Brady, and Cummins, reported owning no property at all. The opportunity to prospect for Granby Mining offered them a new chance to establish their independence. That opportunity also attracted young men from the surrounding countryside who had no mining experience. Missouri-born brothers Benjamin and Thomas Holmes, who were nineteen and sixteen years old, respectively, in 1870, both mined but had no property between them. It made sense that miners with the most skill, experience, and past success gravitated to the developed Granby field. By contrast, the miners who prospected in Minersville lacked the capital, equipment, or connections to gain access at Granby. Instead, they listened to the assurances of Granby Mining that workers with no experience could discover and exploit lucrative new claims. According to company ads, "good laboring men of steady habits soon learn to mine, and will find it profitable as the mineral is readily found in paying quantities." The company had no trouble attracting workers to Minersville but wanted them to deliver more galena. According to a visitor in 1870, "the general opinion prevails, that richer deposits of mineral lie below any depth yet reached." To encourage the pursuit of those deposits, Granby Mining ran a contest that offered a $500 bonus to the miner or miners who produced the most mineral at Minersville in a three-month period. In July 1870, Moffett and Sergeant won the award.[8]

Like many of their counterparts, Moffett and Sergeant wanted to work for themselves. They used the windfall to finance a search for a new, independent claim. In August 1870 they leased two sites along the east bank of Joplin Creek, one from John Cox and the other a few hundred yards to the south on land owned by Oliver H. Picher, a local judge. At first, their prospecting did not go well. According to local legend, Moffett and Sergeant exhausted their supplies and their money with little success before borrowing enough blasting powder "for one more shot before abandoning the effort" on the Cox leasehold. That last shot, the story goes, revealed a rich vein of lead mineral at the thirty-five-foot level. Now they had enough money to keep digging. In the three months that followed, Moffett and Sergeant made $60,000 from what became known as the "discovery shaft." News of their good fortune spread fast.[9]

Hundreds of prospectors followed Moffett and Sergeant to the banks of Joplin Creek. Although rich, the deposits were not continuous, so no one knew where the next discovery might be made. The only way to find out was to dig. By the summer of 1871 miners worked both sides of the valley for three miles from where it joined Turkey Creek. Many of them came from Granby or Minersville, but others came from area farms after hearing "the news of rich strikes of lead near the surface." Like Moffett and Sergeant, they all sought their own paying claim, free of company control. At first, they leased the right to dig from one of the three landowners in the valley: Cox, Picher, or George Porter. There in July 1871, in the midst of trees and fields, observers from Carthage, the Jasper County seat, found tents and "little mounds of earth scattered over the hillsides" where prospectors worked in pits that differed little from those seen in Granby in 1857. Next to each mound, a visitor reported, "there was a hole in the ground just like an ordinary well" surrounded by "a windlass, rope, tubs and picks." Two men worked this mine: one "down in the shaft shoveling around in the mud and water" while "the other man worked the windlass." Miners throughout the valley had discovered galena as shallow as six feet, usually in the fragmentary patterns seen elsewhere in the area. These two "were enthusiastic concerning their prospects for an early discovery of the mineral," the visitor noted, although he confessed "infidelity concerning their bright anticipations."[10]

No one predicted that these anticipations would in fact prove too conservative. Sometime in 1871, Jasper Moon, who had recently relocated from Granby, struck shallow deposits of galena, many of them the size of small boulders, on Cox's land. Over the next two years, miners would take over 5 million pounds of mineral from the Moon Range, as it became known. Many prospectors met similar success. William Swindle, a farmer from nearby Sarcoxie, spent the summer of 1871 digging on Porter's land before he found galena. For almost two years he made an average of $10,000 per month from his mine on what was soon known as Swindle Hill.[11]

When the dismissive visitors from Carthage returned a month after their first trip, they expected to see nothing but the remnants of fizzled dreams, maybe "four or five shafts." "Contrary to our expectations," however, "we found four or five hundred men, and plenty of shafts." Nearly all of the miners were making money, some "from forty to fifty dollars a day." "There was one shaft that had six hundred dollars' worth of ore lying beside it." The camp "has lead in unlimited quantities underneath it," another report stated, and "the sound of the shovel and the pick is heard daily in the bowels of the earth." "Some of the miners are making small fortunes every week." Stories like these attracted hundreds and then thousands more miners to southwest Missouri.[12]

As prospectors dug more mineral from the ground, they created opportunities for upstart companies to challenge Granby Mining. So long as Blow's firm operated the only large-scale smelter, it retained control of the regional lead market. Prospectors might have developed their own independent mines in 1870, but they still had to sell their galena at Granby Mining's price. That changed in 1871 when Moffett and Sergeant opened their own smelter along Joplin Creek. For several months the former prospectors bought all of the galena produced in the valley, "about $1,000 worth of lead per day." Others soon joined them. Patrick Murphy and W. P. Davis, who owned a freight business based in Fort Scott, Kansas, came in early 1871 to sell supplies. They soon made enough money to buy land from Picher and build a second smelter. The completion of the Atlantic & Pacific Railroad to Newton County in December 1870 made these companies viable. Unlike in the 1850s, smelting companies could now ship pig lead on trains that ran twice a day to St. Louis where it sold for over seven cents per pound in 1871. The new smelting companies initially created a more competitive market for galena that benefited miners. Their production and prospecting soon attracted more investors in land and smelting facilities. As other companies emerged to compete with Granby Mining, whether founded by former prospectors or by investors with purely commercial backgrounds, old questions about the relationship between miners on one side and smelting and land companies on the other became more acute.[13]

As in Newton County in the 1850s, the success of prospectors attracted investors who wanted to secure primary mining rights on land in the Joplin Valley. In late 1871, John H. Taylor, a lawyer from Kansas City, bought the land surrounding the "discovery shaft" from Cox. Around the same time, three merchants from Hannibal, Missouri, leased part of Picher's land. Likewise, Samuel Corn, a local merchant, and John Wahl, a St. Louis–based lead trader, took out mining leases on land along Joplin Creek but also elsewhere in Jasper County and near the original Spurgeon mines in Newton County. Unlike Moffett and Sergeant, all of these investors, including Davis and Murphy, were interested primarily in selling pig lead, not mining. They formed leasing and smelting companies, patterned on Granby Mining's contractual practice, to govern production on their holdings. Facing new competitors, Granby Mining expanded by purchasing several hundred more acres near Minersville and at Leadville on Turkey Creek. The sudden growth of mining activity and smelting companies prompted investment in a new railroad that connected the Atlantic & Pacific to Minersville with future construction aimed toward Kansas. The profitability of these companies, as always, depended on the productivity of the miners.[14]

Miners and prospectors eyed the proliferation of leasing and smelting companies with suspicion. In February 1872, eighty "Citizens and Miners of the Joplin Lead Mines assembled in mass meeting" to defend their access to the land and authority over the work. Their declaration focused first on preventing investors from monopolizing land to ensure that new prospectors would have a fair chance: "No Miner shall hold more than one lot, nor company of miners more than one lot to every able-bodied man." The second article stipulated that contract miners with claims must work their lots themselves, not subcontract work to another miner, and that they must employ "one able-bodied hand for each lot so held." This provision not only attempted to exclude speculators who might consolidate control of the field but also pushed successful miners to redistribute some of their earnings to wage hands, who in turn might become prospectors themselves. The assembly agreed that any miner or company of miners who violated the first two articles or left their claims unworked for three weeks should forfeit their claim. The remaining articles called for the formation of a jury and the appointment of a miners' magistrate to enforce the above code and arbitrate any disputes. This attempt at self-government closely resembled similar efforts by miners to maintain democratic principles in the first years of the California gold rush, including the provisions to restrict the number of claims one miner could hold. It reflected the concerns of working miners who intended to preserve the camp's opportunities for men like them, then and in the future.[15]

In addition to these rules, the miners also passed a set of resolutions that proclaimed their independence from the leasing and smelting companies.

The first asserted "that it is the inherent right of the Miner to prescribe the terms and conditions of mining on ground once thrown open to mining work." This concept of "miner's freedom" was widespread among American miners, whether in Missouri lead, Pennsylvania coal, or Nevada silver, and had roots in southwest Missouri that extended back to the wildcat miners who resisted the impositions of Blow & Kennett in the 1850s. It was especially meaningful in the early 1870s for miners and prospectors who, in their effort to escape the authority of Granby Mining, had discovered and developed mineral land in the Joplin Valley only to find themselves facing possible submission to new companies seeking to restrict their ability to sell on the open market. "We hereby wish it distinctly understood," their first resolution concluded strongly, "that we deem ourselves grossly wronged by certain persons claiming the right to buy our mineral at their own prices." They also resented landowners who demanded high royalties. "We deem ten per cent of the price of mineral sufficient rent," another resolution stated. They wanted their work governed by market incentives.[16]

Although some of their demands resembled those made by miners elsewhere, the Joplin miners differed in their emphasis on the profit motive. It was "the duty of the Miner," the second resolution declared, "to sell his mineral to the man who will pay the highest price for it." In the 1870s, every organization of miners made demands about remuneration, whether coal miners paid by the ton or precious metal miners paid by the hour. Few others, though, defined their duty in terms of seeking the maximum proceeds available through competitive trading. In doing so, Joplin miners made a powerful argument about how the hard work of prospectors who sought their own economic benefit had led the discovery and development of the richest mineral deposits in the district. They had assembled to defend and preserve that lucrative system and their "just rights" in it from the impositions of others who did not mine. However, their statements revealed a common interest that worked against any organization more formal than this assembly of individuals: a fair means for men to seek their independence through economic risk. They punctuated these claims with a public assertion of authority as white men by quoting President Andrew Jackson's first State of the Union address: "We will ask nothing but what is right, and submit to nothing wrong."[17]

The new leasing and smelting firms accommodated some of these demands but not all. Most settled on a form of the contractual model pioneered by Granby Mining to harness the ambition of a miner to work "at his own risk and expense." According to an 1873 survey of the district, these companies leased or subleased plots 200 feet square to prospectors working in pairs or small groups. Contrary to the miners' resolutions of 1872, some firms offered a strict contract that required miners to sell their mineral to the company at a discount. Others, however, offered a more flexible contract that allowed miners to sell to another smelter in exchange for a royalty, which varied from one-eighth to one-fourth of the value of the traded mineral. In most cases, companies financed prospecting directly by paying miners a piece-rate wage for opening new ground. To discourage miners who might be tempted to neglect their sublease in order to prospect elsewhere, companies included a provision under which any miner who did not work his contracted claim for a period of time, usually fifteen days, forfeited it. In an attempt to lure prospectors back to its lands from Joplin, Granby Mining offered a substantial $100 bonus to miners who opened new exploratory shafts. Only one company strayed from the subcontracting model in the early 1870s: Corn and Wahl "worked the land they control at their own risk by hired labor." The company employed 50 to 100 wageworkers between 1871 and early 1873. The model apparently did not deliver satisfactory results, however. In late 1872 the firm began offering contracts that paid prospectors on some of its less developed lands fifty dollars per 1,000 pounds of mineral, a startling high price. By the summer of 1873 Corn and Wahl had adopted the subcontracting system on all of its holdings. Whether or not there was consensus on contract terms, the lesson of Granby still held: "Prospecting makes prospects, and good mines cannot be found without it; and in this work no parties in the whole West surpass the Joplin miners and companies."[18]

Joplin's reputation as a poor man's camp grew as its promises repeatedly came true. Between August 1870 and the end of 1873, miners in Jasper County sold over 50 million pounds of galena that yielded pig lead worth an estimated $2.45 million. Meanwhile productivity was continuing to rise — in 1873 alone, miners in Joplin and Minersville produced over 19 million pounds of mineral. Local newspapers filled their pages with stories of success. In May 1873, for example, readers learned about "a couple of young miners named Richardson and Cody" who "at last struck a rich prospect" from which "about 500 pounds of a superior quality of ore was raised in a few hours' time." "The discovery was made," the report stated, "only after much hard work and great perseverance." The following week the same newspaper ran a story about Bradbury and Simpson, "two industrious miners" who struck a piece of galena four feet long and three feet wide in their shaft on the Moon Range. "All day the miners worked hard and faithfully, overcoming one obstacle after another" until "their labor and patience was finally rewarded by seeing the monster safely landed." What did their faithful efforts deliver? "The largest chunk of lead ore which has ever been brought to the surface of the ground," the report declared, "the greatest strike in the mines!"[19] In February 1873, a St. Louis newspaper commended "the pluck and the industry" of Joplin's "poor miners." "While they have lacked the capital they have possessed abundant nerve," the report concluded, "and this has pulled them through." Of course, landowners and smelting companies made a lot of money too. John Taylor reported weekly profits of $1,600 in late 1872; his companies had made over $50,000 in profit the year before.[20]

Aspiring miners arrived by the hundreds in Jasper County every month. When residents along Joplin Creek organized a municipal government in 1873, the new city of Joplin had over 5,000 residents. Most of them sought the success that their first mayor, Elliott Moffett, had achieved. Later that year the residents of Minersville also created a government. They renamed the place Oronogo because, one miner declared, "it's ore or no go." Meanwhile, prospectors pushed their search for mineral into undeveloped areas. In late 1873, neighboring farmers John Webb and William Daugherty began digging on their adjacent lands on either side of Ben's Branch, a small tributary of Center Creek, a few miles southeast of Oronogo. The pair mined for a year, with some promise, but high water levels in the mine led Webb to lease his share to Daugherty, who kept mining with a succession of partners. The Webb diggings, as the camp became known, produced a modest 40,000 pounds of galena in the first year but suggested more. Because of numerous efforts like these, one observer noted in 1874, "gradually the limits of the great mines of the Southwest extend their bounds. New fields are opened. Already it is not possible to keep trace of them."[21]

Local children were also drawn to the logic of the poor man's camp. They were allowed to collect small pieces of mineral for sale to merchants. These scrappers, as they became known, were "of all sizes, black and white, dirty and noisy." "Their occupation is that of picking nuggets of mineral from waste piles at wash places and abandoned shafts," a local newspaper reported, "and their earnings, like that of any other class, vary according to the smartness and industry of the 'scrapper.'" They could make from fifty cents to one dollar a day, which some spent on candy but others saved or contributed to the household earnings. "In this way," the correspondent figured, "thousands of pounds of mineral which otherwise would go to waste, is gathered up and bought and sold, and if we are to judge from appearances, the weekly aggregate of this business in our city amounts to hundreds of dollars." The scrappers of today might become the prospectors and miners of tomorrow.[22]

And so, amid sometimes tense contract negotiations, by 1873 the miners of southwest Missouri fashioned an ethic of entrepreneurial labor that framed the promise of the poor man's camp for the rest of the century. Those who made good encouraged more workers to "invest a few dollars in a pick and shovel, fill a haversack with rations and go for the mines." "It has made fortunes for hundreds of poor men," one miner told a journalist. They sub-leased land from established companies like Granby Mining and from new ones, or they leased it directly from farmers. They covered the length of Joplin Creek with tents and shafts and windlasses, mined the land at Oro-nogo with unprecedented fervor, and made Leadville productive again. They renewed attention to the old field at Granby and other abandoned mines in Newton County. Most important, they explored the spaces between, hunting for surface indications of the wealth that surely lay below. In all their efforts they followed a belief that the Joplin camp, "the head-centre of individual and combined enterprise, the neglected, calumniated creation of poor but genuine miners, the present seat of the most enterprising, well-to-do and persevering companies, partly formed by originally poor men—Joplin, the favorite of fortune, the most rapidly developing lead mine of Missouri—is just now in her infancy." Although the profitability of lead in the early 1870s moderated the potential disagreements between miners and leasing and smelting companies, the growth of the district would reveal new incentives that challenged the peace.[23] Few expected that those new incentives would come from an emerging zinc market.

In the 1850s, G. C. Swallow had predicted that the area's zinc ores would only become profitable with improved smelting techniques and a railroad con-nection. Zinc can be released from its mineral compounds only at tempera-tures over 2,000°F, when it escapes as vapor. Unlike lead smelters, which burned wood, zinc smelters required coal to obtain such high temperatures. Because they burned so much coal—three and a half tons for every ton of ore—zinc companies built their facilities near abundant coalfields. Once charged, the smelting facility used an elaborate series of containers to cap-ture and gradually cool the zinc vapors to the point of liquid condensation at 900°F.

Despite European advances in the 1850s, Americans mastered the tech-nology on a significant scale only in the 1860s. The first commercially viable zinc smelter west of the Appalachians opened in La Salle, Illinois, in 1860. In 1869, the Missouri Zinc Company built a second smelter south of St. Louis to process zinc ore found in eastern Missouri. The arrival of the railroad in southwest Missouri in late 1870, so heralded for what it did for the profitability of lead, presented area miners with possibilities for taking advantage of the nascent zinc market.[24]

Miners who had contracts to dig lead mineral now found themselves in possession of large quantities of zinc ore that outside companies increasingly wanted to buy. In a district organized around galena, working miners and prospectors, not local landowners or smelting companies, would be the first to profit from the rising value of zinc. This development would both strain contractual relationships and give new energy to the already rampant entrepreneurial spirit of the poor man's camp.

At first, miners responded coolly to zinc buyers. The early market prices were simply not high enough for them to shift their focus from galena. In 1871 buyers for the Missouri Zinc Company paid one dollar a ton for zinc ore from the waste piles that sat near every mine. The following year the Illinois Zinc Company, which had just opened a smelter in Peru, Illinois, paid two dollars a ton. Granby Mining began buying ore from miners on its lands for three dollars a ton in early 1873. For all three companies, the still-high cost of shipping large amounts of unprocessed ore to St. Louis dampened what they could pay. Miners gladly sold waste rock at three dollars a ton, but they needed a higher offer to actively pursue zinc ore when galena sold for around thirty dollars per 1,000 pounds. "The mining of it at that price can not be thought of in Joplin," a local correspondent averred. Geologist Adolf Schmidt concurred during his 1873 canvass of the district for the state geological survey. "Zinc blende is abundant at most of the mines," he reported, "but at present prices it scarcely pays the miner to remove it out of the way of the lead."[25]

Rising demand for zinc would within a matter of months make this dismissive account sound absurd. Between 1870 and 1875, the national consumption of zinc tripled. No mines outside of New Jersey could deliver enough ore, and so smelters focused more serious attention on southwest Missouri. In 1873 the farsighted Chicago Zinc and Mining Company built a zinc smelter at Weir, in Cherokee County, Kansas, where extensive, shallow coal beds had been discovered in the 1850s. Buyers for the company began paying eight dollars a ton for ore in Joplin. One correspondent reckoned that the new smelter "will scatter at least $1,000 per week in our midst for an ore that has heretofore been considered worthless." While "Joplin is looked upon as the wonder of the world, almost, on account of her inexhaustible lead deposits," he believed "that five years will develop the fact that the zinc will double discount the lead" and "surpass the wildest theories of men."[26]

When zinc buyers raised bids for ore to ten and then twelve dollars a ton in late 1873, miners and prospectors took advantage. "We are pleased to perceive all over the diggings that the mining of zinc blende has been commenced with great energy," an industry observer noted that October. "Tons after tons are brought up from shafts where, hitherto, no attention was paid to it." "Miners are now more generally turning their attention in this direction," another reporter noted, because of "a market having been opened for it . . . at fair prices." These miners believed that they should be able to operate in two markets: one selling galena to smelters as contracted and another selling zinc ore freely for whatever buyers would pay.[27]

In most cases, at least initially, local smelting companies took a relaxed approach to miners trading zinc ore. None could smelt the ore and so at first "did not claim or ask any royalty on it." As more miners began selling and then mining it, however, some companies insisted on imposing a small royalty. Picher, for example, charged his leaseholders one dollar for each ton of zinc ore sold, a nominal rate, "and no complaint was made by the miners." Other firms added zinc terms to new sublease contracts that required miners to pay a flat percentage royalty, usually 10 percent, of "the cash market price." Granby Mining, under new ownership after Blow sold his majority stake in late 1873, presented the only exception by buying all of the zinc ore its contract miners produced for a discounted flat rate, five dollars a ton in early 1874, that would fluctuate with market prices. Granby Mining's holdings were rich in calamine, a main source of the zinc oxide that St. Louis buyers sought to make zinc white paint.[28]

Miners could extract zinc ore from existing diggings at little additional cost. They mined it from the same shafts and with the same techniques they used to produce lead mineral. Zinc ore required more processing before it could be smelted, however, because its metal content, around 30 percent, was less than that of galena, 70 percent or more of which was pure lead. To isolate zinc ore from the largest pieces of nonmetal material, miners used a process known as concentrating. They crushed the ore with hand-operated rollers and then washed these pieces in a jig, a shallow box two feet by five feet submerged in a bigger water tank that separated ore from waste rock when shaken vertically and sideways in a vigorous pattern that must have resembled dancing. Miners could build and operate their own jigs or, if more successful, hire a wageworker to do it. After jigging, miners had concentrated ore ready for sale.[29]

From 1873, miners responded with a rush for zinc ore that led to new prospecting and the reopening of lead mines thought dead. At the Spurgeon mines, the oldest in the region, miners found "zinc blende of a very superior quality . . . said to be the richest yet discovered." Opening old mines led many miners to new runs of galena, so that "with the zinc mining new lead deposits are being struck." At Granby, "the whole appearance of the mines was changed by this getting of Zinc-ores." Many miners had moved away from the company's land in Newton County, where for nearly thirty years the galena deposits had been picked over, in favor of other more promising lands in Jasper County. The miners who worked at Granby, however, knew that although the galena was fragmented, the calamine ran in wide predictable sheets. By going back for these sheets, "the miners were sure to earn pretty good wages." Working a foot-thick sheet of calamine, they could take out a ton of ore, or five dollars' worth, a day. If they found any galena, they sold that too. "In this way," a state geologist reported, "many good Galena deposits were discovered and mined, and the productions of Lead-ore increased with that of Zinc-ore." Miners and lead-smelting companies both benefited from the arrangement, so long as prices for lead remained high.[30]

Just as miners deepened their investment in lead and zinc, the global economy collapsed. The worst economic catastrophe since the 1830s began with the banking panic of 1873 and then developed into a prolonged period of stagnation and deflation. Because the price of lead remained buoyant and the price of zinc continued to rise, however, miners dodged the long depression for several years. The crisis hit hardest in America's burgeoning industrial centers, cities such as Pittsburgh, Chicago, and St. Louis, and in the coalfields, where workers lost wages, hours, and often their jobs. In these places workers clashed with employers and the state and created more expansive unions, particularly the Knights of Labor, a new national union that became popular among coal miners. Miners in western precious metal camps also suffered from falling prices and demand as their corporate employers cut wages and lengthened workdays. They, too, built stronger unions, led by those in Virginia City and Gold Hill, Nevada, and in Butte, Montana. Miners in southwest Missouri remained on guard against smelting and leasing companies that pushed for stricter leasehold terms, particularly on zinc ore sales. Some protested against the threat of monopoly control. Most did not, at least directly. They either accepted the terms or, more revealing, continued to search for new, independent discoveries. For these miners, the rising price of zinc blunted the worst effects of the economic crisis and sustained faith in the market.[31]

For a time in 1874, some miners seemed poised to form a union similar to those emerging elsewhere. The miners' assembly had reorganized itself on a permanent basis as the Joplin Miners' Union. Their counterparts in Newton County followed suit with the Miners' Union of Granby. Both groups collaborated with the Grange, an agrarian organization gaining support among farmers hit hard by depression. Together, they warned against the growing power of monopolies and government corruption and called for a cheaper currency to ease the burden on debtors. The unions and the Grange planned to run a third-party "people's movement" slate for local office in the November election, in conjunction with the new People's Party, a state-level coalition of Republicans and Grangers. Far from radical, the Missouri People's Party shunned the national Republican Party's Reconstruction policies and appealed to independent producers with an antimonopoly, small-government platform.[32]

Some in the miners' union took a harder line in defense of their working rights. The trouble started in early 1874 when the Lone Elm Mining Company, an outfit owned by Philadelphia investors, terminated several subleases without explanation. Whether legally justified or not, the company's assertion of authority mocked miners' efforts to retain control of their work while under contract. To retaliate, a group of men bombed the company's new steam engine with blasting powder. However legally murky the company's act might have been, no one misunderstood the reason for the attack: Lone Elm "had taken the diggings away from some men and this was to retaliate for it."[33]

The trouble was not over. A few weeks later, W. H. Picher, who had taken over management of the family land from his brother, Oliver H. Picher, tried to raise the royalty that his lessees, which included smelting companies and working miners, paid on zinc ore from 10 to 20 percent. "The miners refused to pay," a local newspaper explained, because they claimed it violated their contracts. The dispute lasted into June, when Picher "notified miners working on his land to raise or dispose of no more zinc ores, until further notice." Some parties gave in. The Jasper Lead and Mining Company, a lead-smelting firm that leased land from Picher, agreed to pay the higher charge on zinc ore, which it in turn demanded of its sublease miners. A month later, in response, masked men bombed the furnace and offices of Jasper Lead. Everyone interpreted the attack as a protest against Picher's decision to raise royalties and the smelting company's collusion with him, but prosecutors could not convict the alleged bombers because of a lack of witness testimony against them. While their acquittal reflected some degree of popular sympathy, local newspapers denounced them as "Communists," dangerous revolutionaries like those behind the recent Paris Commune.[34]

The miners' unions did not survive association with the attacks. Even though no one was held accountable, most district residents, including most miners, seemed to "recognize that no property is safe," including theirs, if such violence continued. Tense election-season debates over Reconstruction and African American rights also heightened scrutiny of the unions and exacerbated partisan divisions among miners. Both Democrats and Republicans blamed those associated with the local "people's movement" for the bombings. Meanwhile, the unions' support for the Republican-backed state People's Party alienated miners loyal to the Democratic Party, which ran a racist campaign against Republicans for supporting the proposed Civil Rights Act. Although the People's Party narrowly carried Jasper County, it lost Newton County and the election; local third-party "people's" candidates were trounced everywhere. Democrats swept Missouri's statewide races and won a majority in the House of Representatives. Newton County party leaders championed the victory as a triumph for white workingmen: "Rough mechanics, miners and farmers have shut down on Republican blarney." Divided by race-baiting partisanship and accusations of being "smelterburners," the union groups soon disbanded.[35]

Rather than back down, landowners and smelting companies tightened contractual terms on mineral and ore in response to the depression. Most companies lowered their price scale for mineral to twenty-five dollars per 1,000 pounds when pig lead sold for seven cents a pound in St. Louis. Miners earned 10 percent less than that as the price of lead fell to 6.3 cents per pound in 1874. Others tried to impose new controls on zinc production. When Jasper Lead relinquished its leasehold, the Picher brothers decided to manage their lands themselves under the auspices of the new Picher Lead and Zinc Company. After the recent violence, the new company abolished zinc royalties altogether and offered a discounted cash rate fixed to a market-tracking sliding scale, like Granby Mining, for all of the zinc ore its lessees produced. Corn and Wahl did likewise.[36]

Many miners accepted these terms in order to maintain access to rich ground where they could still make good money while national economic conditions worsened. An 1875 visitor to Granby Mining land along Joplin Creek found leasehold miners working a number of productive mines. In a single week that summer, miners Cyphert and Williams had hauled up over 21,000 pounds of mineral. Their neighbors, Oliphant and Company, had sold over 3,000 pounds of mineral and a ton of zinc ore the week before, while Hinton, Smith, and Company turned in 1,000 pounds of mineral and a ton of zinc ore. Prospectors Irwin and Haley were still digging without results, but the observer predicted that "before the summer is over these young gents will have a nice pile of greenbacks to start them in business." Despite earlier troubles, miners on Picher land still expected big returns. According to a report in late 1875, the Coyle brothers sold over 5,000 pounds of mineral in five days. Even with the company's restrictions, contract miners continued to invest in their own operations. Burton and Company had a mine that produced 250 pounds of galena and 500 pounds of zinc ore a day. They ran a steam pump to keep water out. Burton had recently sold his lease on another mine for $400 to Dan Wenrich, who considered the prospect and the contract terms offered by Picher Lead and Zinc a good investment. "Nothing venture, nothing gained," he explained. Their entrepreneurial plans were hard to unwind. The Coyle brothers, for example, were "determined in their purpose, notwithstanding the great expense incurred by their undertaking." They not only worked the mine themselves but also gave "employment to several hands." Such a scenario favored experienced miners, of course, and especially privileged white miners, both native and foreign born, but not exclusively. A few African Americans had started mining in the area. An 1875 report included news that Willard and Davis, whom the newspaper identified as "colored," sold 6,000 pounds of mineral in a week.[37]

Many miners, however, left established fields to explore underdeveloped or new lands. Noting a now recognizable pattern, a correspondent from Oronogo reported in late 1874 that miners had left the camp because of the low prices offered by Granby Mining. Some secured leaseholds at the Webb diggings. Daugherty and his new partner, Thomas Davey, a machinist, had introduced a steam pump to drain the tract, opening 360 new acres for lease. With "good inducements to the miner," a phrase that usually meant reasonable royalty rates, Daugherty and Davey attracted over 200 prospectors to the camp they dubbed Carterville. Meanwhile, John Webb had re-leased his adjacent land, now the site of a small camp called Webb City, to the new Center Creek Mining and Smelting Company, which in turn offered good sublease terms for miners to invest their labor "with a certainty of realizing immense profits." During the winter of 1874–75, prospectors on these holdings discovered what many considered "the richest deposit of lead ore in Jasper County." This "new bonanza," a correspondent declared, might rather "become the richest lead discovery yet made in the State." Miners Poundstone and Parker, whose shaft reached twenty-two feet, produced 40,000 pounds of galena in three weeks. Other miners hit "chunks of pure ore weighting 2,000 pounds" each. By the fall of 1875 the new Webb City–Carterville field drew in waves of miners who were fed up with tightened contract terms elsewhere. In a two-day period at the end of August, according to one report, "one hundred miners came in … from Granby and Joplin."[38]

These opportunities, along with rancorous memories of 1874, turned most miners away from insurgent antimonopoly politics in 1876 despite the ongoing depression. Local campaigns were restrained, avoiding the race-baiting divisions of two years earlier. The new Greenback Party, which campaigned nationally for a plentiful paper currency and cheaper credit, was relatively popular in Jasper County, where it claimed a state-best 520 votes, 8 percent of the total. Farmers delivered most of these. Only fifty-five people in Newton County voted Greenback, none of them from Granby. Republican presidential candidate Rutherford B. Hayes won Jasper County and the White House; Democrat Samuel Tilden won Newton County and the state. The nation abandoned Reconstruction after the election of 1876, as white voters, particularly in Missouri, forged a partisan truce around common assertions of white supremacy. That truce did not end all conflict, however, as workers across the country battled on a new scale with corporations over responses to the economic crisis. Those clashes included small coal-mining strikes in Missouri and Illinois in 1876 and a general strike in St. Louis that punctuated the national railroad strike of 1877.[39]

While these events dominated the news, prospectors and miners around Joplin and Granby continued to deliver success stories with pig lead prices still above 6 cents a pound. In April 1877, even as the new boom camps of Webb City and Carterville expanded, prospectors located a new mineral field seven miles west of Joplin along Short Creek, just across the Kansas state line. After miners hit a single surface deposit, subsequent digging by newcomers uncovered a rich and shallow but fragmented array of galena and zinc ore deposits several miles long. In keeping with the pattern of discovery and development in the district, however, news of their success attracted investors from Joplin and nearby Baxter Springs, Kansas, who bought some of the mineral land from area farmers and leased other portions from the Kansas City, Fort Scott & Gulf Railroad. Led by Joplin's Patrick Murphy, these investors created two companies, the South Side Mining and Manufacturing Company and the Galena Mining and Smelting Company, to manage the holdings. Both adopted the district's now standard leasing terms for galena, with discounted prices on a sliding scale, but allowed miners to sell zinc ore for open-market prices minus a royalty. That summer hundreds of prospectors and miners from Joplin, Oronogo, and Granby rushed to Short Creek to take advantage of these "good inducements." They soon named the camp Galena, after its principal product.[40]

The Galena rush rivaled anything contemporary observers had witnessed at Joplin and gave new energy to the ideal of the poor man's camp. Only two months after the first discovery, a reporter from Kansas City, Missouri, noted that miners had sunk over 1,300 shafts. The deepest shaft was twenty-four feet and yielded several tons of mineral and ore a day, while the shallowest paying mine went down only three feet. "Several poor fellows have already been suddenly hoisted from extreme poverty to fortune," one reporter wrote. This was no miracle, he explained, but the result of men willing to work hard for the chance of making money. "One poor fellow who came here some days ago without a cent in his pocket, leased a claim, went to work, and yesterday struck a 'big bonanza' at the depth of four feet from the top of the ground." Within hours, "he was offered $2,000 for his claim." He refused and instead "went to work with renewed energy, and may be found in his shaft from early morning till late at night digging out immense chunks of the shiny ore."[41]

Despite the weakening of the lead market in the late 1870s, the southwest Missouri mining district, now including the Galena camp and increasingly known as the Joplin district, still offered perhaps the best opportunity for poor men with little capital to invest their labor and time in hard, physical work that paid. The rapid development of the district after Granby Mining reopened its lands in 1865 demonstrated that promise to the almost 3,000 miners who worked there in 1880. Despite the proliferation of leasing and smelting companies patterned after Granby Mining, miners seeking greater independence and profits discovered most of the rich mineral deposits. When the companies set less favorable terms or mineral deposits ran thin, they branched out in search of new fields. In hundreds of individual decisions to explore further rather than to fight control, prospectors developed the richest deposits of zinc ore in the country at a time when manufacturers demanded it more than ever. That determination to mine for themselves, even if not always successful, pushed leasing and smelting companies to adopt contractual terms that preserved market incentives for miners, particularly on zinc ore, and thus the promises of the poor man's camp.

Their entrepreneurial efforts turned rudimentary diggings into bustling camps and prosperous towns and cities. In 1880, more than 7,000 people lived in Joplin. A city that had not even existed in 1870 was now the sixth-largest city in the state. Granby was home to more than 1,800 people. The newer camps grew fast: the Census Bureau found more than 2,700 people in Galena and more than 2,000 in Webb City–Carterville. Railroad companies built new lines, one finished with a ceremonial lead spike, that connected Joplin to the smaller camps as well as to the zinc smelter in Kansas and the trunk line near Granby. The miners who lived in the Joplin district shared similar origins and backgrounds with those who came to the area before them. Native-born whites predominated but made up a lower percentage of the mining workforce than of the county population, which was 95 percent native born. The foreign-born miners came mostly from Great Britain and Ireland. Judging by the birthplaces of their children, many miners had lived in Illinois, Missouri, or Wisconsin at some point in the previous two decades, which meant that they likely had metal-mining experience, although rising numbers came from nearby farms. They continued to settle in family groups that provided social cohesion and economic stability. These communities overwhelmingly favored white newcomers but not exclusively so. More African Americans worked as miners in 1880 than ten years earlier, although the numbers were small: twenty-four mined in Jasper and Newton Counties, and twenty mined in Cherokee County. Together, their output pushed these mining camps to the forefront of the base metals industry. From 1847 to 1869, district miners had sold 37,300 tons of mineral and no ore. During the 1870s, by contrast, they sold more than 124,000 tons of lead mineral and 108,000 tons of zinc ore.[42]

By 1880, these miners worked according to an informal, district-wide leasehold system that tied success to the market. They were paid for what they produced at market prices in exchange for conceding certain discounts and rights to smelting companies and landowners. According to an increasingly standardized contract, lessors contracted partnerships of working miners, also known as companies, as they had since before the early 1870s to work specific lots. Most contracts allowed miners and their families to live on the lot free of rent; they provided their own housing, in most cases rudimentary "board shanties." Miners were required to sell all of the lead mineral they produced to the lessor at a discounted price that followed a sliding scale pegged to the price of pig lead in St. Louis, a standard of twenty-five dollars per 1,000 pounds when pig lead sold for seven cents a pound. If the lessor did not buy, miners were usually free to sell elsewhere in exchange for a royalty. The efforts to enforce a discount price on the sale of zinc ore, however, had not succeeded. The most common contract gave lessors a right of first refusal on all zinc produced by a lessee but at "the cash market price" in Joplin, not a company-specific rate, minus a royalty. While many lessors ran lead smelters and so could demand discounts on mineral, no lessors had zinc-smelting capabilities and thus no equivalent leverage when purchasing ore. Offering market prices on zinc ore was also an easy way to attract and keep miners who were not shy about leaving for a better deal. The distinction between discount and cash market prices mattered to miners because the latter gave them a more direct stake in the market. These terms prevailed in most of the camps.[43]

The resolutions of the Joplin Miners' Union lived in these incentives for zinc ore mining, but its resolutions on workplace control did not. Most leasehold agreements gave the smelting or land company or its agents the right "to go and remain upon said lands, at all times to inspect said lands and to see that this contract is complied with." Where miners failed to comply, the company reserved the right to take possession of their lots without notice. These terms assured lead-smelting companies of constant production at predictable prices and allowed miners to profit directly from their success while bearing the expense of developmental work and the risk of loss from accidents, complications, or geological capriciousness. Miners called contracts like this "jug handles," because "the risk is all on" one side, theirs. Yet they readily assumed that risk in "the hope of a big deposit just ahead, to be made accessible by a few more blasts or strokes of the pick."[44]

To outsiders, this method of wide-open exploration made district miners look irresponsible and careless. The unpredictability frustrated government officials. "It is extremely difficult to give the weekly wages of lead miners," W. H. Hilkene, the Missouri commissioner of labor statistics, noted in 1879, when "some weeks miners do not average as much as common day laborers, other times the run of lead may be such as to give them $25 per week." Another commentator reported in 1879 that the district's camps "have a strange, confused appearance. The yawning mouths of the shafts are at every hand, and little, box-like miners' cottages nestle about among the huge piles of debris" that left "very little hope for gardens and flowers to soften the ugliness of the abodes." These conditions raised questions in the minds of observers who held to middle-class notions that white men should behave with prudence to safeguard their dependents at home. "It seems hard to have women and children surrounded by such chaos," this observer sneered. How to make sense of people who chose to live like this? Like drunks, gamblers, and criminals, he averred, "perhaps the men do not care because the quest of 'mineral' becomes all absorbing."[45]

Where these critics saw wanton waste, others saw admirable economic dynamism. "Each miner tries to get as much as possible out of his own lot," F. L. Clerc, reporting on behalf of the U.S. Geological Survey, explained in 1882. To maximize production, companies of working miners "hire laborers to assist them, and by hard labor" were "known to have delivered 100,000 pounds of lead ore in one week," Hilkene admitted. Mine laborers earned between $1.00 and $1.50 a day; many saved their wages to invest in their own prospecting business. Two-thirds of those working in the mines in 1880, however, claimed a direct stake in their product. They were willing to lead precarious lives because they believed that the risks would yield rewards: financial success, economic autonomy, and manly independence. The process might not be pretty, Clerc argued, but it delivered results. "The miners, working on their own account, with hopes of large ultimate gains, have every inducement to work hard and cheaply, and to follow every clew that may lead to the discovery of ore." If not these "enterprising, skillful, well-to-do miners, naturally associated as partners, who have made one or more good strikes, and are always ready to take hold of any new venture that promises well," he asked, "where else could be found capitalists so willing to risk their money in a speculative venture" like lead and zinc mining?[46]

CHAPTER 3
NOTHING BUT HIS LABOR

Belief in the promise of the poor man's camp ethic swelled among the miners of the Joplin district, as it was now known. In early 1881, the *State Line Herald*, which served Joplin and Galena, printed a jesting, psalm-like paean to "the miner" that praised the poor men who had braved this system. Although "none knoweth his nativity nor the dwelling place of his forefathers," the miner could secure a grubstake and "goeth abroad over the land and seeketh a spot wherein he may pitch his tent and dig for the precious metals." Once the miner "secured a pick and likewise a shovel," "he diggeth deep" but "his labor cometh to naught." Dismayed, the miner "murmured against fate, lifting up his voice and crying, 'I am undone, yea, flattened out like unto a pan cake.'" Unable to pay his debts, the miner received scorn from "the inhabitants of the land," who "hold up their hands and say, 'Behold the dead beat he eateth up our substance and payeth not therefor.'" But the miner did not quit, he "worketh day after day and liveth on the husks of the land." And then, "he striketh her big, both of lead and zinc metals striketh he them." Having endured hardship, the miner now set loose "rejoicing on the plains of Joplin," where "the mighty men of the land doth hearken unto his voice and say, 'See what this man has wrought!'" The people, too, "sayeth, 'There is much in the man we knew not of before.'" The miner gained fame and elected office, became "a city dad." From then on, "he liveth a life of virtue, eschewing evil and becometh a mighty leader in the land. Selah."[1]

Despite its absurdist lyrics, the song reflected the remarkable actual events of the preceding decade. Independent miners had discovered the lead and zinc that created the district's richest camps at Joplin, Webb City, Carterville, and most recently, Galena. In small companies and sometimes alone, working miners succeeded in part because the earlier generation of miners had secured favorable leasing terms from those who controlled the land. Those terms allowed men with little property a fair opportunity to enter the metal markets for their own benefit, with some limitations. Thousands of men followed in the belief that they, too, could succeed by investing time and effort in the possibility of striking pay dirt. Most were white English speakers— some with mining experience, others from area farms—who could negotiate the business of mining on a basis of equality with those who came before. Like the miner in the song, they sought economic independence and social respect as workingmen who fulfilled the era's ideal for responsible manhood as husbands, fathers, and community leaders. They viewed entrepreneurial mining as a proper means to achieve manly standing.

As "The Miner" hinted, however, the poor man's camp ethic had nothing to offer white men who remained poor and dependent. Until he succeeded, the song's miner was disdained as a deadbeat, less than a man. This would matter as the national lead market collapsed in the late 1870s. The price of pig lead had held up through four years of depression but fell sharply in late 1877, partly because of surplus stockpiles and partly because of rising production from new silver-bearing lead mines in Nevada, Utah, and Colorado and new discoveries in eastern Missouri. "This has been in a general way one of the most disastrous years we have ever had," one industry executive complained. By 1878 the St. Louis price had plummeted to 3.4 cents per pound, just a fraction above the all-time low set in 1845. Smelting and leasing companies cut prices for mineral by half or more. Some smelters closed for good. After thirty years of favorable markets, miners in the Joplin district now faced parlous conditions that strained the logic of the poor man's camp to the limit.[2]

Rather than reject that logic, like other workers who were mobilizing in opposition to industrial capitalism, they redoubled their commitment to earning independence as white men through economic risk. Miners in the Joplin district explored alternatives, as long-held suspicions of the smelting and leasing companies remained high. Some rallied to the antimonopoly politics of the renamed Greenback-Labor Party and the Knights of Labor, both of which made gains nationally and in the region after 1878. But even these men, who were a minority, still looked for reforms that would privilege small, market-driven producers like those who had made the poor man's camp. The vast majority remained confident in capitalism. That confidence led many miners and prospectors to emphasize the production of zinc ore, which, unlike lead mineral, they had the right to sell for open-market prices that were still rising in the early 1880s.

The new wave of zinc-mining prosperity that these miners unleashed would bring changes that ended the poor man's camp. To profit from zinc, small mining companies, many of which grew from prospecting ventures, increased their scale in subtle ways that required a little more capital than true poor man's operations. Zinc-mining companies in turn hired larger groups of wageworkers. Incremental, gradual, and led by local white men who themselves had mined, these changes seemed at first to reaffirm the district's promise for working miners. By the early 1890s, however, miners in the Joplin district faced worsening chances to become owner-operators. Most of them, inspired by examples of past achievements by other poor men, many of whom were now neighbors or bosses, continued to imagine that the future would offer them the same opportunities as the past. If earlier generations had lived on the "husks of the land" and still prospered, so could they.

The lead slump decimated the earnings of miners under the sliding-scale terms of their "jug handle" contracts. With pig lead selling for four cents in St. Louis in 1879, miners earned only fourteen dollars for 1,000 pounds of mineral. For the first time, successful miners in the camps around Joplin struggled to make headway. "Many miners who have good diggings," a local newspaper reported, did not produce "enough lead to give them a living." The Missouri Bureau of Labor Statistics asked miners that year about their current earnings and, if they had been around long enough, what they had earned in 1872. Thirty-two miners reported average weekly earnings of only $6.53 in 1879, down from $12.27 a week for those who reported earnings from seven years earlier. "A great many are leaving," noted a correspondent in Oronogo. "Those that cannot get away are going to work for anything to keep soul and body together," he continued, citing wages as low as fifty cents a day. For the first time, the miners and prospectors of the poor man's camp endured a depression that challenged their entrepreneurial ambitions.[3]

Many miners blamed the crisis on the smelting and leasing companies. Hit by steep price cuts in early 1878, miners at Granby railed against the Granby Mining and Smelting Company and its new owners, who were fighting the railroad over alleged lease violations. Some went so far as to submit affidavits in support of the railroad's claims against Granby Mining. While the courts resolved the dispute between the companies, the miners got nothing. That November, more miners than before voted for the Greenback-Labor Party, which now supported proposals to protect the rights of workers from corporations in addition to currency and banking reform. Its platform promised to restore the independence of small producers, not foment radical change; the party opposed "strikes, revolutions and all violent measures for the relief of labor." The Jasper County branch appealed directly to the men of the poor man's camp with proposals to cap royalty rates and give miners "the right to sell all ores raised by their labor to the highest bidder." The party's local candidate for Congress carried Newton County with 1,210 votes, 39.4 percent of the total, and finished third in Jasper County with 1,722 votes, 28.2 percent. The Republicans won the seat. The moderate Greenback-Labor appeal remained strong among area farmers and market-minded miners. In

1880, presidential candidate James B. Weaver received 971 votes in Newton County and 1,111 votes in Jasper County, around 28 and 20 percent, respectively. Weaver ran a strong third in Granby, Webb City, and Joplin. But old partisan loyalties again caused problems. The Greenback-Labor Party ran a joint ticket with Republicans at the state level, which hurt Weaver's support among regular Democratic voters. These elections revealed considerable popular worry about the power of leasing and smelting companies but also an enduring faith that poor but hardworking men could again thrive in the market economy.[4]

Not all miners believed they would flourish in the Joplin district. Some left to seek new opportunities in western metal camps, where experienced miners could earn high wages. To Joplin miners, one of the most enticing and best known was Leadville, Colorado, a booming but isolated camp over 10,000 feet above sea level and more than forty miles from the nearest train depot. The St. Louis Smelting and Refining Company discovered silver-lead deposits there in 1876 and built a smelter. Picher Lead and Zinc Company sent D. Baumann, who had been an agent of Henry Blow, to monitor developments. With good business sense, Baumann left the company's employment to manage a Leadville mine, which he named "Joplin." In 1878, he encouraged Missouri miners to come work for him as wageworkers at the relatively high rate of three dollars per day. Many did so. "Southwest Missouri miners (a few black sheep excepted) are recognized here as the best hands for the Leadville deposits," Baumann announced in a late 1878 letter to Jasper County newspapers, "and I never hesitate to state here openly that I prefer them to any others for the two reasons that they are Southwest Missouri men and good workers besides." Other operators hired them too. With lead prices falling, work in the Colorado silver-lead mines compared favorably to the diminished profitability of the lead and zinc operations in Joplin. "They prefer a little silver in their lead," a local paper explained in 1878, "and there is a prospect that two-thirds of the population of Joplin will be in Leadville by the end of next season." They were so numerous in the camp that one hotel operator called his place the Joplin Lodging House.[5]

The efforts of Joplin miners to gain advantage in Leadville collided with the workplace struggles of the miners already there. Leadville began as a poor man's camp in the 1860s, when thousands of prospectors, mostly Irish, Cornish, and native-born white Americans, came to pan the streams around California Gulch for gold. The discovery of lead and silver, however, sparked the rapid development of capital-intensive deep mines by outside investors. In January 1879, a large portion of the camp's miners formed Local Assembly (LA) 1005 of the Knights of Labor, a growing national labor union that appealed to both moderate antimonopoly reformers and wageworkers who favored more militant responses to their employers. However unfocused on strategy, the Knights advocated for the autonomy of all workers, regardless of skill or craft, against the rising corporate powers of industrial capitalism. Coal miners led the expansion of the Knights in the mid-1870s, first in Pennsylvania, then across the midwestern coalfields of Ohio, Indiana, Illinois, and Kansas and into Colorado in 1878. Although Leadville's metal miners were the first miners outside of the coal industry to join the Knights, they had connections to independent metal miners' unions that had formed in Nevada in the late 1860s and in Butte, Montana, in 1878. Like these unions, the Leadville Knights demanded higher wages, particularly a daily rate of four dollars that offset the high cost of living in the mountaintop camp. In May 1880, they went on strike.[6]

Confronted with direct demands for worker solidarity, the Joplin miners in Leadville continued to work. They thought conditions were fair. "As a general thing," a correspondent reported, "the Joplin miners are opposed to the strike, claiming that three dollars is enough for eight hours' work." When the strike collapsed after the Colorado state government dispatched the militia, their loyalty seemed wise. Mine owners rewarded them. A visitor from Jasper County in August 1880 met fifty Joplin men employed in various Leadville mines. Most worked in managerial positions and reported good earnings. "It makes me feel proud of Joplin," the visitor declared, "to know that a Joplin miner can always secure work when there is work to be had; he is at a premium, for Leadvillians realize that he has skill born of experience."[7]

Most Joplin miners in Leadville did not want to become permanent wage laborers, no matter how well paid. Many were frustrated by the lack of prospecting chances there. According to miner F. F. Smith, who returned to Joplin as soon as he could, "Leadville is a fraud for the laborer." All of the prospecting "claims that are worth having are all taken up," he explained to a Jasper County newspaper, "and if you have a claim it takes a small fortune to even commence prospecting on it" due to the camp's rough terrain and inflated prices. Worse still, Smith said, "the mines are principally controlled by wealthy corporations, thus leaving the poor man the one alternative of working for what he can get."[8]

On the other hand, the strike seemed to show that Leadville's union miners also made work there unreliable. "Every ore-producing mine in the vicinity of Leadville has been utterly stopped from operations by the striking element among the miners," one writer noted during the strike, "vowing vengeance on" anyone who went back to work. "So you see," this Missourian concluded, "the devil is to pay in Leadville." Caught between big corporations and a forceful union of wageworkers, the men from the Joplin district identified with neither. Smith and many others returned home, where they still saw the best chance for small producers. "The day for the poor man is past in Leadville, and he stands no better show there now than he does anywhere else. If a man is doing well—making a good living—he had better be contented, and not run off on every tangent that strike up—it don't pay." The question remained, however, whether the Joplin district would offer anything better.[9]

In the 1880s, many Joplin miners turned decisively toward zinc production to make their living. Conditions in the lead market remained turgid as prices bounced between 4.00 and 4.75 cents before falling again in 1884 to 3.50 cents. By contrast, the market for zinc ore was growing along with demand for brass, galvanized metals, and corrosion-resistant paint. New land companies, such as the Joplin Zinc Company, founded by Picher and several other investors in 1881, and established firms, including Granby Mining, constructed a series of larger, more advanced zinc smelters in the region: at Pittsburg, Kansas, in 1878; at Joplin in 1881; and at Rich Hill, Missouri, and additionally at Pittsburg in 1882. The increased demand raised market prices for zinc ore, from $14.00 a ton in 1876 to $21.50 a ton in 1886. During that period, district miners increased their annual ore production sevenfold to more than 75,000 tons, which made the Joplin district the principal source of zinc ore in the United States.[10]

Miners in the field around Webb City and Carterville were the first to focus intensively on zinc. In the late 1870s, they discovered that the field's shallow deposits of galena and sphalerite extended down only forty to sixty feet before giving way to a thick layer of limestone. Some stopped digging because miners rarely found rich galena deposits below such bedrock formations. Others, however, explored cracks in the limestone. In 1877, according to one story, a miner subleasing a claim from the Center Creek Mining and Smelting Company in Webb City "broke through and discovered a rich zone of zinc blende ore." His success "induced others to sink deeper shafts, and the limestone was soon found to be the cap-rock or roof of the zinc ore deposits." According to mining engineer F. L. Clerc, writing in 1882, miners followed this "immense deposit of zinc blende ... continuously for over half a mile." Their digging revived work in the field and fashioned a new model for profitable mining. "The zinc bodies of ore are so much more extensive that the lead ore is not sought for," Clerc observed. By 1882, he reported, miners in Webb City and Carterville "worked principally for zinc ore" and produced "more than half of the zinc ore raised" in the district.[11]

Small companies of working miners in other camps followed their lead. Market conditions informed the new emphasis as much as geology. Miners who began work in 1878 or after, during the lead slump, had little positive experience with mineral under sliding-scale contracts. They saw greater promise, however, in the zinc provisions of contracts that allowed them to sell ore at prevailing market prices, if not to their lessor then to other buyers, minus a royalty payment. By 1882 seven regional smelters offered prices that continued to climb, up to twenty-four dollars a ton at the end of the decade. Although zinc ore was still worth less than an equal weight of lead mineral, the gap was closing. Many companies reckoned that they could make up the difference and more by exploiting the greater quantities of zinc ore. For the first time in 1883, for example, leaseholders on the South Side Mining and Manufacturing Company's land in Galena produced more zinc ore than lead mineral, 1,762 tons to 1,184 tons. Three years later, they produced 7,237 tons of zinc and only 835 tons of lead. In all cases, working miners who had assumed the risks of leasehold mining made the decision to pursue zinc, a strategy that emerged, almost literally, from the bottom up.[12]

Small companies could achieve a marginal increase in zinc production with old methods. In Granby companies of two to five miners predominated in the 1880s. When M. L. Wolfe, the Missouri state mine inspector, reported on the camp in 1887, he found thirty-two companies working shafts mostly between fifty-five and eighty-five feet deep: twelve producing lead and zinc, eleven producing zinc only, and nine producing lead only. Most hoisted men and material with animal-powered whims, although some still used windlasses and a few employed more expensive steam engine hoists. Similar operations predominated in many parts of Joplin. When the inspector visited Granby Mining holdings in Oronogo in 1889, he found that "most of the mining done on the land is by small companies. The ore is usually found near the surface, therefore it does not require much capital to open a mine." These companies produced from two to four times as much zinc ore as lead mineral "on a very cheap plan." Such small-scale operations bore the legacy of the earliest days of the district, but profitable zinc production called for more elaborate means.[13]

Some companies began increasing the scale of mining. At first that required digging deeper. "There is not a single instance," Clerc noted in 1887, "where the supply of zinc ore has been exhausted downward, in working a mine, and the depth of the deposit is therefore unknown." While some

Poor man's mine, ca. 1880s. R620, 2-015. Courtesy State Historical Society of Missouri Research Center, Rolla.

mines in Joplin went below 100 feet by 1890, most of the mines in Webb City and Carterville went below 140 feet. The deepest mine, worked by miners Allen and Sheldon on land subleased from Center Creek Mining, went down 180 feet. Deeper mining enabled companies to exploit greater quantities of zinc ore that required enhanced handling and processing capabilities. In Galena, for example, companies working the holdings of South Side Mining earned two and a half times as much from zinc ore as lead mineral, $379,074 to $148,891, between 1885 and 1888 but only by producing six and a half times as much material, 19,989 to 3,043 tons. In addition, miners found that the deep, more plentiful ore deposits usually occurred in hard formations of cemented chert, a silica-rich sedimentary rock. Unlike in the soft ground, where pick work often sufficed, miners working these hard ground formations needed blasting powder to dislodge seams of ore. As a result, miners generated not only more zinc ore the deeper they went but also more rock waste, all of which increased the quantities of material that had to be handled and hoisted for milling.[14]

These consequences required companies to use more machine power. Those working in Webb City and Carterville were most likely to adopt steam-powered hoists, although a majority continued to use horse-powered whims in 1890, even in mines 160 to 170 feet deep. Of the eleven companies mining on William Daugherty and Thomas Davey's land in Carterville, for example, four used steam hoists. The mine inspector observed "new machinery being put in" the other mines. The trend was toward mechanization. Between 1887 and 1890, the mine inspector's survey recorded an increase in steam hoists from 35 to 117. To counter water in deeper shafts, most lessors provided pumping services in exchange for ground rent, while many mining companies installed their own pumps.[15]

The use of machines for concentrating zinc ore provided another measure of how practice was changing. Methods of hand crushing and jigging developed in the 1870s could not cope with the unprecedented quantity of material that needed milling in the 1880s, so companies increasingly used steam-powered equipment. Most lessors of holdings rich in zinc ore, particularly those in Webb City and Carterville, provided this equipment to leaseholders in exchange for an additional ground rent. As with pumps, however, some mining companies invested in their own machines. For example, the Webb City Lead and Zinc Company, which subleased two plots from Center Creek Mining, used steam-powered jigs, crushers, and rollers at both mines. Machines not only handled far greater quantities of material but also improved the quality and purity of the concentrated ores. According to Thomas Davey, one of the first to mine in Carterville, the use of these machines would usher in "an era of prosperity for land owners, operators and miners that will make the past exceedingly insignificant."[16]

Mining companies constructed new buildings to house the machines, which changed the way their operations looked. These clapboard structures, which became known widely as "Joplin mills," consisted of a squat, two-to-three-story barn to cover the hoisting derrick and the mine shaft; a longer, single-story room for concentrating the ore; and one or more narrow towers, in some cases as high as eighty feet tall, where elevators raised and then discharged waste rock, known locally as chat, onto ever-larger dump piles. These mills were relatively cheap to build and move. Still considered "crude" and "peculiar" by experts from highly mechanized mining districts, the Joplin mills nonetheless marked an important development in the district: the replacement of the open-air whims, wash troughs, and hand-jigging installations of small-scale operations with machines driven by steam power housed on small factory-like sites.[17]

Companies using more intense operations hired more wageworkers. The initial increase was modest and made up mainly of the sons of local miners and farmers. Most zinc producers employed from eight to twelve wageworkers in 1890, about twice as many as their smaller, mixed-production

A growing zinc mine, Prosperity, Missouri, ca. 1890. Historical Mining Photographs Collection. Courtesy Joplin History and Mineral Museum.

counterparts. The largest among them in Webb City employed twenty-four to twenty-eight workers each. In most cases, a majority of these workers labored above ground, either tending the hoist and pumps, operating the concentrating mill, or moving ore and waste rock by hand. Rising production figures, however, reflected an increase in men who worked for wages in the mine. In 1890, the state mine inspector counted 1,631 miners, although without specifying how many owned a share in the claim and how many worked for wages, and 1,872 other mine employees in the Missouri portion of the district. Most worked nine-hour days, the district standard. The increase per mine was inconspicuous, though, as mining companies proliferated during the zinc boom, up to 485 independent firms in 1890.[18]

These changes opened a new, albeit still narrow divide between the owners of small mining companies and their employees. The forty-one mining companies operating on Center Creek Mining land employed an average of five workers above ground and four below ground. The partners who owned the smallest of these companies continued to work in the mine in the late 1880s alongside a few mine laborers to help dig and handle material. This followed a traditional, and still most common, pattern. By contrast, the owners of slightly larger companies, generally those with more than ten employees, had stopped working in the ground or had at least hired others to oversee parts of the operation. By the end of the decade, these outfits were more likely than smaller companies to list wage rates for "pit bosses" and "engineers," which ranged from two to three dollars per day. These managers made hiring decisions and choices about where to dig and oversaw the miners who did the actual work as well as the larger crews of mine laborers. Most of them had been miners themselves, some very recently, and likely lived near the men they supervised. These close social ties softened, and likely obscured, the slight hierarchies that were emerging.[19]

The "black jack" zinc boom of the 1880s brought a new wave of prosperity to the district but also ushered in a more elaborate system of wage labor. In 1891, the area's mining companies, led by those in Webb City and Carterville, produced over 120,000 tons of zinc ore that sold for a total of $2.6 million and lead mineral worth an additional $713,000. Successful companies raised wages and hired more mine laborers. That year jig operators, the men in charge of concentrating ore, earned the most of all nonmanagerial workers, from $2.00 to $2.50 per day. Their daily wage matched or exceeded the earnings of most skilled urban workers nationwide. The men who labored for wages in the mines, whether they dug ore or shoveled it, earned from $1.75 to $2.00 a day, rates that exceeded those of semiskilled workers elsewhere but were not as high as those of metal miners in the West. General laborers on the surface earned the least, $1.50 to $1.75 per day. Although production stoppages in small mines were frequent, whether due to cash flow problems or other disruptions, mine laborers in the Joplin district made more money than they had in 1882, when Clerc reported that all mine laborers earned from $1.00 to $1.50 per day. As more companies produced more zinc ore with expanded operations year on year, everyone expected opportunities to expand and wages to rise. Many workers looked forward to using their wages to fund prospecting ventures. Pay in Joplin mines might have been lower than that in Leadville and other western camps, but the money went further. In 1889, for example, according to the new state mine inspector, C. C. Woodson, "wages are fair and the opportunity for owning a home good." Indeed, Joplin, Webb City, and Galena grew and prospered with businesses, schools, churches, and fraternal organizations, all linked by a new interurban trolley system. Joplin was a remarkable mining center, another observer claimed, because "it is within the civilized world" with "all of the advantages of any city." Although more stratified than before, these communities cohered along racial and national lines that reinforced a sense of democratic fairness between white men; Jasper County was 96.5 percent native born and 98.2 percent white. They did not exclude African Americans from the mines yet; as many as 100 worked in the district in 1890. By then, over 32,000 people lived there, including 9,900 in Joplin, 7,600 in Carterville and Webb City, and 3,500 in Galena. Prospectors had made Joplin the "Lead Metropolis of the World" in the 1870s; now their successors turned it into "the town that Jack built."[20]

As the district changed, albeit gradually, miners and mine laborers in the 1880s still believed that they could also become owner-operators. Despite the collapse of lead prices, the productivity and profitability of the mines continued to rise with the turn to zinc. In every camp, the hopeful saw handsome sums made by men who had been prospectors themselves not many years before. Charles DeGraff was one. In 1883, he left his leasehold in Galena to seek a better opportunity near Joplin. He was among those who located a rich deposit of zinc ore near the old Turkey Creek mines that would become known as Zincite. By 1890, he and his partner controlled a steam-powered mine where they employed eleven miners and laborers. Others discovered new ore at Aurora, in Lawrence County, thirty miles east of Granby. Few doubted that more deposits would be found and that prospectors would make the discoveries, as they always had. Many aspiring miners considered the expansion of wage work a help rather than a hindrance by providing a new, more ready means of finance. They considered their hardships a necessary sufferance in order to establish economic and social independence. Most did not notice the incremental industrial changes that threatened their chances. They still easily imagined succeeding like the men they knew and worked for had done, as their common background and culture rendered doubt into evidence of individual inadequacy, an unmanly failure of nerve. At a time when many men might have questioned their future in the industry, aspiring miners instead took greater risks, both financial and physical, in ever more desperate efforts to become prosperous themselves.[21]

In the 1880s, prospectors worked in all parts of the district, in fields old and new, large and small. "There is a large body of keen, hard-working prospectors," Clerc reported in 1882, "who during the season wander from place to place, live in wagons, under tents, or in the open air, and carefully observe and follow every real or supposed indication of ore." Although some were as footloose as this depiction suggests, many prospectors worked on undevel-

Prospecting at Hell's Neck, Missouri, ca. 1890. R620, 1-036.
Courtesy State Historical Society of Missouri Research Center, Rolla.

oped land near existing mines, often under contract with the same leasing company. In some ways, the growth of small mining companies helped aspiring miners. Prospectors took wage work in developed mines as a means to finance their own digging. By the 1880s, mining companies, particularly those digging deeper for zinc, paid higher wages than ever before and had little interest in exploring the shallow ground that best suited small-scale methods. Consider a group of prospectors, described in an 1881 report by the Missouri Bureau of Labor Statistics, who funded themselves by "working for wages in the day time" and then "club together and sink a shaft by working four and five hours every night until the work is accomplished, and thus, by hard work, they try to better their condition." Eight years later, C. C. Woodson described an identical system. The laborers he met "save their wages, and when a hundred or so of dollars have been accumulated, sink it into some prospecting shaft." If unsuccessful, he noted, the prospector "takes his rope and windlass and bucket to another spot of land, digs another hole and prospects again, keeping on thus until ore is discovered." In 1891, the Missouri mine inspector counted 1,054 prospectors in Jasper County, 163 in Newton County, and 168 near Aurora in Lawrence County. The latter camp, where a farmer had discovered deposits of zinc ore in late 1885, re-created the excitement of the 1870s and in turn, again, seemed to validate expectations of future success. In 1888, a local miner boasted that now Aurora "is pre-eminently the poor man's mining camp of the world. Many have made money from the first day's diggings."[22]

Yet the economics of zinc ore mining limited what was possible for even the most successful prospectors. Their small companies, equipped only with human and animal power, could not produce enough volume to make much headway. Companies in Aurora, for example, each earned an annual average of $2,429 for zinc and $352 for lead in 1890. That left only modest incomes for company partners after they paid wages and royalties. Companies that did not have the equipment necessary to reach the deeper, richer deposits lost out to those that did. Already by 1888, according to H. R. Ruby, an Aurora miner, many "had to stop and abandon good claims, either on account of inability to purchase machinery for draining, or because they were unable to get down through the rock to where the best ores lay." Ruby admitted that only the companies with the means to go further "will make a pile of money," not the prospectors or small operators.[23]

Inspector Woodson, far more skeptical than most observers, nevertheless marveled at how miners maintained their faith in the poor man's camp. For thirty years, miners had worked the shallow ground deposits from Granby to Galena. The absence of new major discoveries between 1877 and 1885 threw miners back into the established camps, where they explored the ground around and between existing mines. Only the Zincite and Aurora discoveries offered evidence that new camps might emerge. Still, at Granby Woodson met an African American prospector who had spent fifteen months digging a seventy-foot-deep shaft "without making a dollar." Every day he could be seen "working away as faithfully as he did the day he commenced." He was "now beginning to get a few 'shiners,'" or evidence of ore, and so had renewed hope. Woodson might have dismissed this man's behavior in racist terms, but white miners acted no differently. Another miner had invested twelve years of savings in a failed prospect. Although "as poor as when he began," Woodson noted, he was still "rich in anticipations." "He will save his wages," he explained, "expecting that at last the next prospecting will prove a bonanza." He interviewed another miner, who along with three partners worked a leasehold near Carterville. They "operate the mine by themselves without any outside help, each man sharing alike." The man earned $546 the previous year and saved eighty-two dollars. He lived with his wife and one-year-old daughter in a three-room house that they rented for five dollars a month. Woodson noted that the house had "a very poverty-stricken appearance, no carpet on any of the floors, poorly furnished and lacking in cleanliness." Rather than save his money to improve these living conditions, however, the man reinvested the surplus in the mine. "Owing to this gambling spirit," Woodson concluded, "but few miners seem prosperous or own their own homes." The promise of the poor man's camp provided logic that encouraged miners to endure its poverty.[24]

In interviews, Woodson learned that miners and their families often invested all of their resources, no matter how meager, in the hope of discovering ore or mineral. In Carterville, he met a family of five that included a miner and his wife, her mother, and her two younger sisters. The husband earned $1.75 a day working in a mine but only for nine months during the year. He spent the other three months prospecting. To cover his lost earnings, the wife and mother took in washing. Altogether they earned $657 in 1889: $379 from the husband's wages, $186 from the women's washing work, and $92 from prospecting. This meant they could afford to rent only a two-room house, with all five sharing the single bedroom. Similarly, a man and son who worked as mine engineers earned $751 a year, combined, to support a household of six. They spent any leftover money on prospecting efforts. Others fared worse. Nearby, Woodson visited a prospector, his wife, and their five children, the oldest of which was ten. The father worked for wages in a mine and prospected on the side. The family owned the two-room house they lived in but not the lot on which they raised vegetables for household use. Because prospecting reduced the father's availability for wage work, he earned $310 that year, an average of $5.96 per week. Although the wife sometimes took in sewing in exchange for butter and some of the children did odd jobs for neighbors, the family was "poverty-stricken and dirty." The house had "no books, newspapers nor pictures; children have scarcely any clothes on and very dirty looking; wife can't read." What surprised Woodson was that this family differed little in its outlook and ambitions from neighbors who earned more. All risked, consensually or not, their future security and present comfort to search for undiscovered metal, a quest that demanded the work of whole households. Women provided crucial labor and earnings to the effort. Such precarious striving stretched the ideal of manly independence awfully thin.[25]

The highest-paid wageworkers also invested in prospecting; poor man's camp aspirations were still widely shared. In Carterville, Woodson interviewed a father who worked as a ground boss and his two sons, the older a prospector, the younger a mine laborer. The father made $2.50 a day; he earned $480 working eight months in the year. The youngest son earned $2.25 a day. Together they prospected a promising claim that yielded ore worth $250 in 1888. The older son and his wife still lived with his parents, however, in a four-room house along with the other brother and two grown sisters, one of whom was also married. To provide some visual privacy, they partitioned the rooms with a board wall that went halfway to the ceiling. The father told Woodson that he had been mining in the district since 1872. He "had money when he started; has none now." Still, Woodson reported, they ate well, read newspapers, and sent their children for music lessons. He interviewed another Carterville man who worked as a ground boss and prospected on the side, with earnings of $469 and $250, respectively. This man did not want to follow his older neighbor's example. "Growing tired of mining," Woodson noted, but "thinks he will be compelled to remain at mining until he strikes a lead and makes sufficient to go into some other business." "The only hope for a miner," the man explained, "is to strike a good lead and quit mining." This man was stuck in the logic of the poor man's camp: success was the only imaginable means of escape.[26]

District miners steadied themselves on stories of poor men who hit pay dirt. "The successful few are heard of not once, but time and time again," an exasperated Woodson reported. "The story of how one man digging a well for water in his backyard found, instead of water, a vein of ore from which he took out $100,000, is told around almost a hundred thousand times." Most observers relished the remarkable tale of how the district had come to be. "It is true we have no millionaires," Thomas Davey admitted, "but the money made is more diffused among the masses. We have many men who have become worth from fifty to two hundred and fifty thousand dollars, but they started with almost nothing but their hands only a few years hence." It was true, and many believed it would continue to be so. To Clerc, in 1887, "it is a settled fact now that the lead and zinc deposits of this locality are permanent and inexhaustible." Anyone could tap them, he concluded, because "the mines have always been and are yet being worked with the capital which the mines themselves have produced, and all our mine operators, with but few exceptions, have acquired all their wealth from the mines." No one remembered, however, those who failed. "The trouble is," Woodson explained, "the savings which could certainly buy a comfortable home are dumped into holes in the ground, which possibly may result in gains of thousands, but which probably and generally result in the loss of all." Still, the miners and prospectors piled up evidence of their belief that past glories could be re-created. Throughout the district, Woodson reported, "the earth is upturned—holes and crevices, piles of debris, rickety scaffoldings, abandoned shafts, masses of rock and ore—a result of the prevailing system of mining."[27]

In the context of new structural limits, however, the system's enduring promise encouraged miners to take greater risks, not just with their family finances but also with their own bodies. With less experience running paying mines, narrow operating margins, and little means to develop zinc mines for the long term, they worked faster, with little regard for safety. "Each miner," Clerc said, "tries to get as much as possible out of his own lot, is only interested in it as long as he expects to work it." Few miners built timber supports in the shafts or inspected the roofs for loose rocks. "The roof and pillars are badly trimmed," Clerc concluded, "and in many cases dangerous, fatal accidents being distressingly common." Woodson observed similar conditions. He counted twelve fatal accidents and twenty-five serious accidents in 1889, most a result of rock falls. These accidents, he said, resulted from men "working the mine for immediate profit, regardless of the future."[28]

Neither the Knights of Labor, the nation's largest, most expansive union by the mid-1880s, nor their allies in government could convince Joplin miners to see their interests differently. The Knights built strong local assemblies in Missouri, particularly in St. Louis, as well as in the coal-mining counties in the central part of the state. The Missouri Knights coupled robust organizing efforts, including in the Joplin district, with successful campaigns to enact new legislation based on the national body's 1878 declaration of principles, primarily the creation of a state Bureau of Labor Statistics and mine safety laws. In 1879 Missouri joined Massachusetts, Pennsylvania, and Ohio as the only states with agencies dedicated to documenting the lives and labors of workers. The new commissioner of labor statistics, W. H. Hilkene, began surveying workplaces that summer. In 1880, the Knights supported a Missouri law that for the first time required county courts to oversee the inspection of mines, although only those that produced coal. In 1887, a year after a wave of strikes across the state's coalfields, the legislature created a new statewide mine inspector's office with jurisdiction over all mines, regardless of product. Miners in the Joplin district could not escape the influence of the organization and its political efforts to protect and empower workers. Some of them were attracted to the Knights, a sign of ongoing interest in antimonopoly ideas, but most of these men were not prepared to abandon their commitments to capitalism.[29]

The Knights of Labor established several local assemblies among workers of all kinds, including some miners, in the Joplin district in the 1880s. The first organizer in the area, John Loftus, was the leader of an assembly of coal miners in Stilson, soon to be renamed Scammon, in Cherokee County, Kansas, whose work was controlled by smelting companies in nearby Pittsburg. The Irish-born Loftus organized the district's first assemblies in Galena, Webb City, and Granby in late 1879 in the midst of the lead crisis. Although these assemblies lasted less than two years, the Knights continued to send organizers to the region, most of them from St. Louis. Their appeals led to the formation and re-formation of unstable assemblies in Joplin, Webb City, Carterville, Galena, Zincite, and Oronogo throughout the decade.[30] Most were "mixed" assemblies, meaning workers of all occupations belonged. Miners and mine laborers predominated in Webb City, Carterville, Zincite, and Galena. At Webb City, for example, some of the most "untiring and active" local leaders, R. J. Davis, Edward Armstrong, and P. McEntee, mined metal. Other local leaders were carpenters, blacksmiths, and stonemasons. The assemblies were never large, usually with no more than a few dozen members, and were prone to disorganization.[31]

In contrast to the striking Leadville Knights, the order's members in the Joplin district favored moderate reform. They boasted some local public influence. Knight J. G. W. Hunt, for example, was both a miner and a real estate developer in Carterville, where he was elected justice of the peace in 1882. Some in the order campaigned for the Greenback-Labor Party, which continued to appeal to farmers and workers with calls for antimonopoly measures, a cheaper currency to help debtors, and a protective tariff. They wanted to preserve the competitive economy, not overthrow it. Nothing made that clearer than the tendency of the Missouri Greenback-Labor Party to run joint tickets with Republicans, as in the 1884 presidential election. They did well in that race, but otherwise independent Greenback-Labor candidates failed to get more than 15 percent of the vote in southwest Missouri. Meanwhile, the Knights organized the first public observance of Labor Day in Joplin in September 1886 with a parade that featured bands, civic leaders, and city officials. Webb City's Knights requested badges featuring the words "Miners Assembly" for the following year's parade. Local miners not only had ample opportunity to join and support the Knights, but it was even respectable to do so.[32]

Some miners were drawn to the Knights' vision of the "cooperative commonwealth." The order heralded worker cooperation as a primary means to counteract the "aggressiveness of great capitalists and corporations" and escape the vicissitudes of wage labor. Cooperative endeavor, the Knights argued, would help workers share in the wealth they produced. In late 1879, Loftus reported that members of LA 1373 in Webb City "are going in the cooperative business with vigorous measures." They sold stock, at five dollars a share, to fund two ventures: a general merchandise store and a plan "to lease 400 acres of ground to prospect on." Although LA 1373 did not become a lessor of mining ground, it did open a cooperative store with twenty-six members and $400 in operating capital. The store failed within the year, however. Other members, perhaps, recognized the ideals of the Knights in their mining partnerships. The ground boss father and his sons who prospected together in Carterville, discussed above, all belonged to the Knights as late as 1889. Their prospecting effort, with its emphasis on productive labor and self-responsibility, in many ways realized the highest aims of the cooperative ethic; many in the Knights were sympathetic to small capitalists like them.[33]

As these miners interpreted the cooperative ethos of the Knights to suit their aspirations, just like other members did across the country, their aims strained the dominant "noble and holy" culture. The order valued the achievement of individual workers, even those who were petty capitalists, so long as they remained within the confines of a republican ideal of true manhood that emphasized obedience to community and family, temperance, and mutual assistance. While Joplin miners aspired to be respectable men, their often reckless pursuit of commercial gain never quite fit with this valorization of sober self-restraint. Even Loftus, who led Stilson LA 535, one of the Knights' oldest assemblies, seemed tempted by the poor man's camp. Within months of arriving in Joplin, he began prospecting with a convert's zeal. "I have been prospecting all summer," he informed Terence V. Powderly, the order's national leader. "I have the best prospect around here." As a skilled coal miner, Loftus had high hopes of making money because the leasing system allowed miners to "go to work & prospect & if you strike up lead or zinc ore you can pay your 10 per cent royalty out of that." He wrote to Powderly to urge the Knights to develop cooperative mines in the district, like the one LA 1373 proposed, and "make this the *great central point* of cooperation in our order." The Knights did so elsewhere, with investments in several midwestern coal mines. Loftus hoped the endeavor would provide fair conditions and security for workers. But he also wanted to make money himself. He punctuated his report with a claim that clashed with most of what the Knights stood for: "A man may be poor here one day & rich the next."[34]

Many of the Joplin miners who joined the Knights could not fully accept the ethic of solidarity that bound the union together, especially when asked to support wageworkers elsewhere who did not share their faith in capitalism. Beginning in 1884, the Knights required all assemblies to pay into an "assistance fund" to support strikes. Following the order's 1878 declaration of principles, its national leaders opposed strikes in all but the most justified and winnable situations, but as hundreds of thousands of new members swelled the organizational ranks in 1885 and 1886, they struggled to maintain discipline. Local assemblies across the country launched a wave of strikes, most spectacularly against Jay Gould's southwestern railroad system, which called for greater contributions to the assistance fund. Miners in Webb City LA 6240 and Zincite LA 7278 did not want to pay because they did not see the benefit for themselves. In 1887, LA 6240 requested "exoneration from paying Assistance Fund." They explained that they were "a Lead Miners' Assembly, never had a strike, and are not likely to require assistance." Although their request was denied, the assembly survived. The men in LA 6240 did not necessarily oppose the strikers, since they continued to advocate for the "industrial masses," particularly in the elections of 1888 and 1890 when LA 6240 campaigned for the Union Labor Party, the latest guise of the Greenback-Labor movement. But they could not imagine themselves ever going on strike, because doing so would mean directly opposing their employers, some of whom were also members, such as the father and son pit bosses in Carterville, and thus risking their own prospecting dreams. Their ideal of independent manhood rested on economic risk taking, not reliance on assistance from other men. Moreover, with more time and a little luck, they might soon find themselves employing wage laborers in their own mines, the end goal that informed their opposition to monopoly in the first place. The Knights no longer seemed right for those miners who had joined. After 1888 only one assembly remained active in the district: LA 6240, which folded in 1891 amid a dispute about its financial accounts.[35]

The majority of miners and prospectors showed even less affinity for the state labor reforms the Knights helped create. At the time of Woodson's visit, most miners, regardless of the size or method of their operation, violated the state's mine safety regulations. The 1887 law not only created the office of state mine inspector but also required protective measures in all mines, including safety catches on hoisting equipment, covered cages for lowering men into the mine, escape and ventilation shafts, and adequate roof supports. When M. L. Wolfe, the first inspector, visited Joplin in the summer of 1887, he found most mines "being worked in a desultory and primitive manner" by miners who did not know that state law now applied to them. "The violation was not intentional," he noted. Maybe not, but they should have known because political supporters of the Knights of Labor had pushed for it.[36]

Many miners, especially those wringing what they could from small, nonmechanized operations, considered the law a risk to their livelihoods. Wolfe thought that most would comply where "practicable" but also accepted arguments that small-scale miners, especially prospectors, "cannot comply literally" because of the expense involved. Woodson found that nothing had changed when he inspected district mines two years later. Rather than enforce the law, he, too, resorted to the logic of the poor man's camp. "As the principal mining done is by miners or small operators, who have limited means with which to open up their mines," Woodson explained, "they would be compelled to abandon their mines if the law was enforced requiring them to put in covered cages, with safety catches, appliances, etc." The burden of safety equipment would fall hardest on cash-strapped prospectors and small-time operators, he added, who simply could not afford it. Despite his misgivings about their "gambling spirits," aired in the same report, Woodson concluded that the strict application of a law meant to safeguard the health and safety of miners "would be death to the lead and zinc industries of Southwest Missouri." Both he and Wolfe called for the state legislature to develop a new, separate law to govern lead and zinc mining "that did not hinder or injure our poorer class of miners." The consequences were real: ten men died in Jasper County mines the following year.[37]

And so, paradoxically, the men charged with enforcing a law meant to protect miners argued instead that the miners of the poor man's camp should ignore labor union and government efforts to rein in their entrepreneurial risk taking. The logic and power of poor man's mining had impressed Wolfe and Woodson, particularly because of the apparent willingness of Joplin miners to accept the risks of industrial capitalism in the hope of hitting pay dirt. "The miners," Wolfe reported in 1887, "are intelligent and seemingly contented. Strikes are unknown." That Webb City's Knights, the strongest union voice among district miners in the 1880s, said much the same thing seemed to confirm his analysis. Wolfe credited the system of prospecting and leasehold mining for the harmonious labor relations. It offered miners "better opportunities for acquiring wealth," he explained, and in doing so "makes them free and independent."[38]

The continued expansion of zinc-mining operations in the 1890s forever closed those opportunities for most miners in the district. The zinc boom energized the sale of leaseholds and land, which now traded at the Kansas City Mining Exchange and the new Joplin Mining Exchange. "Jack is up," one newspaper reported in 1892, as the price of zinc ore went above twenty-six dollars per ton. Investors, many from the East and abroad, formed new land companies that consolidated control of prime mining tracts. This was similar to contemporary developments in coal, particularly in Appalachia, and in eastern Missouri lead, but rather than run the mines themselves, these investors maintained the customary leasehold system. They replaced human prospectors, however, with steam-powered churn drills to quickly survey the quality and extent of their holdings before leasing. Their leases, in turn, favored bigger, more mechanized mining companies. Poor men were steadily pushed out. For most miners in the Joplin district, remaining free and independent had never been more difficult.[39]

A small number of investors took control of large tracts of mining land in response to the rising profitability of zinc production. Local land and smelting companies with longtime holdings led the way. Granby Mining, which owned over 15,000 acres in Jasper and Newton Counties, and Center Creek Mining, which leased 200 acres between Webb City and Carterville, both demonstrated the profitability of leasing land to zinc-mining companies. Center Creek Mining was so confident that it bought 160 acres of its leasehold from the Webb family in 1890 for $315,000. Others followed. In early 1891, the Picher brothers and three other local investors bought 1,000 acres on the eastern edge of Joplin that prospectors had explored only at shallow depths. They chartered the Rex Mining and Smelting Company with $250,000 in capital stock to lease and manage the land. Outside investors also consolidated large landholdings. In the summer of 1891, for example, Richard Heckscher and August Heckscher, cousins who owned anthracite mines in Pennsylvania, formed the Empire Zinc Company, capitalized at $750,000, to purchase more than 200 acres near Joplin. By the summer of 1894, 186 of the 332 mines in Jasper County were on land controlled by one of seven such leasing companies. Their holdings, all leased or subleased to local mining companies, yielded 71 percent of all zinc ore mined in the county that year.[40]

With higher stakes, mining companies accelerated the intensification of methods. While visitors might still marvel at how "the surface of the ground is completely covered with old dump piles," engineer John Holibaugh reported in 1894, "these are only the relics of the early days of mining for shallow deposits of lead and zinc." The present and the future belonged to companies mining deeper ore. "To-day," he continued, "we find modern ore dressing and concentrating plants in full operation among these old dump piles" because of the rising productivity of companies "producing the ore from a depth of 150 to 200 ft." In order to handle greater quantities of ore and waste rock, mining companies invested in more steam-powered hoists, pumps, and jigs. The largest companies on Rex Mining and Smelting land, for example, "all have good steam hoisting and pump plants" as well as steam-powered concentrating plants. Although many companies continued to mine in the old way, with animal and human power, Holibaugh concluded, "the greatest and most marked improvements have been made within the past six years and the next few years will see even greater changes in the mining and handling of the ore."[41]

New, more accurate prospecting methods supported these investments. Land companies replaced the traditional means of prospecting with crews using steam-powered drilling rigs to survey geological formations. Drill crews could sink a prospecting hole up to 400 feet deep in two weeks. Rex Mining and Smelting was the first land company to rely on drills, which quickly "proved large deposits of lead and zinc ore." The company leased mining lots based on the drill record. Accurate readings of the location, depth, and quality of ore deposits attracted the best-equipped mining companies, allowed land companies to more accurately assess royalty charges, and lent assurance of profits for all concerned. The other large land companies soon adopted the same practice. They employed independent drilling firms, such as P. L. Crossman and Brothers, to do the work on a contract basis. A drill survey proved especially valuable at depths below 100 feet, where the richest and thickest zinc deposits lay. In 1899, P. L. Crossman and Brothers reported it drilled "more holes from 200 to 250 ft. deep than any other depth." While the basic leasehold pattern of mining did not change, the adoption of drill-hole prospecting favored the largest, most mechanized mining companies.[42]

As opportunities for traditional prospecting dwindled, most miners entered an increasingly permanent state of wage labor. Between 1891 and 1893, the number of prospectors at work in Jasper County fell from 1,054 to 500, in Lawrence County from 168 to 86, and in Newton County from 163 to 70, most of them seeking shallow deposits that the larger zinc-mining companies did not want to pursue. Those companies hired bigger crews of mine laborers. Mining firms on land controlled by Center Creek Mining, Chatham Mining Company, Eleventh Hour Mining Company, and Rex Mining and Smelting employed ten to fifteen wage miners each, on average, in 1894. The J. J. Luck Mining Company, meanwhile, employed between twenty-five and thirty-five workers at its operation in Galena. These companies were still small compared to other mining firms in the United States. In the eastern Missouri lead field, for example, the St. Joseph Lead Company employed 583 miners to operate eight shafts. Still, more and more miners worked for wages in southwest Missouri each year: up from 3,578 in 1891 to 4,117 in 1893. Many of these additional mine laborers came from area farms with little or no mining experience but hoped to make a good wage and perhaps somehow realize the promise of the poor man's camp. By 1892, however, Granby Mining offered the only realistic opportunities for them on its shallow holdings in Newton County, where in an attempt to re-create the magic of the late 1860s, it provided small-scale miners all of the equipment they needed in exchange for half the proceeds of their efforts. Still, this model was the proverbial exception, which even Granby Mining admitted when touting itself as "one corporation at least, that gives every laboring man who chooses to work on its lands a chance to make a fortune."[43]

For most miners, the consequences of this transformation would have been difficult to perceive in the midst of the unprecedented national economic crisis that began in January 1893. By May, the bankruptcy of several national railroads and other companies had led to the nation's worst financial crisis yet. At least 3 million workers were unemployed by the end of the first year of a depression that would last four years. For the first time since the early 1870s, the price of zinc ore plummeted, from $26.00 per ton in August 1892 to $16.50 in June 1893. It continued to fall that summer, down below fifteen dollars per ton, the lowest price since 1877. As mining companies failed to find buyers, most simply stopped production. "Many of the large operators closed down," Holibaugh noted, "while others worked only a small force in prospecting and developing, so that zinc mining was almost at a standstill." After producing over 8 million pounds of zinc a week in April 1893, district companies mined less than 2 million pounds a week in August. Operators told the Missouri state mine inspector that they would not reopen their mines "until such time as they can be made to pay." Miners had weathered past economic crises by prospecting new lands and developing zinc production, but they were more dependent on wage labor in 1893 than ever before. Although many still believed in the possibilities of the poor man's camp, they were now forced to consider its limits and perhaps its demise.[44]

The first tremors of the depression were felt in the district when union coal miners in southeast Kansas went on strike. In May 1893, thousands of miners from Scammon, Weir City, and Pittsburg protested a wage cut by stopping all coal production in Cherokee and Crawford Counties. These miners still belonged to the Knights of Labor but had recently also affiliated with the new United Mine Workers of America (UMW), a national union of coal miners formed in 1890. Their strike, supported by the UMW and soon joined by coal miners in central Missouri, deprived both the largest mining companies and the area's zinc smelters of fuel. The district's most mechanized mines felt the coal shortage first. Within a week the Joplin press reported, "The effects of the strike are already becoming manifest in the closing down of some producing mines in Joplin, Carterville, and Webb City." More closed in the weeks that followed for want of coal but also because area smelters stopped buying ore. The potential long-term closure of the smelters presented a greater threat. Due to the complicated process for smelting zinc ore, smelting companies had to keep furnaces charged in order to avoid costly damage that would need repair. Most had enough coal reserves for a few weeks. Smelter closure, a Joplin reporter feared, would mean "a failure of a large portion of the demand for zinc ore, a fall in the price of that, until it would not pay to mine it, and at least 5,000 men in the zinc region would be without work until the smelteries would again be started." He hoped the union would stop the strike before its effects decimated the Joplin district. "This would mean the making of a scene of suffering which the miners' organization can scarcely afford to assist in causing" and "would do their cause more injury than any other scheme that could be devised."[45]

As the strike wore on into June, many in Joplin blamed the UMW entirely for the falling price of zinc ore, misinterpreting a symptom of the depression as its cause. According to the editor of the *Joplin Morning Herald*, a Democratically affiliated paper, the union's "demand is made in the name of Labor, but examination discloses that it is essentially selfish, because only a particular branch of labor would be benefitted, and that at the expense of other branches." He targeted UMW "walking delegates," union representatives who coordinated regional action, as "agitators" whose influence would ensure "a continuance of the strike to the bitter end." Two weeks later a contributor stated simply, "The coal miners' strike is causing mines to close down." Business leaders agreed with the faulty analysis. "In consequence of the strike in the coal mine the business this month will be very small," reported William F. Sapp, a Galena-based mining company boss.[46]

Antiunion animosity peaked when striking coal miners arrived in the Joplin district looking for work, reportedly for wages as low as $1.25 a day, far below the standard wage. "Ye men of Joplin who depend on your labor to support yourselves and families and are used to good wages, what do you think of that?" a *Morning Herald* correspondent asked. Rather than "sympathize with strikers who came here to lower the price of your wages in order that they may barely make a living whilst they bring about a better price for their own labor," the writer advised, "we should look out for our own interests first." The paper's editor lauded the absence of labor unions among the metal miners by comparison. "Labor is better paid and is better contented in the Joplin district than in any other mining district in the country, and it is a noteworthy fact that there is an absence of labor organizations. The miners are independent workers." According to his analysis, the traditions of the poor man's camp focused the minds of Joplin miners on new discoveries as the way to prosperity, even if they worked as mine laborers. "Strikes in the lead and zinc district are quite different from strikes in the coal district," he quipped, "and much better for everybody."[47]

Despite these tidy explanations, miners and others soon realized that the metal industry faced bigger problems. The price of zinc ore continued to fall as smelters in Illinois also stopped purchases. "The bottom seems to have fallen out of the lead and zinc market and prices can hardly be worse than they now are or have been for several weeks," the *Morning Herald* reported in late June. "The coal miners' strike is a factor in the matter," the paper still argued but now admitted the concurrent effects of "the prevailing business depression over the country." In early August, as the UMW strike collapsed after the coal operators began employing African American strikebreakers, observers turned their full attention to the broader crisis. One account declared the second week of August the district's dullest for business in ten years. A mine owner explained that he would not reopen his mine "until the price of jack gets back to $24 to $26 per ton." Soon, all of the large mining companies in the district, including all of the largest zinc producers, had ceased operations. Everywhere else, companies cut wages.[48]

Some miners considered organized action in response to these developments. In Webb City, a small group revived LA 6240 that summer. "I think that next report that you get from us will be very encouraging," miner A. A. Phillips informed the Knights' national office. "I never saw so many that were anxious to join when approached in the right way," he explained. "Times is very hard here and work very slack and people are investigating the causes of it." The assembly gained only thirty members, however—fewer than had joined in the 1880s. It folded in early 1894.[49]

The crisis prompted others into public debates not seen since the early 1870s. In late August, 300 miners from Webb City and Carterville met in a series of open meetings to discuss what to do when "the pumps were stopped, wages lowered and the miners and their families were suffering." Some wanted to strike. At a meeting in Prosperity, a new camp south of Carterville, most called for moderation, no doubt influenced by the collapse of the UMW strike the week before. They agreed that the "pressure of earnest moral suasion be brought to bear on the lease holders by personal application and petition to have them" open the mines. Eight hundred miners and mine laborers signed.[50]

In doing so, they appealed to a poor man's camp tradition of fairness and equality of opportunity between men who came from the same social class and shared common interests. Their petition included a statement accepting lower wages for the duration of the crisis, as long as operations restarted. They justified the concession by asserting their manly responsibilities as heads of households. "The drift of nearly all the speeches was that the men needed work to support their families," according to one report, "and that they were willing to take reduced wages until the price of ore came up." This paternalistic explanation not only ignored their reliance on women's labor but also belied a sense of crisis among men who expected to be self-made. They remained committed to capitalism, even if only out of desperation. The miners "were willing to suffer some loss on account of the low prices of ore." Some spoke like men who still expected to become mine owners. A miner named Cox said that while "he did not advocate low wages" because of pride in his craft, "it was the sheerest nonsense to expect men to run their plants at a loss, just to give the miners work at full pay. Men had better work for a dollar a day than not work at all." A miner named Patton denounced talk of strikes as "just the kind of talk that drove capital away, tied up money in bonds and left such fellows to go hungry." He wanted work, "and it took men with money to give work during such times." "Treat your employer right," Patton concluded. These men still accepted poor man's camp logic. "The miners are in perfect accord with the operators," one newspaper concluded, "and are willing to work at reduced wages for a time, just so they get work."[51]

Some miners looked with hope to the smaller, owner-operator companies that seemed to fare better than the larger outfits. Due to the closure of the mechanized firms, district miners produced only 86,800 tons of zinc ore in 1894, down from the all-time high of 128,200 tons in 1892. Most of that production came from companies working shallow deposits. "The production was confined entirely to subleasers and small concerns who were satisfied to work their mines if they could only make miners' wages," Holibaugh reported. From Galena, Sapp noted that the only companies that remained active were "those where the miner is obliged to labor for his daily bread." Desperation was still seen as an important source of virtue. According to Francis LaGrave, the Missouri state mine inspector, "the small producer, frequently employing only the labor of himself and family, can always make a living at prices which would mean ruin to larger operators."[52]

Meanwhile, the manly appeals at Prosperity seemed to work, as landholders opened land to unemployed miners for prospecting. This was done mainly to forestall protests. For mine laborers who had not seen the roaring events of the 1870s, however, these developments suggested that the poor man's camp was not dead. Even LaGrave, who should have known better, agreed in his annual report. "The closing of these mines," he observed, "seems only to have changed the manner of occupation of miners, as there are comparatively few men out of employment around the mines." Some of these prospectors turned unemployment into pay dirt. They "have made several valuable discoveries of ore," LaGrave reported. A poor man could not only survive but still thrive in the Joplin district, another industry expert reasoned, because he "invests nothing but his labor and if he can secure the food to renew his physical energies he can accomplish more than capital." Miners did not object at the ballot box in November 1894, despite the chance to vote for the insurgent People's Party, which inherited the Greenback-Labor legacy. The winning Republican congressional candidate carried both Jasper and Newton Counties; the People's Party candidate finished a distant third with less than 15 percent of the vote.[53]

When larger companies resumed production in late 1894, district watchers saw only resilience, not augurs of change, despite the ongoing national depression. Prices remained low; zinc ore sold for an average of $16.86 per ton in 1895, 25 percent below the all-time high average of $22.51 in 1890. "In spite of this," LaGrave reported, "the mines are in a prosperous condition, and while not worked to their full capacity, are steadily pushing developments, expecting that the near future will bring greater demand and better prices." Larger firms resumed buying land and leases, especially where prospectors had recently made new discoveries, such as Duenweg. These companies hired more and more mine workers. LaGrave counted 4,366 mine laborers in 1895, up from 3,341 a year before, a jump of 30 percent. Land companies also resumed systematic prospecting, albeit with drill crews. "I may judge from the amount of drilling and the large number of shafts being sunk in what may be termed new territory," he added the following year, "the probability of an increased production for several years to come turns to an assured fact." Market conditions in the first half of 1896, when ore neared twenty dollars per ton, seemed to prove him right.[54]

Yet the contours of power in the district had changed a great deal: the largest land and mining companies emerged from the depths of the depression with more control over production than ever before and a realistic new sensitivity to downward swings in the metal markets. Economic uncertainty in the summer of 1896 exposed the new dynamic. In July, the Democratic Party's presidential nomination of William Jennings Bryan, a former congressman from Nebraska with strong ties to the People's Party, spooked the nation's manufacturers. Some feared that Bryan's demands for inflation and cheaper credit would spark a new, more severe economic crisis. As many companies slowed production to await the outcome of the election, commodity prices slumped. Zinc prices fell too. "There is an uneasy feeling among the mine operators in regard to future prices of ore until after election," an observer in Joplin noted in August. Many mining and land companies ceased production altogether, while "those who are working their mines are reducing the number of men employed and are making only enough ore to pay expenses and keep the men employed." Unlike during the recent crisis, however, these companies did not allow unemployed miners to prospect on their holdings. For example, when Scott McCollum and four other miners approached Dan Dwyer, the superintendent of the Rex Mining and Smelting Company, to lease a plot of land, he turned them away. Dwyer "said he would not lease any until after the election, and if McKinley wasn't elected, didn't know as he would open up any ground at all." "Every place a man went," McCollum explained, "it was just about the same—nothing 'till after the election."[55]

Prospectors such as McCollum, following the path of so many capital-poor but ambitious miners before him, now had nowhere to turn. "When mines are closed down and land owners will not lease to men to prospect," he asked, "what can they do?" The question did not yield easy answers.[56]

CHAPTER 4
THE JOPLIN MAN SIMPLY
TAKES HIS CHANCES

While Joplin's miners looked backward, they became more and more entangled in the dynamic web of a rapidly maturing industrial economy. The
market variables affecting the lead and zinc industry were increasingly inter-
connected in multiple new ways. The price of lead now depended on the
price of silver, because many western mining districts, such as Leadville,
Colorado, produced large quantities of lead as a byproduct of more lucrative silver mining. The price of zinc followed the fortunes of coal, as the 1893
Kansas strike demonstrated. Continued railroad expansion throughout the
western United States meant that changes in any one mining district were
quickly felt in them all. By the 1890s, Joplin miners faced a growing local
system of wage labor in an industry regularly whipsawed by the dynamic and
complex relationships of modern capitalism.

They could not navigate these relationships without also reacting to the
efforts of miners' unions in the 1890s to challenge the power of corporate
employers. Until then, most Joplin miners stood aloof from the national
labor movement, even though they remained distrustful of monopolies.
Some had rejected the Knights of Labor in Leadville. Others had joined
small, short-lived assemblies around Joplin in the 1880s. Viewing themselves as future owner-operators, most did not see the relevance of big national unions premised on permanent class divisions. The Kansas and Missouri coal miners' strike of 1893 rattled their sense of exceptionalism, but
its outcome seemed to validate a turn away from the Knights and solidarity
with outsiders, in spite of the depression.

At the same time, however, metal miners in other districts cast their lot
with a new union movement of wageworkers who understood their interests
in opposition to those of the mine operators. In May 1893, diverse groups
of former Knights in western gold, silver, and lead mining camps organized
a new union, the Western Federation of Miners (WFM), to achieve better
pay, win workplace rights, and reduce the hazards of underground labor.
Two years later, miners in the iron and copper ranges of Minnesota, Wisconsin, and Michigan's Upper Peninsula formed the Northern Mineral Mine

Workers' Union (hereafter referred to as Northern Mineral) to achieve similar goals. Although neither union's initial regional focus included the Joplin district, their confrontational actions, particularly the WFM's strikes, would soon pull Joplin miners into the heart of the struggle but not as allies. In the face of emboldened workers, many of them immigrants, mining corporations looked to the Joplin district for miners who might work in place of strikers. The unions, in response, belatedly sought to organize Joplin miners into a labor movement that split over political philosophy, ethnicity, and ultimately, how to grapple with the future of capitalism. Forced to choose amid the structural changes that recast the practice and promise of work in the mines, their decisions would reverberate throughout the industry, affect the strategies and aims of metal miner unions into the new century, and give rise to a racist, nativist, and often violent masculine culture that would, in time, transform their view of the possibilities of wage labor.[1]

Joplin miners such as Scott McCollum were not the only American metal miners grappling with the hard imperatives of industrial capitalism in the summer of 1896. After tenuous regional origins, their new unions seemed close to uniting in a national metal miners' organization with the support of the American Federation of Labor (AFL), which had overtaken the Knights as the nation's largest and strongest federation of workers. Northern Mineral affiliated with the AFL in December 1895; the WFM affiliated in June 1896. AFL president Samuel Gompers hoped to include the Joplin district in this nascent movement. He dispatched E. J. Smith, an Indiana-based organizer for the Cigar Makers' Union, to southwest Missouri in June 1896 to speak to the miners. The possibilities of this constructive effort were soon dashed by crisis. Elsewhere, the WFM and Northern Mineral both plunged into bitter and costly confrontations with wealthy mining corporations. The WFM's local in Leadville began the largest and most consequential of those strikes on June 19, the same day that Smith visited Joplin. Desperate to counter the effects of the depression, the union launched into a fight relying on a sense of solidarity that had not yet been built. Neither the WFM, Northern Mineral, nor the AFL would send another organizer to Joplin until 1899. In the meantime, Leadville's mining corporations offered men such as McCollum a solution to the problems they faced in the summer of 1896: lucrative work as strikebreakers against a union they had never before encountered.[2]

In Leadville, the WFM local demanded the restoration of a wage cut it had taken in the depths of the depression three years earlier. Bent on destroying the union, the operators of Leadville's mines refused to negotiate. John Campion, the owner of the Ibex, the camp's richest mine, hired the

Thiel Detective Agency to infiltrate the WFM local and soon learned that the union was divided along ethnic lines on strategy. Its Irish-born leaders and rank-and-file majority remained committed to the strike, while the Cornish and American-born minority wanted to return to work. Such divisions were common throughout western mining camps and in the ethnically diverse WFM, although its first and strongest local, in Butte, Montana, was overwhelmingly Irish. Campion and his fellow operators hoped to exploit ethnic divisions by hiring non-Irish miners to return to work, but the union convinced most local workers to stay away. Leadville's operators realized they needed outsiders to defy the WFM. Joseph Gazzam, superintendent of the Small Hopes Mining Company, later recalled that he "suggested that, if they decided to import labor, they bring in miners from the Joplin, Mo., district, as they were native-born Americans and would not be intimidated." Both Gazzam and his boss, Seeley W. Mudd, knew Missouri mining. They had trained together in the mine engineering program at Washington University in St. Louis in the early 1880s. Mudd proposed the idea to Campion, his former boss, who instructed the St. Louis-based Thiel Detective Agency to dispatch a recruiting agent to Joplin.[3]

Missourians had a history in Leadville. Missouri-trained mining experts developed the district's silver-lead mines in the 1870s. Many Joplin miners sought work there after the lead market collapsed in 1878, and some had helped break the strike of the Leadville Knights in 1880. Many of these miners went on to local management positions as mine superintendents and pit bosses and could have still been in Leadville in 1896. Regardless, Mudd and Gazzam would have been well aware that Joplin miners were skilled and overwhelmingly native born and also probably that the WFM had no presence in southwest Missouri. The state mine inspector's annual published reports made this information available.[4]

Thiel agent T. Z. Pickers attracted keen interest from miners when he arrived in the district in mid-August. After meetings in Webb City, Carterville, and Joplin, he reported that over 300 men had agreed to go to Leadville. "This is a good field to work in as the mines are shutting down every day and throwing morem [sic] men out of employment," Pickers explained. "Miners here only get from two to four days work a week," he noted, "and are anxious to go where they can get steady work." Although Leadville operators sought skilled miners, the district's mine laborers, now more plentiful than ever before, also showed interest. "There are a great many men here who have worked in and about these lead and zinc mines that are not, what operative calls, experienced miners," reported John F. Farley, head of Thiel's Denver office and Pickers's boss. "They are what they call here, shovelers," he explained, and "they can do anything in and about the mines, shoveling, wheeling and working at gigs." "These men are good, able bodied young men and anxious to learn to use the hammer, and are ready to go." Whether experienced miners or mine laborers, they were eager to take advantage of Pickers's offer.[5]

Not simply desperate, these miners and shovelers were acting on an entrepreneurial ethic with deep roots in the Joplin district. Despite their need, prospective strikebreakers took time negotiating assurances from Pickers: that the Thiel agency would pay their transportation; that they would make three dollars a day, the wage the WFM was striking to achieve; and that they would be able to retain work after the strike ended. Pickers guaranteed these conditions. Beyond that, he reported, none of them cared about the WFM or the strike. "The fact that there is a strike there cuts no figure with them," he explained, "providing they can get steady work." "Work is what they are after," he added.[6]

Yet when Campion hesitated, perhaps in doubt about their loyalty, Farley reframed Pickers's reports to appeal to the mine owner's obsession with foreign-born labor radicals. "Operative has circulated the truth about the Leadville strike among his miners," Farley assured Campion, "and they are anxious to go. The miners in this country do not believe in Unionism (Labor Unions)." "They have tried to form Labor Unions here, but failed," he concluded. Then Farley added, for emphasis, "They are all Americans." Here he embellished the facts because he knew Campion wanted to stoke the ethnic conflict that already plagued the WFM in Leadville. Pickers did not mention nationality in any of his dispatches from the field. Indeed, no contemporary observer or commentator had associated the lack of labor organization in Joplin with nativity or nationality before he did, at least not in any prominent publication. Nor had anyone, local or otherwise, referred to nationality during the 1893 United Mine Workers of America (UMW) strike, when a significant minority of UMW miners had been born abroad. Still, Farley's point happened to be true, since more than 95 percent of district miners were native born in the 1890s, an increased proportion since the 1870s. Farley's assertion did not forge a causal link, but it did create a new way of understanding the minds of Joplin miners and, importantly, a new way for Joplin miners to understand themselves.[7]

The men who considered going to Leadville had no shortage of information about the controversy or danger involved. Despite a divisive presidential election campaign, Joplin newspapers affiliated with both parties lined up to oppose Pickers and strikebreaking on principle as a dishonorable intrusion into the affairs of others that violated prevailing expectations of respectability and restraint between white men. "Miners will do well to remember first, that if they got any work at Leadville it would be scab work," the *Joplin Daily Globe* counseled. "Scabbing in Colorado is a dangerous business, and it is a despicable business anywhere. Stay at home," the paper's editor advised, "and give your fellow miners in Leadville a chance to win their fights against the Leadville bosses. That is the fair and right thing to do." The editor of the *Joplin Morning Herald*, who had criticized UMW strikers who came to Joplin looking for work in 1893, likewise reckoned that only "thugs and bums" would now go from Joplin to Leadville. Men willing to threaten the work of other men, he declared, "are a menace to any community." The *Joplin Mining News*, meanwhile, denounced the men talking to Pickers as "consumptives and bums," a reference that undermined their status as real men by invoking new eugenic theories of physical weakness. These denunciations also obliquely raised the common association of strikebreaking with African Americans, many of whom had broken the UMW strike in 1893. No paper provided any validation for strikebreaking on the basis of nationality or race. To get more information about Leadville, Scott McCollum wrote to his brother, a telegraph operator in Colorado. "You would be taking ten chances to one of being shot should you attempt to come here to take the places of the strikers," his sibling replied. "It's alright to go to mining but these strikers have too much dynamite and Winchesters."[8]

Weighing the need for money against the threat of death, hundreds of Joplin miners went to Leadville to break the WFM strike. McCollum was among them. In a November letter to the *Daily Globe*, he reported that he had no complaints "as the work goes—can work seven shifts each week, at $3 a shift," exactly what they had negotiated from Pickers. Although McCollum did not like the place, he wrote, "it is the money I'm after, and I can speak for the rest of the boys." Despite the high pay, he felt compelled to explain further. The miners who went to Leadville, McCollum wrote, did so because they could not find work in Joplin. Like those miners who had accepted lower wages in 1893, he justified strikebreaking with an appeal to paternal responsibility. "When they can't make a living at home they must go somewhere else, for they can't let their families starve," he declared. "If we could have got work at home every one of us would be in Joplin today." McCollum seemed to acknowledge their transgression of the manly code. "The men who came here are the ones who had to have work at once," he explained a second time. Such special pleading betrayed an unease about the social role of white men as prospecting gave way to wage work, an unease that resounded in McCollum's denial of the important role of women, and the family economies they ran, in the making of the poor man's camp. This interpretation allowed him to assert that Joplin strikebreakers actually deserved respect. "Joplin ought not to blame the men for leaving," McCollum argued, "for if a man is a man he will do the best he can by his wife and little ones." For miners like him, the decision to break a strike in lieu of prospecting followed the well-worn risk-and-reward logic of the poor man's camp, with an emphasis on risky work as the responsibility of individual men as men.[9]

The first group of strikebreakers arrived to an armed camp on the verge of open war. Five days before, strikers had attacked two mines with dynamite and guns, as McCollum's brother had warned. The Colorado National Guard, sent to protect the other mines, escorted the Missourians to their new place of employment. Crowds of strike supporters, led by women, jeered, denounced, and threatened to kill them. "The hissing and hooting did not cease for a moment," the *Denver Republican* reported. "Oh, the scabs, the hungry Missouri scabs," shouted one woman. Another told her son, "See the scabs in the middle, darling; when you grow up, get a gun and shoot them." A man urged the strikers to act now. "If the miners were any good," he said, "they'd get their guns and wipe the —— scabs off the earth."[10]

While the military presence maintained an uneasy peace that allowed the Joplin miners to work, the WFM supporters continued to demonize them in terms that challenged their manhood and racial standing. "Missourians are in bad repute here and about the worst epithet you can apply to a fellow is 'flat-footed Missourian,'" one soldier wrote. This denunciation implied both physical deformity and mental incompetence, the basis for a new phrase that strikers coined to describe the Joplin men: "Show me." "An expression much used is, 'I am from Joplin: you'll have to show me,'" he continued. In other words, according to strike supporters, Joplin miners would do anything the boss asked but required remedial instruction, a shame for any skilled worker. A Congregationalist minister in Denver, meanwhile, called them "an army of cowards." These taunts, however, belied a sense of fear. The strikers, one observer noted, "are seeing their old jobs flit away from them and this causes them to be very irritable."[11]

Backed by military force, the strikebreakers answered this vitriol with pugnacious swagger. Some Joplin miners, goaded by insult, attacked a group of WFM miners with knives in a Leadville saloon. "Knives, bottles, chairs, and pokers were brought in requisition," the *New York Times* reported, "and a terrible battle raged for several minutes, the long dirks of the Missourians bringing streams of blood with every slash." Deepening rancor between strikebreakers and strikers left little room for union appeals for solidarity. According to a reporter, when some WFM miners tried to persuade three strikebreakers who came into town for supplies to join the union side, "the Missourians turned a cold ear to these inducements." Far from ashamed, the Joplin men began to take pride in what they were doing. "We are all good miners and are accustomed to working on harder rock than this," one said. Although a few did leave, most expressed no regrets. This miner felt vindicated as a white man: "A man couldn't be treated whiter anywhere than we have been treated here."[12]

Their labor broke the strike. Despite public support from the AFL, which appealed in its main journal for members to aid the strikers, the WFM could not dislodge the Joplin miners. By the end of 1896, all of Leadville's major mines were "working with non-union men," an industry observer reported, "the Missourians who have been imported during the past few weeks." The union's Irish leaders refused to back down after the arrival of the Missourians, even as the mines reopened. Many Cornish and American-born union members defied the strike leaders, however, and returned to work alongside the miners from Joplin. By February 1897 the mine owners announced they had "all the men they need," including over 400 former union members. The remaining strikers abandoned the effort in March. In defeat, the WFM local collapsed. The most strident union miners struggled to find work in Leadville. Mine owners welcomed back "many of the old workmen who were not prominently identified with the strike" but refused "to discharge their non-union men to make room for union miners," an industry correspondent reported in April. Some strikebreakers returned to Missouri with their earnings. Scott McCollum, for example, came back with enough money to buy a house and go into business as an electrician. Others, such as George Dreiman, decided to stay. They did not cut all ties with Joplin, however. That spring Dreiman returned for a month of vacation, a luxury that surely reflected well on the decision to defy the union.[13]

The defeat shook the whole WFM. Before Leadville, the union had pursued moderate goals that included employer recognition, collective bargaining, and broad working-class electoral politics that reflected the influence of the Knights of Labor and their largest and richest local, the Butte Miners' Union. The union affiliated with the AFL in 1896 to gain allies in its struggle toward these goals. After Leadville, however, radical voices in the WFM blamed their defeat in part on this approach and on the AFL, which they argued had not done enough to help, particularly in terms of money. While the AFL could have done more, neither it nor the WFM was strong enough to win in Leadville. Reading the strike defeat in the context of William Jennings Bryan's loss in the 1896 presidential election, WFM president Ed Boyce concluded that the workers of the West were more militant and attuned to the realities of industrial capitalism than those of the East, who, he asserted, lacked the "manhood to get out and fight with the sword or use the ballot with intelligence." Sensing "little sympathy existing between the laboring men of the West and their Eastern brothers," Boyce informed Samuel Gompers that he had lost faith in the AFL and was "strongly in favor of a Western organization." In May 1897, the WFM effectively cut ties with the AFL. A year later, the WFM led the formation of the Western Labor Union (WLU), an organization for the "unification of all labor unions and assemblies east of the Pacific ocean and west of the Mississippi river," that would replace the conservative AFL and lead a new anticapitalist movement of all workers, regardless of craft or skill. The union's clash with Joplin strikebreakers informed the WLU strategy. According to one WFM member, the "Western Federation learned from experiences with the Joplin field that their only safeguard in the West was a thorough organization of all classes of labor." Despite these claims, neither the WFM nor the WLU rushed to send organizers to southwest Missouri.[14]

Meanwhile, miners in the Joplin district discovered that their willingness to oppose the WFM created new demand for their labor from other companies. In Ouray County, Colorado, the Caroline Mining Company responded to a brief strike by WFM miners, most of them Italian, in December 1896 by recruiting Joplin miners through agents based in Leadville. Observers believed that the operators wanted to provoke trouble. According to the *Emporia Daily Gazette*, "it appears that Ouray and San Miguel counties are on the verge of a miners' strike that may surpass the one now in progress in Leadville, caused by the importation of non-union miners from Missouri." When a special train arrived in January 1897 "bearing a large number of zinc miners from Joplin," however, the union miners remained peaceful to forestall another military deployment. That failed too. Another group from Joplin, at least 100 strong, arrived two weeks later. Although many left the camp due to altitude sickness, the miners who remained helped break the union in Ouray. Again, the opponents of the WFM emphasized the nativity of the strikebreakers. "We got rid of the foreigners," the county attorney boasted in February. "The citizens of Ouray are feeling very kindly towards the new arrivals," who seemed like good, trustworthy Americans, he said.[15]

More than 400 Joplin miners went to Colorado in 1896 and 1897. In the conflicts at Leadville and Ouray, they earned a reputation as a group of non-union, native-born miners with the skill and disposition to defy the WFM, a reputation that created new, good-paying opportunities for work wherever and whenever the increasingly militant union went on strike—a cycle of escalation that would yield more violence and hatred on both sides. Farley's nativist assertions to Campion from the previous summer were becoming true. In 1897, George Quinby, the Missouri state mine inspector, made similar claims in his official report. The Joplin district would prosper again, he declared, because of an "entire absence of labor troubles of any kind" and the "exceptionally high character of the miners, who are almost exclusively of American birth."[16]

As the Colorado clash concluded, however, miners in Joplin enjoyed a resurgence of prosperity that brought the economic crisis of the 1890s to a spectacular halt. In the spring of 1898, as the nation mobilized for war against Spain, armament manufacturers' demand for brass sent zinc ore prices surging above twenty-three dollars per ton in March to a stunning thirty-seven dollars per ton in December. Alongside war industries, other manufacturers also demanded more metal following President McKinley's electoral victory in 1896, the subsequent passage of a high protective tariff, and the 1897 discovery of rich new gold mines in the Klondike. Ore prices continued to rise, from an annual average of twenty-eight dollars a ton in 1898 to thirty-eight dollars a ton in 1899. That April, mining companies in Joplin reported sales over fifty-one dollars per ton. Those companies hired more mine laborers at higher wages than ever before in 1899, the richest year yet as district mines produced over 255,000 tons of zinc ore worth $9.5 million and 23,000 tons of lead mineral worth $1.3 million.[17]

As the new century dawned, more than 7,000 men worked in district mines, twice as many as in 1895, for an average daily wage over two dollars. The populations of Joplin, Webb City–Carterville, and Galena surged above 26,000, 13,400, and 13,000, respectively. At a time when immigrants were entering the United States in larger numbers than ever before, these places, by contrast, were even more dominated by native-born whites, who in 1900 comprised 94 percent of the population in Joplin, the most diverse place, and 97 percent of the population in Webb City. More than 100 African Americans still worked in the mines, but their opportunities were closing. Seventy-six of them lived in Galena; none lived in Webb City or Carterville. After five years of depression, good times had returned to the Joplin district, now an icon of what native-born white Americans could achieve under capitalism.[18]

Prosperity attracted new investors, who financed further mechanization in the district that consolidated the transition to a permanent system of wage labor. Observers reported that firms invested over $10 million during the 1898–99 year, over $2 million in land titles and leases and the remainder in prospecting drills and upgrades to machinery. These efforts rejuvenated old mining fields and opened new ones at Chitwood Hollow and Central City, west of Joplin; at Hell's Neck, north of Oronogo; and near Peoria, just south of Galena, in Indian Territory. Of the new land companies, only one, the Boston-based American Zinc, Lead, and Smelting Company, attempted to run actual mining operations on its holdings. The rest relied on the traditional leasehold system that favored local mining companies, which now numbered more than 500.[19]

Industry commentators lauded the district's native-born miners and its poor man's camp tradition, portraying the place as a safe bet for further investment. "The relations between operators and their employees are close and agreeable," a correspondent to the industry-leading *Engineering and Mining Journal* reported in early 1899, "and the district is noted for the entire absence of all labor disturbances." To help account for the company-miner accord, which was certainly unusual in the mining industry in the 1890s, he claimed that the system of subleasing made it possible that "the miner of to-day may be the operator of to-morrow" and fostered "a spirit of good fellowship peculiar to this district." Miners trusted that fellowship, he explained, because they "are almost exclusively native born Americans." Quinby repeated a similar explanation in his official 1899 report. Subleasing "has the greatest advantage of making the miner independent of control except so far as he is bound by the conditions of his lease," he claimed, "thereby utterly eliminating any danger of labor disturbances on a large scale." He called "the mining population of the Missouri zinc fields the most intelligent of any in the world," an unmistakable assertion that their nativity and race enabled them to appreciate and accept the logic of capitalism. These gospel-like narrations not only rejuvenated the poor man's camp tradition despite its demise in reality but also reassured native-born white men that they were favored under the prevailing system. Miners responded with renewed support for the governing Republican Party. After William Jennings Bryan swept Jasper and Newton Counties with 59 percent of the vote in 1896, Republican candidates surged in 1898; while barely losing both counties, they won eleven of the twenty precincts where miners predominated.[20]

Critics warned miners that the poor man's camp was dead. Writers for the *Appeal to Reason*, a socialist newspaper published since 1897 in Girard, Kansas, forty miles northwest of Joplin, predicted that "the lead and zinc mining business will be monopolized, the little fellows frozen out of their properties without any return, just as in other fields of industry." While admitting that the industry "has made fortunes for many men," the *Appeal* observed that high ore prices had now attracted big investors, who would leave "no more pickings in this business for those who lease and sub-rent."[21]

The district's most powerful producers soon seemed to prove that analysis. In December 1898, the largest land and mining companies organized the Missouri and Kansas Zinc Miners' Association to protect the new price levels against the efforts of smelters to lower the benchmark. Smelting companies that used coal insisted on price cuts because they faced pressure from new Kansas competitors using cheaper natural gas fuel and from the UMW, which was striking for higher wages. In late June 1899, the association's members, who collectively controlled 80 percent of the district's ore production, closed operations for two weeks to enforce their preferred price scale. The Associated Press reported that "over ten thousand laborers were thrown out of employment" as mines, mills, machine shops, and rail yards closed. Yet according to the *Joplin Daily News*, the "shut-down of the mines has been received with universal favor by operators and miners." Not all, surely, but most miners understood the market provisions of zinc leaseholds and widely accepted the connection between the price of ore and wages, as they had in 1893. "The wage earner of the district was equally interested with the producer, as a cut in price of ore meant a cut in wages—their interests were identical," a miner named Daniels claimed. The association's strategy, however, prompted coal-fired smelters to shut down in retaliation. While some mining companies resumed operations in July, the association called for periodic closures that disrupted production, until the two sides reached an agreement in November. Despite the year's prosperity, district miners again faced precarious employment in the summer and fall of 1899.[22]

Now, however, Joplin miners could look to the West, where their reputation as nonunion workers created opportunities for highly paid work. Western mine owners looked to them, too, as the WFM increased its campaign across the region. These interests aligned again in the summer of 1899 when the WFM clashed with mine owners in the silver and lead district of Coeur d'Alene, Idaho. In April 1899, union miners requested a union contract and pay scale from the Bunker Hill and Sullivan Mining and Concentrating Company, the only firm in the district that did not employ union labor. Rather than submit, the company fired all known or suspected union members. In response, the WFM went on strike and blockaded the mine, which prompted the company to ask Governor Frank Steunenberg to send soldiers to protect its property. In the violent clashes that followed, someone, perhaps from the WFM, dynamited Bunker Hill's main mill. After the attack, Steunenberg declared martial law and asked President McKinley to dispatch federal soldiers to keep the peace. With help from the mine operators, the army detained all known union miners and their sympathizers in the district, more than 1,000 men, without counsel or charge in a rough stockade known as the bullpen. The state government also instituted a permit system that required anyone who wished to work in the mines to register their name, country of birth, recent employment, and history of union membership and "to renounce and forever abjure all allegiance" to the WFM. Some union miners who remained free vowed to resist, but most fled the area.[23]

Unable to find enough workers who qualified for permits, Amasa Campbell, co-owner of the Standard Mining Company, dispatched an agent to Joplin in June with instructions to hire nonunion miners. And so, just as the zinc miners' association prepared for the latest district-wide shutdown, Campbell's man, J. R. Smith, placed a tantalizing advertisement in local newspapers offering "steady work the year round" in Coeur d'Alene for up to 1,000 miners at $3.50 per day.[24]

With the shutdown looming, Joplin miners accepted the offer of work in Coeur d'Alene more quickly than they had in Leadville. More than 100 "first class miners," "a husky looking lot," joined the first contingent to depart for Idaho. Upon arrival, S. B. Willis, whom the press described as "one of the leaders," explained that "we are perfectly satisfied with the situation.... Before coming we were told that old miners had blown up the Bunker Hill mine, and that we were only needed because the former hands were either in the bull pen or would not be employed again." Unlike in 1896, the Joplin press did not denounce them, a sign that public opinion was moving in favor of the strikebreakers. The *Joplin Daily Globe* had printed a pared-down tale, similar to Willis's, that omitted key events, such as the initial strike, martial law, and the permit, but Joplin miners found the explanation sufficient. Although wary of calling himself a strikebreaker, Willis clearly understood the economic value of being nonunion. "There are plenty more miners, and good ones, around Joplin, who will start as soon as they hear from us," he promised a local reporter. "There are no unions there." Whether Willis was against unions or simply saying what he thought mine owners wanted to hear, he clearly sought to take advantage of the absence of labor organization in Joplin.[25]

Their primary goal was good-paying work. "The mines near Joplin are being worked on a modest scale," Willis explained, but the real "trouble there is the uncertainty of employment, for small shutdowns are continuous." Willis's colleagues "expressed determination to remain in the Coeur d'Alenes and hold the jobs." Notably, Willis did not justify the decision with appeals to family security as McCollum and others had in 1896. For him, the quest for work with high pay was its own justification. Most of the others agreed. Within days they sent letters and telegrams home with reports of ample work, good wages, and no strike. "I found everything in much better shape than I expected," John Maddy informed his nephew. "If a man comes out to have a good time this is not the place he wants to find," he counseled, "but if he wants to work and to make money, here is the place." "This is the best company I ever worked for," Maddy declared. "They want experienced miners who want work," Lyon Hopkins explained. Ira Esry assured his mother that the "Joplin boys are all well pleased." Indeed, hundreds more followed them to Idaho that summer.[26]

Some of the miners, however, found sympathy with the union men in Idaho and rejected employment. Despite military efforts to keep the union and nonunion miners apart, WFM organizers managed "to make a representation of the case" to a group of new arrivals, some of whom "tore up their permits and refused to go to work." This group of eleven men sent a statement to the Joplin newspapers that claimed they had been deceived about the conflict in Coeur d'Alene and warned others to stay away. "The condition of things here has been grossly misrepresented to us," they wrote. Unlike those who stayed, they voiced concern for the suffering families of union miners. "We have witnessed the heart-rending sight of women and children whose husbands and fathers have been torn from them and cast into the second Andersonville (the bull pen)," they explained, "for no other crime than that of being good union men and adhering to the principle of the Declaration of Independence which makes us free American citizens." They were coming home and wanted to "warn innocent men from being beguiled into this state where a scab is beneath the notice of a bootblack," both the contemporary name for shoe shiners and a derogatory term for African Americans, who were often employed as strikebreakers. Their argument, influenced by the WFM, used the terms of respectable manhood and white American nationalism to make a case for union solidarity. The union made a similar argument in a notice sent to Joplin newspapers urging men to stay away. The WFM explained that the union was not on strike but had been locked out because it would not abide a permit system that asked men "to swear away their rights in this un-American and servile manner." While the effect of this appeal was unclear, the union had created a way for at least some Joplin miners to respect its cause.[27]

Yet as in Leadville, most of the Joplin miners in Idaho were learning that native-born white men who were also nonunion enjoyed a special advantage with mining companies against a majority foreign-born, unionized workforce. The operators stated their demand for workers in language that placed a premium on linked markers of race, nativity, and trustworthiness. This was made clearest in the terms of the permit system. The Spokane *Spokesman-*

Review reported that of the first group to arrive, "all but two of them are native-born Americans, and typical Anglo-Saxons at that," the latter a term that reflected the burgeoning national politics of racism and xenophobia. When presented with the permits, the Joplin miners "signed the applications with alacrity," the paper reported, "well satisfied with the situation." The permit not only allowed miners to work but also marked them as good, loyal white men. Whereas before the April strike more than 80 percent of miners in the district were foreign born, the state afterward issued nearly all of its permits to American-born miners. John Maddy was not bothered at all about signing the permit. "We had to sign a paper," he informed his uncle, "but all there was in that is that we are 'American-born and belong to no union'" and played no part in the April bombing. He was proud to answer these affirmatively. "Now if that takes away a man's liberty mine is gone." As for the men held in the bullpen, he said, "only 20 can speak good English." The detention of trouble-making foreigners meant more work for Americans. Frank Meyers likewise appreciated the opportunity. Martial law had been declared, he told his father, because union men blew up the Bunker Hill mill and declared that "no white man (or in other words, no non-unions) could live in this canyon." That Joplin miners now considered nonunion status a positive marker of racial and gendered standing was significant. The WFM had called for a closed shop, but Meyers read beyond this, aided by the terms of the permit system, to frame the struggle as one of hardworking, nonunion American men against work-shy, unionized foreign radicals.[28]

Unlike the small group that sided with the WFM, many strikebreakers articulated this emerging ideal of patriotic, native-born antiunionism through public aggression and even violence against the union. During a Fourth of July celebration in Wallace, Meyers bragged, "the Joplin boys had about fifty fights with the union men, and they never lost a fight." "We go where we please, and if a man calls us 'scab' we knock him down." These comments, all printed in Joplin newspapers, not only overwhelmed the WFM's appeals for solidarity but also showed the miners redefining strikebreaking in new terms that emphasized the bellicose prerogatives of native-born white men acting in their own self-interest. Men like Meyers did not justify strikebreaking as a reluctant defense of family but now bragged about their power to smash the faces of foreign strikers and get paid for it.[29]

WFM miners again fueled the animosity, which further complicated the union's efforts to reason with nonunion men. Some union members directly confronted the Missourians, but martial law made that dangerous. Doe Isbell recounted how two Italian miners who accosted him were quickly arrested and imprisoned. Without physical means, the union resorted to moral attacks. Like "show me" in Leadville, "Joplin" and "Missourian" became synonyms in Coeur d'Alene slang for "scab" or "cut-rate laborer." The union men wrote a song, "Strike Breaker's Lament," inspired by the poem "Bingen on the Rhine," that imagined the dying words of a greedy Missouri strikebreaker:

> They told us that our wages would be three to four a day,
> And that, you know, in Joplin is more than double pay;
> The thought of such great riches, it made my heart to glow,
> For I'd felt the rack of poverty in Joplin, Joplin, Mo....
> Just then his voice it faltered, he ceased to murmur low,
> His soul it went a-scooting to Joplin, Joplin, Mo.
> His partner wept above him, and sadly fell his tears,
> Then tried to drown his sorrow by drinking many beers;
> He boxed the stiff and shipped him, as fast as he could go,
> To the land of scabbing miners in Joplin, Joplin, Mo.[30]

Harsh moralizing shored up union confidence but did little to soften the hearts of strikebreakers toward the WFM or toward unions more broadly.

Again, as in Leadville and Ouray, Missouri strikebreakers proved essential to the antiunion campaign of Idaho's mine owners. More than 1,000 Joplin miners went to Idaho that summer. The operators vowed to "make no concession whatever to union miners." As production increased, state authorities closed the bullpen in December 1899. Hundreds of miners from Joplin remained in the camp; the WFM was broken in northern Idaho. Many other strikebreakers, however, brought their gains and their experiences home to Joplin.[31]

Yet just as it seemed that Joplin miners had turned decisively against unions, labor organizers found a more complex situation in the district. In May 1899, an AFL organizer named Baxter helped a group of masons, bricklayers, and printers in Joplin form a laborers' protective union, an institutional form that gave workers direct affiliation with the AFL in the absence of national union representation. This was the first AFL-affiliated union in the area. Baxter's next goal was to organize the "lead and zinc miners." In August, amid shutdowns and the Idaho conflict, miners in Oronogo, led by zinc miner S. G. Dodson, formed a local union for zinc and lead miners, also directly affiliated with the AFL. This group was small—AFL rules required a minimum of seven paying members in these local bodies—but its formation suggested that some miners had not accepted the district's new system of wage labor with as much equanimity as most observers thought, perhaps influenced by the UMW's recent victory in the 1899 coal strike. The new union also suggested that Joplin miners might find agreement with the policies of the AFL, as an alternative to the WFM, if offered a means of affiliation. The AFL pursued a pragmatic strategy that focused on organizing the nation's most skilled workers into well-funded national unions that provided benefits for members and their families, particularly financial security in case of injury, unemployment, and death, while pursuing conciliatory negotiations with employers over pay and hours, especially the eight-hour day. The AFL sought to avoid strikes that it could not win, particularly those launched by local unions, a lesson influenced by the crushing defeats of the Knights of Labor and the WFM, particularly in Leadville. With this approach, the AFL explicitly appealed to the country's most privileged white workers, both native born and of northern and western European origin, often backed by calls from Gompers and other national leaders for immigration restriction. If Joplin's native-born white miners would join a labor organization in the late 1890s, surely this would be it.[32]

With similar inclinations, and perhaps influenced by AFL organizers, miners in Joplin formed an independent union that August that paralleled developments in Oronogo. Claiming 350 charter members, the Joplin Miners' Union declared a pragmatic set of aims: to "protect the interests of the craft of mining," "to protect members against dishonest men who attempt to cheat employees out of pay for their labor," "to assist in having passed and enforced laws intended to protect the lives of men working in the ground," to provide sickness and death benefits for members, and "to co-operate with laboring men of all callings for the advancement of the general interests of the laboring man." The union also announced commitments to find accord with local mining companies: to promote "the interests of the mining industry in general," "to protect just and honorable operators from dishonest and incompetent workmen," and "to encourage the principle of conciliation and arbitration in the settlement of differences between employers and employees." With these resolutions, the Joplin Miners' Union recognized the possibility, if not the inevitability, of disputes with employers, even while advocating against strikes, as miners in the district had always done. While the origins of the independent union are unclear, its resolutions tracked the aims of the AFL, although with more generosity toward members' employers. In September, for example, the Joplin Miners' Union publicly supported another round of mine closures by the Missouri and Kansas Zinc Miners' Association to resist efforts by the smelters to reduce ore prices.[33]

The movement to form local AFL unions soon subsumed this independent group. In October 1899, AFL members in Oronogo organized new local miners' unions in Hell's Neck and Webb City, "towns where there has never been a labor union of any kind before," organizer J. A. Burkett reported. In Joplin, meanwhile, the first laborers' protective union had split as national craft unions established constituent locals among sheet metal workers, carpenters and joiners, bricklayers and stonemasons, cigar makers, electrical workers, iron molders, machinists, musicians, printers, and painters. Representatives of these unions chartered a Central Labor Union (CLU) governing body with the AFL to coordinate activity and to continue organizing "the lead and zinc miners as well as all other wage earners." The Joplin Miners' Union, meanwhile, left no trace of continued activity; presumably, its members joined the ranks of the AFL.[34]

These developments caught the attention of Samuel Gompers. He wanted to build a national AFL union of metal miners that would incorporate or replace the WFM. Gompers believed that the only way to counter the power of organized capital was "for the combined forces of labor in this country to unite more thoroughly than ever before" on a national, not a regional, basis. In late 1899, he proposed "that the mineral mine workers ought to be organized under one national head, upon a comprehensive, broad basis, where the interests of one would be promoted in the interests of all." Although relations with the WFM had collapsed after the formation of the WLU in 1898, Gompers continued to urge the WFM to rejoin the AFL. In the summer of 1899, the AFL publicly supported the WFM's struggle in Coeur d'Alene and even sent money. Gompers refused to recognize the WLU, however, and began preparing for the possibility that the WFM would never abandon its regional strategy. He planned to use Northern Mineral, which was badly weakened by strike defeats, as a foundation for the new national union, which would, if successful, provide a stronger institutional home for miners like those in Joplin than the small, isolated local unions. Gompers dispatched two national organizers, Frank Weber and Robert Askew, both from Northern Mineral, to connect the nascent union movement in Joplin to this new national organizing plan.[35]

Gompers's commitment of two trusted organizers at a time when AFL resources were scarce testified to how much he valued the mission. While no doubt looking to steal a march on the WFM, Gompers also seemed to legitimately care about establishing the AFL in Joplin. "You understand," he informed Weber, "that I shall accept nothing but success at the hands of yourself and brother Askew. The workers of Joplin must be organized and in affiliation with the American Federation of Labor." He did not directly acknowledge the district's strikebreakers but advised Askew that the AFL's conservative strategy "presents the most comprehensible platform upon which all may stand, united in heart and hand, mind and spirit, to secure justice for the workers," especially in contrast to the radicalism of the WFM. "There is no reason in the world why the workers, miners included, of Joplin as well as in any other part of the country," Gompers insisted, "should not be in full affiliation with the" AFL.[36]

Some Joplin miners seemed to agree. Weber and Askew organized new local unions in Duenweg, Zincite, and Central City in November and December 1899. "These men until recently have not deemed it necessary to organize their trade," Weber explained to Gompers, "and have scoffed at the idea of ever being compelled to unite for mutual protection." Weber, like so many others, blamed the promise of the poor man's camp tradition. Yet "the change of conditions in mining" away from prospecting and small operations, he argued, "has caused them to realize the danger awaiting them." Now, Weber reported, they "are beginning to understand that the wornout cry 'every man has an equal chance to become rich' is a delusion and a snare." He was confident in his ability to convince the rest of their errant thinking. Both cocksure and condescending, Weber reported that "the zinc and lead miners and mine workers are now thoroughly aroused from their mental lethargy."[37]

Joplin's AFL miners led the chorus in support of Gompers's plan at the federation's convention in Detroit in December 1899. Burkett proposed a resolution to create a new national union that would give local metal miners' groups regular affiliation with the AFL. "The Zinc and Lead Miners and Mine Workers are not satisfied with their mode of organization and request that an International Federation of all mineral unions and mine workers be formed," he wrote, preferably "to be known as Federation of Mineral Miners and Mine Workers of America." The AFL's committee on organization approved Burkett's resolution with the recommendation "that the incoming Executive Officers be instructed to use their best endeavors to bring about an amalgamation of the Mineral Mine Workers of America." It also urged Gompers to continue to appeal to the WFM to rejoin the AFL fold. Organized labor finally had some momentum in the Joplin district, propelled in part by the AFL's strategy to outflank the WFM on the right. "The cry that is going forth to-day" from Joplin, Weber wrote, "is the organization of all the mineral miners into an international union to be known as the American Federation of Mineral Miners."[38]

The AFL's focus on Joplin prompted the WFM to send its first organizer to the district later that month. The WFM dispatched John Lewis, a Welsh silver miner from Colorado, who, at least early on, "was working hard to show those people the error of their way and meeting with good success." After a month, however, Lewis gave up and returned to Colorado. He reported "that another man who was more familiar with the situation could do better" in Joplin. Lewis died in an avalanche soon after, without leaving further explanation. His struggles possibly stemmed from a lack of experience outside of the capital-intensive, deep mines of the Rocky Mountains, where he had worked since arriving in the United States in 1892. Lewis would have known little about the Joplin district or why miners there might have favored the AFL's approach. Of course, the WFM's failure to send an organizer to Joplin for more than three years after the first strikebreakers arrived in Leadville also made his task more difficult. Lewis not only encountered anti-WFM sentiment but also had to contend with the AFL's proposed alternative organization.[39]

There were some Joplin miners who were not completely opposed to the WFM. Organizer Solon Cress, Lewis's successor, formed WFM Local 88 in Joplin in March 1900. In contrast to Lewis, Cress claimed several points of identification with miners in Joplin. Born in California, he had worked as a lead miner in Colorado since the 1870s, including in Leadville in 1881. His reports to the WFM displayed a keen understanding of the old prospecting and leasehold system and its recent demise. "The advent of labor saving machinery and improved methods of mining and treating ores have doubtless made mining more profitable for operators," Cress explained, "but for the man with his bare hands it becomes increasingly harder for him either to find leases or employment." He reported enthusiasm for Local 88, with "new members at every meeting." Cress even hoped to cooperate with the local AFL movement. This stance set him apart from many in the WFM but aligned with recent AFL unity overtures and so probably aided his efforts in Joplin. Cress commented favorably on the city's CLU, which, he explained, was "beginning to create a healthy sentiment in favor of unionism." He did not believe that the WFM would sweep Joplin soon but voiced cautious optimism. Miners were "beginning to join the unions," he informed *Miners Magazine* and, much like competing organizers in the AFL, pledged that "we will ere long be able to give a good account of ourselves in the Joplin district." But, like Weber and Lewis, Cress could not resist scolding the miners for past transgressions. "We shall do our best to see that this shameful thing"— strikebreaking—"shall not again occur."[40]

Joplin miners entered the new century with considerable confusion about the best way forward. Conditions in the district seemed good. Mining companies produced around 250,000 tons of zinc ore a year on average from 1900 to 1903, as well as over 30,000 tons of lead mineral each year. While the price of zinc ore fell back from the average high of thirty-eight dollars per ton in 1899, companies achieved prices that bettered those in any year before 1898 and were rising: twenty-four dollars per ton in 1901, thirty in 1902, and thirty-four in 1903. For the thousands of miners who now relied on wage labor, however, conditions remained uncertain. Wage scales varied considerably across the district. The biggest companies in Webb City, Carterville, and parts of Joplin paid top wages, between $2.25 and $2.50 per day. Smaller companies, particularly those in Aurora, Granby, or Galena, paid less, between $1.75 and $2.00 per day. Miners could not rely on steady wages, especially if they worked for one of the big companies. Those firms led continued efforts through the producers' association to defend ore prices by shutting down operations. But miners had choices. Western mine operators offered frequent strikebreaking opportunities as their conflict with the WFM sharpened and spread across the region. The WFM and AFL, meanwhile, presented miners with related but rival visions of collective action. Although more Joplin miners came to favor the possibilities of unionism, most found little of interest in either appeal.[41]

In 1900, conditions in the district seemed to provide an opening for AFL and WFM organizers. As prices retreated from the all-time high set in 1899, the producers' association closed many mines for six weeks. According to Cress, more than 3,000 miners were thrown out of work. He noted that most received lower wages when the producers' association resumed operations. They "are still wondering where their part of the unexampled prosperity is to come in," Cress reported. Weber, meanwhile, recognized that "the individual prospectors and the small producers are being driven out of the field" and with them any hope of the survival of the poor man's camp tradition.[42]

The unions, however, struggled. The AFL's plans to create a national metal miners' union, so grand in late 1899, faded away. In January 1900, Gompers reassigned Weber to campaigns in New York, Pennsylvania, and Ohio. Robert Askew remained in the district and organized another laborers' protective union, which probably included some miners, in Galena, Kansas. Despite little support from the national office, some Oronogo miners launched a wildcat strike in June to protest the recent wage cut. The ill-advised strike failed within the week. None of the district's seven AFL miners' locals survived into 1901. As much as he wanted to supersede the WFM, Gompers believed foremost in consolidating organizations in places where union members could achieve the most leverage at work and at the ballot box, such as Buffalo or Cleveland. As AFL membership in these places surged in 1900, the executive council devoted all available resources and national organizers, such as Weber, to ensure effective institution building, often to the detriment of new organizing drives. When the federation sent Edwin Trappe to Joplin in 1901, he reported that it was "the hardest town to organize he ever went into." Trappe hoped to reinforce the city's CLU but noted nothing about union sentiment among the miners.[43]

The WFM likewise failed to build on Local 88. After a promising start in early 1900, Cress seemed to speak past local miners. A Socialist, like many others in the WFM, he gave a series of speeches that emphasized political action to overthrow capitalism. "Labor cannot hope to participate in the advancing gains of civilization under a competitive system," Cress argued. However true this statement might have seemed, many Joplin miners would have struggled to reconcile it with the potent traditions of the poor man's camp that supported the opposite view. Cress also challenged their motivations during recent strikebreaking episodes. "All men have not liberty to work, but only liberty to hunt for work," he argued, "and they have not even that liberty in Idaho, where one must obtain a permit from a state official who is a paid hireling of the Standard Oil Company." Cress's claim that the men who worked as strikebreakers had no liberty made for a risky appeal to men who likely had been strikebreakers. His effort soon faded. That autumn, Cress ran as a Social Democrat to represent western Jasper County in the Missouri House of Representatives. He received 241 votes, a sad showing compared to 7,162 for the victorious Democrat and 6,178 for the Republican runner-up. Many miners were willing to vote for Democrats in response to new economic uncertainties, but few were willing to go further. Local 88 collapsed soon after.[44]

While both unions staggered, Joplin miners saw fresh evidence of the advantages of strikebreaking as conflict flared again in Idaho. After breaking the WFM in Coeur d'Alene in 1899, the mine owners ended martial law and the permit system in April 1901. Union miners believed that many nonunion Missourians would leave the camp rather than work without armed protection. The mine owners, however, planned to hire more Joplin miners to ensure that the union could not reorganize. J. R. Smith returned to Joplin in February 1901. He praised the productivity of the miners who went to Idaho in 1899 and again offered high wages, from $3.50 to $5.00 per day. Although miners would have to pay their own transportation costs, Smith guaranteed them a job upon arrival. He claimed that over 100 miners signed up in a few days. Smith "has no trouble in securing men," the *Engineering and Mining Journal* reported, "as those who went before indorse the methods of the company."[45]

Once in Idaho, the Missourians met renewed resistance from covert union miners. The WFM diehards concluded that the Joplin miners would never join the union and would instead give the owners undeniable leverage to maintain open-shop mines. According to company informants who infiltrated their ranks, many union miners reacted "in a disheartened sort of way and act as if they were very much discouraged concerning the future of unionism in the Coeur de Alenes." Some called for violent resistance, but most realized that state authorities would crush such efforts. As an alternative, some hoped that rudeness and social isolation would convince the Missourians to leave. One union miner advised colleagues who shared a boardinghouse with them not to "speak to the Missourians also not to pass them any thing at the table." He aimed to make them "sick and tired of the place as soon as possible." None of these strategies had much effect. One miner told the company informant "that it hurt the old-timers to see new men come in here in bunches as they could not now get up a mob to beat them off as they once could." Others, according to the informant, "remarked that it seemed that the only proper thing for union men to do was to leave the camp, which many of them are doing."[46]

A few strategic thinkers tried to convince the Missourians to join the WFM. In contrast to Smith's claims, union miners told the company informant that several of the Joplin miners were not happy with conditions in Coeur d'Alene. Apparently some companies paid less than promised and occasionally closed operations for days or weeks at a time, just like in Joplin. "These men say that a good many of the Missourians are good men, when they understand how things are here, and they claim that Agent Smith misrepresented things in Joplin." According to one report, some of the Missourians even joined the union in Idaho. But most could not countenance that, no matter how much they came to dislike the Idaho mine operators. Many Joplin miners "are willing to set in sympathy with the Union," the informant noted, "but are unwilling to join as the Union has such a bad name." Other Missourians did not even extend sympathy to union miners. They relished the opportunity to work. While "most of the Missourians were good men," one union miner said, "some of them were scab at heart."[47]

The WFM's leadership denounced them all as bad men in terms that framed the union as the defender of manly fairness and self-restraint. In May 1901, Ed Boyce, who as president of the union also edited its new official journal, *Miners Magazine*, published a piece that cast all Joplin miners beyond the pale of honorable workers. "For a number of years," he reminded readers, the Joplin district "has been the recruiting station for scabs to take the places of miners struggling for their rights throughout the mining regions of the West." Despite union efforts to slow it, "this insufferable influx of scabs has not abated in the least," as the arrival of hundreds of Missourians in Idaho made clear. According to Boyce, miners from the Joplin district were so corrupt that they should no longer be considered full men according to existing ideals. "It is strange," he continued, "how degraded some men can become, when, for a miserable job in a cold, damp mine, they will sell their honor and manhood and try to deceive others so they may follow in their footsteps." In this view, Joplin miners were irredeemable, perhaps congenitally so, and should be shunned. "They are a dangerous class of men and not to be relied upon," Boyce explained. "They will seek admission into unions when they find it is to their advantage, but union men should not tolerate them because they are a disgrace to themselves and to any organization that harbors them." "No wonder," he concluded, "that the few decent men who came from Joplin are ashamed of the name and deny that they ever worked in such a scab hole."[48]

As the WFM became more militant and confrontational following the defeats in Leadville and Coeur d'Alene, it launched more strikes that multiplied the potential points of conflict with nonunion miners from Joplin. Those conflicts came thick and fast after July 1901 when the WFM struck the copper mines and the smelter of the Rossland Great Western Company in Rossland, Washington, and Northport, British Columbia. The WFM objected to the dismissal of union members at the Northport smelter and to wage cuts at the mines in Rossland. Many of the union miners had recently fled there from Coeur d'Alene. They were in no mood to compromise with the powerful Idaho-based capitalists who controlled the Rossland Western holdings. Company directors, in turn, sent agents to Joplin to recruit strikebreakers. Over 200 miners accepted the offer of work "in spite of the warnings coming from there to the effect that there is a strike on and warning the men to stay away." This lot was only the most visible contingent of nonunion Missourians to go west in 1901. That summer, in the months after the AFL and WFM locals in Joplin failed, the Missouri Pacific Railroad office in Joplin reported selling nearly 2,000 tickets for journeys to metal-mining camps in Utah, Idaho, and Colorado.[49]

These clashes, which reprised scenes from Leadville and Coeur d'Alene, deepened the chasm that separated union and nonunion miners. In Northport, the first trainload of strikebreakers ran a gauntlet of armed strikers. This time, the strikebreakers armed themselves. "Since the Missourians arrived," one report stated, "the most of them have carried firearms, as have also the striking smelter men." According to the local press, "a crisis should be imminent." The first battle erupted when "a number of Joplin men who were in the saloon drinking" took offense when a group of union men began "singing a song which made some reference to 'scabs,'" likely one of the songs from the Idaho struggle, probably "Strike Breaker's Lament." After exchanging threats, according to the bartender, "the Joplin men went into the restaurant next door, and came back with their coats off. They said they were from Missouri, and they were ready for business." A brawl ensued and someone fired shots—"bullets for the chorus," the *Los Angeles Times* quipped—one of which killed a prominent union miner.[50]

Afterward, the strikebreakers became more aggressive in public. "The Missourians have been inclined to carry things with a very high hand of late," the local press reported. "Things are in a very critical and dangerous state, and unless something is done in the near future, it may be necessary to declare martial law." The sheriff disarmed both sides, but that worked to the advantage of the mine owners and strikebreakers, who defeated the strike and the WFM in both camps. In response, WFM miners stepped up pressure against nonunion miners across the region. In Cripple Creek, Colorado, the WFM local posted signs that autumn warning that any nonunion miners, particularly those "from Missouri," caught in the camp "will be considered a scab and an enemy to us, himself, and the community at large, and will be treated as such." "You are for us or against us," the Cripple Creek WFM declared. "There is no middle ground."[51]

The WFM's *Miners Magazine* responded with more intense denunciations of all Joplin miners, whether they broke strikes or not, that denied their manhood, patriotism, and even basic humanity. The Joplin district, Boyce declared, was a "scab incubator" full of men who, "dead to all sense of honor and manhood, have never attempted to improve their condition, financially, morally or intellectually." Rather, he claimed, Joplin miners "take peculiar pleasure in hindering the advancement of other workingmen in their efforts to better their condition." They possessed a "brute spirit" and "will hesitate at nothing, not even robbery and murder." This account recycled epithets widely used at the time to vilify foreign-born radicals and African Americans to attack Missouri strikebreakers who had justified their own actions in Leadville and Coeur d'Alene in patriotic, nativist terms. Now, the WFM pointed out, these hypocritical, native-born Americans had done what they claimed to despise by going to British Columbia as strikebreakers. "What a splendid sight it is," the journal concluded, "to see those 'free born' American citizens of whom we hear so much in these days of flag worshipping crossing the Canadian line, armed with a six shooter and bowie knife," as "hired thugs" to undercut "workingmen struggling for their rights." A letter to *Miners Magazine* from a former Local 88 member confirmed this interpretation. "I have always done what I could in my humble way to point out to the miners in Joplin the error of their ways in not organizing for their own protection in place of going to the Coeur d'Alenes and Northport at the solicitation of the mine owners," he explained. He interpreted their refusal to respect the WFM as evidence of an almost pathological lack of solidarity. He and some others who had belonged to Local 88 "have had many round ups with them, but it doesn't seem to do them any good; it is bred in them and they can't help it." He saw little future for any union in the district, although some diehards remained. "The few of us who are here," this miner concluded, "hope the time will come when men from Joplin can associate with other working men and not be looked upon as scabs wherever we go."[52]

These views heaped bitterness upon the already tenuous politics of metal miner unionism. The WFM now blamed the AFL for destroying unionism in Joplin. *Miners Magazine* printed a letter from the Cripple Creek local that blamed "the pernicious work of the A. F. of L. agents at Joplin, Missouri, two years ago" for the failure of Local 88. The letter claimed that "the paid wreckers of the A. F. of L. appeared upon the scene to spread discord" after the WFM had already organized a local, an untrue account that reversed the actual order of events. To make matters worse, the editorial continued, Weber and Askew, "the wreckers, having accomplished their work of destruction, made no effort to restore the organization under their own sovereignty." It was true that Gompers reassigned Weber from Joplin and that his plan for a national metal miners' union vanished. But was it true, as an editorial in the next issue asserted, that the AFL's withdrawal from Joplin "allowed unionism to die"? Some WFM leaders thought so, including Boyce, who still blamed Gompers and the AFL for what had happened in Leadville. "From that dead sea," *Miners Magazine* stated, "have come thousands of 'scab' miners who aided the mine owners in the Coeur d'Alenes and every place the Mine Owners' Association has attempted to defeat unionism." Because of the AFL, the Cripple Creek writer concluded, "Joplin remains the menace that it has ever been to the W. F. of M." These conclusions, whether right or wrong, pushed the WFM and AFL even further apart. In 1902, the WFM reorganized the WLU as the American Labor Union, a new national radical alternative to the AFL that would, with leading Socialist allies, found the Industrial Workers of the World in 1905. AFL leaders, meanwhile, vowed to destroy the American Labor Union. Any future effort to organize Joplin miners would have to navigate this yawning divide.[53]

AFL organizers with the Joplin CLU tried to restart the defunct local miners' unions in late 1901. That summer, S. G. Dodson reported to the AFL national office that a new Missouri law mandating an eight-hour workday in all mines had started to awaken interest in unionization in the district. Democrats backed by coal miners in the northern part of the state had created the bill, and a large majority of the general assembly, including the two representatives from Jasper County, approved it in March 1901. In Joplin, Dodson explained, many companies refused to obey the law and challenged it in court. The miners he talked to "are determined that the 8-hour law shall be enforced everywhere that it applies." Although Dodson made little organizational headway, the CLU persisted. In the summer of 1902, while Boyce and others in the WFM directed a torrent of invective against Joplin miners, F. N. Ford, who represented Joplin carpenters, reported that he had organized a local union of miners in Chitwood and that "prospects are good for organizing the whole district of Miners." Whatever opportunity Ford saw soon vanished. The AFL never registered the Chitwood local. Dodson left Joplin. At the 1903 AFL convention, Ford requested a national organizer to replace him, but the AFL focused resources elsewhere. Once again, without national support, the Joplin CLU could not sustain its efforts to organize area miners.[54]

The WFM, meanwhile, restarted its push to organize Joplin miners in early 1903 under the leadership of new president Charles Moyer, a South Dakota miner. In contrast to Boyce, who retired in 1902, Moyer pushed "a vigorous campaign of organization" among metal miners in the East. He courted the leaders of Northern Mineral, which suffered after the AFL's shift away from the industry; it would merge with the WFM in early 1904. Moyer believed organizing Joplin's nonunion miners, even those who had worked as strikebreakers, was essential for the union's survival. In the spring of 1903, he sent organizer D. C. Copley, who hailed from Cripple Creek, to Missouri. Copley had success in eastern Missouri, where Northern Mineral had organized locals the year before. He then held a series of meetings in the Joplin district that slowly but steadily attracted support. Despite finding "quite an opposition to organization," Copley organized three new locals in April and May: Local 186 in Chitwood, where the AFL had made some headway; Local 195 in Joplin; and Local 205 in Webb City. In August, the WFM sent William M. Burns, a miner from Ouray, to carry the work forward. By the end of the year he organized two additional locals: 207 in Neck City and 210 in Aurora. The locals were small, around fifty members each, but showed promise for growth. To signal the WFM's commitment to the district, Moyer visited the new locals himself in November.[55]

The arrival of labor agents seeking more strikebreakers that fall added new urgency to the WFM organizing campaign. Throughout September 1903, agents placed help wanted advertisements in Joplin newspapers offering work at high wages ranging from three to four dollars a day to replace union labor in mines in Prescott, Arizona; Tonopah, Nevada; Randsburg, California; and Cripple Creek and Telluride, Colorado. The most important fight for the WFM was at Cripple Creek, a gold camp where union miners sustained a strike against a well-financed group of mine owners backed by the Colorado National Guard for almost a year in 1903 and 1904. Determined to break the strike and the union, the Cripple Creek mine owners deliberately copied the model that Idaho's mine owners had used in 1899: secure mine property with an armed force, recruit nonunion replacement miners, and institute an open-shop system using permits. As in Coeur d'Alene, the Cripple Creek owners dispatched labor agents to Joplin. Now, those agents appealed to the most violent forms of masculinity that had emerged in Idaho and later conflicts. The Cripple Creek owners explained that they wanted Joplin miners because of their reputation for being "the toughest in the whole country" and their eagerness "to undertake whatever hardship will be incurred in putting the first blows at organized labor." In Leadville and Idaho, one owner explained, nonunion Missourians had "distinguished themselves by making a 'rough house' with union men whenever the two factions met." Joplin miners were now sought not just for their lack of union organization but more so for their aggressive antipathy toward the WFM. The Cripple Creek owners hoped that the ensuing conflict would give them the excuse they needed to crush the union for good. The WFM recognized the trap but was not sure it could be avoided. As one representative admitted, "Trouble will ensue as soon as the new men strike the camp."[56]

WFM leaders understood that the union's fate in Cripple Creek was linked to the fortunes of the nascent union movement in Joplin. The union worked hard to frustrate the recruitment of strikebreakers, often in conjunction with the Joplin CLU. Copley and Burns realized that collaboration with the AFL was the only way to overcome local animosity toward the WFM. They held rallies to educate miners about the strikes. The Joplin WFM local distributed handbills that countered the claims of labor agents. The CLU circulated these handbills among AFL members in other trades and raised donations to aid the strikers in Colorado. "May success crown your efforts," Thomas Sheridan, the president of the CLU, wrote to the WFM executive council in late 1903. They achieved some success. Sheridan heralded the WFM's newfound commitment in Joplin and predicted that the district would "take its place in the industrial army of labor which will bring about freedom and peace." Despite the past record of strikebreaking, he argued "that there is just as good material here to make union men out of, as anywhere else in the world, and we will demonstrate it in the future." Likewise, Copley reported with confidence to the executive council upon his return to Colorado that "the seeds of unionism sown there will grow and Joplin shall redeem itself of the odium that has so long attached to her, and finally become one of the strongholds of the Federation." Neck City Local 207 concurred with this sentiment. "We have a nice little union here," a representative informed *Miners Magazine* in early 1904. "I think if we win the Colorado strike we will have as strong a union in Missouri as there is in the West."[57]

Despite this rare example of cooperation between the WFM and the AFL, however, hundreds of Joplin miners went west in late 1903. At least 400 Missourians went to Cripple Creek as strikebreakers, around 100 of them by way of Coeur d'Alene. Some of these, particularly those from Idaho, were veteran enemies of the WFM. Others, however, took the opportunity to earn high wages after the lead and zinc miners' association again shut many large mines that fall.[58]

The local movement could not escape the poisonous legacy of violence and ethnic animosity between the district's strikebreakers and the WFM. Cripple Creek mine owners, backed by the Colorado state militia, had pushed the WFM to the brink over the winter by declaring martial law, imprisoning union leaders, including Copley and Moyer, and enforcing a new permit system. On June 6, while the WFM held its annual convention in Denver, where delegates read Copley's hopeful report from the previous fall about prospects in Joplin, a bomb exploded among a large group of nonunion miners waiting to board a train in Cripple Creek. Thirteen men died and six sustained severe injuries, including the loss of limbs. In Joplin, the bombing opened old wounds. Local newspapers carried reports of the attack that blamed the violence on the WFM and deceitful foreigners. "The blowing up of a railroad depot and the killing of nearly a score of miners was an act of cowardly and hellish malice," the *Webb City Register* declared. The culprits, the paper continued, could not be "a product of this free and manly country" because "the American spirit hates assassination and will not endure such foul means of revenge." The WFM's organizer in the Joplin district, Matt Wasley, a former president of Northern Mineral, reported that the bombing undercut support. "The prevailing impression with many people was that the miners of Cripple Creek, Colo., were responsible for the atrocities committed in that section," he stated. Although no records indicate any Missourians among the casualties, Joplin miners, union and nonunion alike, still sympathized with nonunion victims of WFM violence.[59]

Joplin miners soon abandoned the WFM outright. In August 1904, Wasley reported that all of the locals were in disarray, with most members several months delinquent in the payment of dues. In Joplin, he discovered that only a dozen or so members still attended meetings. The Chitwood local had stopped meeting altogether. In Webb City, Wasley could not find the local officers and spoke with some former members "who thought that they could get along without organization." He reported that although more than 2,000 people marched under the CLU's banner in the annual Joplin Labor Day parade, only three were miners, including himself. With Sheridan and the CLU he arranged a public meeting to rebuild Joplin Local 195, "but no one came." Wasley was about to leave when he received a letter from Moyer, fresh out of jail, with orders "not to give up the fight in this district." He held more meetings "with but little success."[60]

The Cripple Creek violence destroyed whatever basis for union solidarity had been built in the Joplin district. As the locals weakened, a duplicitous local officer ran away with remaining union funds, which utterly decimated the WFM's reputation. While Wasley acknowledged that the theft of money made it hard to rebuild any of the locals, he blamed the collapse of the WFM locals on the inability of most Joplin miners to think and act in solidarity with union members elsewhere. Wasley concluded that too many miners in the district had given themselves over to an aggressive, selfish perspective that the union's version of respectable manhood could not overcome. By early 1905 none of the locals survived. "I must say," Wasley declared, "that I attribute the present deplorable condition of the unions there to a lack of dignity and utter disregard for their welfare and that of their fellows." He concluded, "There was no chance to reorganize."[61]

By 1905, many miners in the Joplin district held a powerful animus against the WFM. That animus stemmed, for the most part, from conflicts between Joplin strikebreakers and union miners in the western mining camps. In those conflicts, Joplin miners learned to define their interests, once thought insulated and separate, in direct opposition to those of union miners as the national metal economy tangled their fates together. Joplin's strikebreakers began with defensive intentions aimed at family survival amid a national depression. Over time, however, they came to claim a special economic advantage, a privilege, as native-born white men who were willing to crush mostly foreign-born union miners, who were, in their view, lesser men. When the WFM fought back, with words and fists, Joplin strikebreakers claimed defense of those privileges as justification for further aggression. Across the West, they learned to assert racial privilege through a newly violent masculinity, often backed by state force. The cascade of conflict was impossible to escape, whether one was in a strike zone or in Missouri, where the WFM grappled with the AFL to organize the miners it fought against in places like Cripple Creek. Miners in Joplin were also learning to seek power with violence against perceived racial enemies in their midst. In Joplin in April 1903, a mob lynched Thomas Gilyard, an African American laborer, for allegedly murdering a policeman. In the aftermath, the mob burned the small black section of town. The *Joplin Globe* assured readers that no "honest toilers" from the mines were involved, but a trial revealed that many miners took part, a few as leaders of the mob. Their attacks would drive the few remaining black miners out of the district.[62]

Joplin strikebreakers brought their aggressive, racist sense of masculine self-interest home, where it set "the Joplin man" against the WFM and even the more conservative AFL. In an era of rampant racism and nativism across the country, that sense of self-interest would also give Joplin miners a positive way to understand their potential power as white men who worked for wages. They displayed it in the 1904 presidential election by giving a substantial majority to Republican Theodore Roosevelt, who championed both the new ideal of rugged white masculinity and the need to restore the democratic possibilities of capitalism. As an industry observer declared, "Joplin has the best American spirit" because "Joplin has no union." What that meant was plain. "The Joplin man simply takes his chances—often they are big chances—puts in an honest day's work, and gets on in the world if there is anything in him at all."[63]

CHAPTER 5
THE AMERICAN BOY
HAS HELD HIS OWN

Americans in the first decade of the new century witnessed the nation's emergence as a global economic and military power. Nothing symbolized the transformation more than the U.S. Navy's new fleet of sixteen battleships steaming out of Hampton Roads, Virginia, in December 1907. President Theodore Roosevelt dispatched this "Great White Fleet," so dubbed because of its snowy paintwork, to display American might around the world and embolden American nationalism at home. Beginning with the wars of 1898, the American economy prospered in part because of military spending and the influence a newly powerful military won in Hawaii, China, the Philippines, Cuba, and elsewhere in Central and South America. Most of the fruits of that prosperity went to a small number of large corporations that had consolidated control of entire industries in new monopolistic cartels called trusts. These industries, selling at home and abroad, demanded more raw materials, particularly metals, than ever before. Between 1898 and 1910, American metal production boomed: pig iron up 131 percent, copper up 105 percent, and lead up 71 percent. The companies that mined and smelted zinc, however, expanded production most, 133 percent. Manufacturers used greater quantities of zinc in familiar ways—to galvanize iron and steel, to make brass, and to fashion decorative sheet metal—but also for components in the new electrical industry. Many manufacturers, particularly in shipbuilding, found zinc-based pigments best at holding color—so it was that Roosevelt's new ships came to be painted a memorable zinc white.[1]

Mining companies in the Joplin district stretched operations to meet growing demand. With no new ore discoveries since the 1890s, firms increased the scale of old mines by investing in engines to power hoists, pumps, and jigs and by hiring more wage laborers; while dozens of companies still used horse hoisters in 1904, all used either steam or electric hoisters in 1914. The need for more scale prompted further consolidation, as larger firms acquired smaller competitors. The number of district companies shrank from over 500 in 1904 to 150 a decade later. Some leasing companies started mining for themselves in this period, including, most important, the American Zinc, Lead, and Smelting Company, which emerged as the district's largest firm.[2]

Compared to the monopolistic structure of other mining areas, however, small-scale production continued to define the Joplin district, despite the trend toward intensification and consolidation. Though American Zinc came to dominate, scores of small companies pursued scattered ore deposits with modest, flexible methods but now with steam engines in place of horses and with a few more employees. Distinctive Joplin mills dotted the landscape. Industry observer T. Lane Carter reckoned in 1910 that a prospective operator "may do a great deal" with an investment of $25,000. Such relatively rudimentary conditions deterred industry giants from takeovers. The American Smelting and Refining Company, which owned copper mines and smelters in Mexico, Arizona, and Texas, abandoned an effort to take control of Joplin's zinc production in 1902. Similarly, the National Lead Company, a trust that controlled most of the lead industry, including interests in the eastern Missouri field, dropped its bid to acquire Joplin's leading producers in 1905. Neither conglomerate believed the area's unpredictable, dispersed ore deposits were profitable enough to justify the trouble of wrangling agreement from scores of producers. The district's mining and milling practices seemed "crude and wasteful to those accustomed to more elaborate methods used elsewhere," engineer Clarence Wright explained in 1913. However backward these methods seemed to outsiders, Carter believed that such practices allowed small and medium-sized companies to flourish in a competitive and democratic environment where "a larger number of men are successful, and the mineral wealth is more evenly distributed among the members of the community than when a mining district is gobbled up by a trust or syndicate and all the profits go to swell the fortunes of a few." Their competition for that wealth was fierce. Even the district's producers' association could not hold together; it disbanded in 1906.[3]

With "crude," partly mechanized methods, Joplin's mining companies relied ever more on the muscle power of underground workers to make profits. As the shallow deposits dwindled, companies began working deeper deposits in the band of Jasper County mines that ran continuously through Oronogo, Webb City, Carterville, and Duenweg, known locally as the "sheet ground." By 1910, these four camps produced over 40 percent of all zinc ore in the district. These deposits were found more than 150 feet deep in a great sheet that was five to twenty feet thick and up to five miles wide and twenty miles long. The problem was that the sheet ground deposits, once milled, yielded a metal content of 4 percent or less, a much lower grade of ore than anyone had tried to mine before. Sheet ground companies had to hoist more material than ever: twenty-five tons or more of raw ore to produce one ton of concentrated ore for sale. To do that, they needed "increased capacity at small cost," Wright observed. While companies invested in steam power to raise hoisting and milling capacities, they also hired larger, cheaper crews of laborers to increase underground production, particularly in the job of hand loading ore for transportation to the surface. To get as much capacity as possible for the smallest cost, companies began paying these shovelers, as they became known, by piece rate rather than the customary daily wage for unskilled workers.[4]

With no future as owner-operators, Joplin miners transformed shoveling into a new form of poor man's entrepreneurialism that sought individual aggrandizement through brutal labor. Like those who broke strikes, they abandoned the manly ethics of the skilled prospector and union miner in favor of a rough masculinity that associated physical strength and recklessness with the risk-and-reward logic of capitalism. Providing the muscle on which district mine production relied, shovelers took advantage of demand for their labor. They gained power and used it to command higher and higher piece rates. Soon, the shoveler—calculating, Herculean, and formidable—embodied the new ideal of aggressive white American masculinity that President Roosevelt and others championed.[5]

By embracing rough masculinity as the means to entrepreneurial opportunity, the Joplin shovelers validated the decisions of the district's strikebreakers and created a new way for district miners to understand their position as wageworkers in opposition both to the expanding but still myriad local mining companies and to foreign-born miners in radical unions. Yet even if they saw themselves as a group apart from other workers, Joplin miners could not escape the labor movement in American mining. Their ferocious work gave them sway but did not make a strategy. Union organizers and strikebreaker recruiters alike continued to seek influence in the district after 1905. The oppositional power of the shovelers not only created new opportunities that the Western Federation of Miners (WFM) and American Federation of Labor (AFL) hoped to exploit but also exacerbated white nationalist animosities that complicated and undermined those efforts. What Joplin miners wanted for themselves above all else after 1900 was money and status as white men free from control. As the zinc market soared during the Great War, they believed both were in reach.

The partially mechanized methods used in the sheet ground mines created new opportunities for underground workers that allowed entrepreneurial, market-based ambition to survive, albeit on much narrower terms, in the district's new wage-labor regime. These hard, low-grade deposits required extensive dynamite blasting that dislodged great amounts of zinc ore, known locally as "dirt." Companies invested in air-powered, steel-tipped machine drills to set charges for blasting and hired drill operators to run the machinery. To load the massive quantities of ore that was dislodged, companies enlisted the district's plentiful and relatively cheap ranks of wageworkers. By 1912, three-fourths of all mine workers labored in the ground; 80 percent of these miners performed one of two tasks, drilling or shoveling. Shovelers were the largest group, accounting for roughly 40 percent of all miners in the ground. Long considered lowly common laborers, shovelers were now essential; companies competed for their services. By 1912, those who excelled were among the district's highest-paid workers; only the most skilled miners, the engineers and the men who set the charges, the powdermen, earned more than the strongest and fastest shovelers. In less than a decade, men with few assets other than their muscle and endurance had become, according to one report, "one of the factors in the district" whose services "are in big demand."[6]

Shovelers quietly became indispensable. Local companies, like their counterparts across the country, had employed common wage laborers to perform unskilled tasks in and around the mines for decades but particularly after the acceleration of zinc production in the 1880s. While skilled miners and prospectors enjoyed the highest status and best opportunities, common laborers went unheralded and badly paid. T. Z. Pickers, the private detective who came to Joplin in 1896 to recruit strikebreakers for Leadville, was among the first to report their emerging usefulness. He recommended Joplin "shovelers" as good strikebreaking material: "Good, able bodied young men ... anxious to learn to use the hammer, and are ready to go." As Joplin companies produced more and more ore, local observers also noticed the shovelers' growing importance. In 1901, the *Joplin News Herald* reported on "a day with the shovellers," who were "the least spoken of." The sheet ground expansion lent their work new importance. Companies that mined more ore needed to move it out of the ground faster than ever. On a daily basis, the writer noted, the average mine hoisted 350 cans, each weighing 500 pounds, a total of 175,000 pounds of dirt. In some mines, companies hoisted one 500-pound can a minute. To fill those cans, companies turned to the power of men armed only with a standard, number-two-size shovel. "He is the brawn and muscle of the lower ground," this observer explained. As unskilled laborers, however, shovelers still earned wages commensurate with low status, on average $1.75 per day in 1901, about as much as common

Shovelers, Webb City, Missouri, ca. 1900s. Historical Mining Photographs Collection. Courtesy Joplin History and Mineral Museum.

mine laborers had earned a decade earlier and roughly in line with what un-skilled laborers earned elsewhere at the time.[7]

Despite meager wages, young men in the district and surrounding region sought work as shovelers because the job provided a way to enter the indus-try now that prospecting was generally closed. "Working in the big mines around Joplin there are hundreds of young men who have in recent years drifted in from country places," the 1901 observer noted. Many of them came from nearby counties, others from rural areas across the lower Midwest and Upper South. These men benefited from the demand for shovelers because of their youth, since older men could rarely perform the sustained, hard work of moving tens of thousands of pounds of ore a day, as employers in-creasingly demanded. "Personal acquaintance with ground men reveals that the shoveler is usually a strong, robust young man," the report continued. The countryside was full of such workers, most of whom had no particular skill or experience outside of routine farm tasks. Once they began shoveling, however, many took defiant pride in their hard work. In 1901, for example, one shoveler expressed disdain for a crowd of people he overheard worrying about a tired road crew. "Why they ain't in it with a spade hand," he said. Give any of them "a Number Two and tell him to hike 'er, he would throw up his job after the first hour." In his view, no other worker, skilled or otherwise, worked as hard as a Joplin shoveler.[8]

Shovelers started to receive higher pay to ensure increased production. When ore prices went above thirty dollars per ton in early 1903, many companies began paying shovelers twenty-five cents per hour, or $2.25 for a nine-hour day, to help ensure full work crews and perhaps to weaken the appeal of western labor agents and union organizers. When zinc prices soared to new highs above fifty dollars per ton in early 1905, companies offered shovelers $2.50 a day. Shovelers no longer performed common labor; they now had an essential task that paid among the best wages in the district. Their rising fortunes helped explain the contemporaneous collapse of the district's WFM locals. The men who told organizer Matt Wasley in 1904 that "they could get along without organization" knew that shoveling was another means, with strikebreaking, of doing so.[9]

Joplin miners had to get along, however, amid unprecedented workplace dangers. This was especially the case in the sheet ground mines, where companies moved more ore than ever before. Since the opening of these mines in the 1880s, miners knew that the dry, silica-rich formations gave off lots of dust, a problem that worsened with expansion. Miners in Webb City and Carterville reported rising rates of "miner's consumption," a term used by coal and metal miners elsewhere to describe a variety of tuberculosis-like lung diseases. Afflicted miners slowly lost their ability to breathe over a period of months until they could no longer walk. Most ultimately died of suffocation. Local papers regularly reported deaths from miner's consumption as early as 1903. Although commonly confused with tuberculosis, which physicians knew was a bacterial infection, miners insisted that this affliction stemmed from work in the mines. In 1905, for example, a miner told the *Webb City Register* that "so many men in this mining district die from what is called miner's consumption" because of "smoke from them dad blamed old lights they wear on their caps and from the dust in the mine." Two years later, the paper reported that Ben Peppers, a forty-four-year-old miner who had been a Kansas farmhand, was near death from miner's consumption he contracted "while doing under ground work." He died three months later. Ben's younger brother William, also a miner, had died of "consumption" in 1903. At the time, however, few observers beyond the sheet ground camps noted the problem. The state mining law, for example, stipulated no measures for ventilation in zinc and lead mines until 1907.[10]

The rising number of miners injured or killed in accidents captured far more attention. In 1903, thirteen miners died in Jasper, Newton, and Lawrence Counties, most as a result of hoisting accidents, premature explosions, or rock slides and falls, the latter causing about half of all fatalities. The problem worsened as miners moved more rock. In 1904, twenty-four miners died in southwest Missouri. In 1905, the number had grown to thirty-nine, and in 1906, fifty-one miners perished on the job. Dozens more suffered serious injury. The majority of those killed or injured that year—fifty-three men—worked as shovelers. The 1904 death of Harvey Dunlap, a twenty-seven-year old shoveler from the Missouri Ozarks, epitomized the banality of the dangers the shovelers faced: "The deceased, in a stooping posture, was loading his tub with a shovel, when a small boulder fell from the roof, striking him on the back, the wound causing death by the time his body reached the surface."[11]

Industry reformers tried to stop the carnage by enforcing safety regulations. Since the early 1880s, at the urging of coal miners, the state of Missouri had passed a series of laws to regulate the operation of mines, first coal but also lead and zinc: regular inspections, the adoption of safety standards and equipment, and measures to improve working conditions, including a 1901 law mandating an eight-hour day for all miners in the state. From the outset, Joplin's small mining companies argued that they could not afford to comply and made little effort to do so, even after some grew larger. In 1905, according to Otto Ruhl, a local engineer, no mining company provided a safety cage for hoisting men into and out of the mine. Companies instead used open, free-swinging tubs, little changed from the 1850s. "The general run of operators would strenuously oppose any law compelling them to install cages and safety appliances to thus protect their men," he explained. Ruhl also blamed complacency among the miners, who, he claimed, readily accepted company arguments about the threat of excessive regulation, a holdover, he believed, from the logic of the poor man's camp. "Habit and custom holds fearful sway," Ruhl concluded. The state mine inspector accused companies of using old justifications to excuse inaction. "Conditions which at one time justified the excuse referred to, in the minds of many, are rapidly vanishing," he wrote in 1906. "The poor operator and shallow shafts are overshadowed by the great majority of those otherwise situated."[12]

As they had in the past, however, Joplin miners continued to accept the risks of their work. They ignored WFM organizing efforts in the midst of rising fatalities despite the union's long-standing demand for more stringent safety regulations. Although miners could not ignore the dangers of the job, they did not naturally accept union arguments that explained those dangers as the result of systemic capitalist injustice. Some miners relied on old forms of private insurance through fraternal orders or private insurance companies, such as the Modern Woodmen of America or the Aetna Liability Insurance Company, to cope with the risks. On average, one in four miners killed or seriously injured in accidents between 1903 and 1906 carried some form of insurance. These men were the most mindful of family security, the main purpose of fraternal and corporate insurance plans alike. However, the vast majority of miners—three of every four miners killed or seriously injured in these years—carried no insurance at all. Some sought spiritual remedies; in 1904, hundreds of people, including many miners, helped turn Charles Parham's middle-class healing ministry into a popular religious revival that became modern Pentecostalism. Most locals soon cooled on Parham but continued to believe in the miracles of capitalism.[13]

While miners showed little support for safety regulations, they expressed strong concern that the state's new eight-hour law might limit their pay. After a series of court challenges, the U.S. Supreme Court upheld the law in 1905 in *Cantwell v. Missouri*, despite a ruling earlier that year in *Lochner v. New York* that had overturned a state law mandating a ten-hour workday for bakers. Although the justices filed no opinions, their decision justified the law as a legitimate exercise of the state's police power. Respondents in the Joplin district claimed that no one liked the law. With zinc ore prices near fifty dollars per ton, companies were in no mood to risk curtailing production by reducing the length of the workday. Meanwhile, many miners complained that a shorter workday would at the same hourly rate result in a pay cut. According to an editorial in Joplin's *Lead and Zinc News*, "both operators and miners resent the fact that they are being robbed by a law that prevents the full exercise of production in the operator's mine and brings about the arbitrary cutting down of the miner's day to a point where his return is for eight hours instead of nine."[14]

Miners wanted to keep their daily wage, and some threatened to organize a union to do it. Joplin's Central Labor Union led that effort in late 1905, although with little to show. Still, some observers concluded that although the threat "did not amount to much," it prompted employers to find a solution. Some companies began exploring alternative ways to maintain or increase production under the law. The editor of the *Lead and Zinc News* correctly reported that a "change from a day wage to a piece wage will be made in order to evade the law."[15]

Larger companies, led by American Zinc, began paying shovelers on a piece-rate basis in early 1906. They adopted the piece rate as a way to protect their low-yield, high-volume operations from constraints imposed by the eight-hour law. The shortened workday made it harder for companies, especially in the sheet ground, to increase production. Companies could hire more shovelers or buy mechanical loaders, like the sprawling St. Joseph Lead Company in eastern Missouri did, but that required further investment to expand the size of the underground workings and perhaps sink additional shafts. Although many companies made moves in these directions after 1905, most could not afford large capital investments, at least not in such short order. Instead, with ore prices near all-time highs, firms turned first to the cheapest solution: getting more work per hour out of their shovelers by paying them for every can they filled. Under the initial rates, shovelers earned between four and six cents a can, depending on size. Companies moved only shovelers to the piece rate; other workers, including drillers, or machine men, as they became known, continued to earn a daily wage. At the time, piece rates were common in coal mining and some manufacturing industries as a way to increase productivity but not in metal mining. Most western metal miners fiercely resisted piece rates. The WFM argued that the system forced workers to do more for less pay and encouraged reckless practices that made work more dangerous. Many large-scale companies that mined base metals, meanwhile, such as National Lead and St. Joseph Lead, hired recent immigrants to shovel for low daily wages. In the Joplin district, however, piece rates made sense for companies looking to raise production without heavy capital investment and for miners who associated masculine risks with remunerative rewards.[16]

Shovelers accepted the piece-rate system with an enthusiasm that shocked many observers. With brutally hard work, they registered staggering increases in production and in pay. By early 1907, district residents marveled at the "remarkable feats performed by shovelers" and began tracking the record for highest earnings in a week. In March, twenty-eight-year-old Ed McAuliffe set the record at fifty-one dollars. The following month, Jack Fox, a twenty-seven-year-old Webb City shoveler who hailed from Michigan, beat it by earning fifty-three dollars in a week. He did so by filling 653 ore cans at a rate of eight cents per can. Consider for a moment the prodigious physical effort his achievement required. Filling cans that held 800 pounds of ore, Fox shoveled a total of over 522,000 pounds, or 260 tons, that week, an average of 87,000 pounds, or forty-three tons, per day. To do that in an eight-hour shift, Fox lifted his shovel, which held about twenty-one pounds of ore per full scoop, over 4,140 times a day, 518 times per hour, 8.6 times per minute, or once every seven seconds. He maintained that pace for six consecutive days. Not all shovelers earned these sums, of course. But across the district in 1907 many shovelers regularly earned five dollars a day, or thirty dollars a week. Eager to boost production with zinc prices still rising, companies responded with new financial incentives to attract the most ambitious workers. Some offered bonus rates that, for example, paid shovelers six cents per can for the first fifty filled in a day and then ten cents a can for any additional cans filled. Shovelers on piece rates could often earn double what they had on an hourly wage.[17]

Workers like McAuliffe and Fox saw in piece-rate shoveling a new means of poor man's entrepreneurialism. Despite the sharp decline of owner-operator mines after 1900, many commentators in the district continued to trumpet its "opportunities for poor men." One newspaper, for example, claimed in 1907 that "the opportunities for creating wealth in the Joplin district today are no less than they were ten years ago," but rather those "opportunities are multiplying, and the man who comes here now stands a better chance to gain wealth than did the men who have already created their fortunes." For men with no capital other than their physical labor and their determination, the likes of whom had once been the engine of prospecting, their only available opportunity for earning more than a standard daily wage was in shoveling. Some men left middle-class jobs to work as shovelers. Fox, for example, had previously worked as a clerk but quit because he preferred the pay of shoveling. As another observer explained, "The economic prize of the shoveler's wages attracts and recruits the beginners." Stories of record earnings like Fox's, told with rhetorical flourishes once used to describe successful prospectors, encouraged more young men to try their hand—and back.[18]

Shovelers turned the piece-rate logic to their advantage to become among the highest-paid workers in the district. "The shoveler, depending on his strength and experience, can make ... considerably more than the skilled laborer," visitors to the district noted in 1914, "a condition probably not found in many other mining camps." The best shovelers honed their technique to make themselves into "fancy" shovelers, whose strength, endurance, and willpower commanded the highest wages. "They say they have created a new profession here," another commentator explained, "that of skilled shoveler." Some shovelers adopted a subcontractor gang-labor model. Like leasehold miners in the 1870s, these shovelers hired helpers, often single, younger men, for a set daily wage or a share of the day's total earnings. This model enabled them to take full advantage of production bonuses by filling 100 or more cans a day. Yet however much shoveling recalled the entrepreneurialism of jug-handle leasehold mining, shovelers ran far different risks than prospectors who invested time and money. If prospectors took chances with their livelihoods, shovelers risked life and limb. Observers emphasized the astounding physical exertion that shovelers delivered. As a doctor explained in 1914, "One can hardly realize the severity of this work without seeing it." Make no mistake, another report concluded, the shoveler "earned his money."[19]

To do so, shovelers embraced a form of working-class masculinity that heralded physical aggression and dangerous action in pursuit of financial incentives. Whereas prospectors had prized the diligence and sustained effort required to locate and develop a paying mine, shovelers celebrated physical power, strength of will, and heedless action under the piece rate. Early on, shovelers were more likely to be young and single and were celebrated for their youthful virility. "They are stout," the *Joplin News Herald* reported in 1901, "and are possessed of that reckless nature characteristic of the country boy." The shoveler had "no fear for the morrow," the paper concluded; he "knows no fear of toil and often I have known men to go to dances, stay up all night, perhaps, twice or three times a week, and never miss a shift." Over time, these men married and started families; as the job gained in status, slightly older men with families also took up the work. According to the 1910 census returns, 53.6 percent of men who worked as shovelers in Jasper County were married. They were less likely to be married than male workers in the local zinc-mining industry overall (64.8 percent), and a lot less likely to be married than men who farmed in Jasper County (80 percent). Still, the best shovelers tended to be younger than other workers. "It was said of a shoveler," according to a local historian, "that rarely could a mature man enter this line of work and make a go of it, that the men who followed this occupation were developed and this development extended over quite some time. In fact, all good ore shovelers began as youngsters." The arrogance of youth, however, led shovelers toward short-term imperatives. Men exploited their youthful strength and energy as if infinite, which added a sense of competition over male prowess. While "a husky shoveler who works with a will" but "does not overstrain himself" could sustain high earnings, an industry observer reported, "the striving of the best men in daily rivalry to show a large tally" caused "their early breakdown." Restraint was difficult, however, when demonstrations of power in pursuit of money garnered respect and status.[20]

Shovelers gained so much acclaim that the *Joplin Globe* held a contest in 1907 to find the best of all. Lem Smith won the fifty-dollar prize, and over thirteen dollars in wages, by filling 303 cans weighing an astonishing total of 257,550 pounds, or just over 128 tons, during a single eight-hour shift. He did it by working "with the regularity of a machine, and with a strength as steady as that of steel." Although he tired toward the end, the twenty-seven-year-old Smith "gritted his teeth as nature rebelled against the awful demands that were being made" and continued to shovel in pursuit of the bounty. Asked how he felt the following day, Smith replied, "Pretty near all in, to tell it square. I loafed the next day. But I reckoned I could afford to take a vacation. I felt tolerably certain that that record would get the fifty." If anyone wanted to challenge him, Smith was ready. "I can out-shovel any man in the district," he told reporters and was "willing to back that opinion with a liberal chunk of cash." Reporters marveled at his achievement but also at his confidence, calm, and crucially, self-possession. "Here is the kind of man you would choose for your expedition if you had an appointment to march across the Sahara and storm the gates of perdition on the other side," the *Globe* declared. As this imperialistic motif suggested, the same thing that excited western mine owners also excited these observers: Joplin shovelers were willing to risk their bodies in pursuit of individual economic gain.[21]

They pressed the individual incentive of the piece rate to extremes that anticipated the findings of scientific management theorists. In his 1911 *Principles of Scientific Management*, Frederick Taylor described how managers could increase the efficiency of unskilled workers through close instruction, physical regimentation, and individual monetary inducements, a set of practices that became known as Taylorism. He had based his theory on experiments conducted with loaders and shovelers, the archetypical unskilled industrial workers. In one example, he trained a man named Schmidt, a "mentally sluggish" pig iron handler at Pittsburgh's Bethlehem Steel, to increase his handling rate from twelve and a half tons to forty-seven and a half tons per day. In another, Taylor recalled how a team of Bethlehem shovelers raised their individual production from sixteen to fifty-nine tons per day. In both cases, Taylor argued that immigrant workers like Schmidt, whom he deemed inferior, needed personal lures and total managerial oversight to achieve such gains.[22] Joplin shovelers, by contrast, applied the logic of individual incentives to themselves long before anyone in the district had heard of Taylorism. Some observers offered them as the model to be emulated in other districts. "The marvelous energy with which the work is carried on, and especially the remarkably high efficiency of labor," a report from the American Mining Congress declared, "are things to be seriously considered by operators in the Far West." The shoveler who earned forty-two dollars per week "at the regulation contract price," he concluded, "is an object lesson in himself."[23]

Joplin's shovelers were now celebrated as exemplars of working-class masculinity. "The shoveler is a typical man, an American," the *Globe* exclaimed, who used "brain, nerve and muscle" to perform heroic work. Although Lem Smith stood only five feet, eight inches, tall and weighed 165 pounds, the press reported, his "will power and endurance" typified "the genius" of the shoveler, "a man that literally moves mountains." "There is an extraordinary mainspring hid in that fellow somewhere that we don't get to see," the *Globe* declared.[24]

Many believed that the hidden power stemmed from an inherent racial and national superiority. According to one report, the district's miners "are in a class to themselves" because "they are Americans." Driven by pursuit of personal prosperity in the tradition of the poor man's camp, this writer continued, "the American boy has held his own and kept away the foreigner from the field which produces lead and zinc in such quantities as to almost supply the world's demand for these metals." They were intelligent, industrious, and independent—in other words, the opposite of unionized or foreign workers, like Taylor's slow-witted Schmidt. "There is no comparison between the personnel of the miners in this district," this writer explained, "and those of the Pennsylvania coal fields. The latter are made up of Hungarians, Italians and different divisions of the Slavonic race, who live in squalor and are debased in their habits to a degree impossible for the American boy who works as a miner in this district." As the *Webb City Register* boasted, "Our shovelers are intelligent Anglo-Saxons."[25]

The shovelers gained status and power because they seemed to embody the ideal image of white American manhood championed by President Roosevelt and other white nationalists. Fearful that native-born white men had become overcivilized and soft and thus vulnerable to domination by other races, Roosevelt advocated that American white men seek a "strenuous life" to hone their racial advantage in the arts of civilization through aggressive physical competition and even violence. Other white Americans joined in this obsession to conflate masculine dominance with assertions of racial superiority, whether they cheered the doomed white boxers who challenged black champion Jack Johnson or Roosevelt's "Great White Fleet" that promised victories over nonwhite people overseas. To observers in the Joplin district, meanwhile, the shoveler seemed to fulfill these great white hopes already. Engineer T. Lane Carter challenged anyone who worried "that the white man is losing the art of hard work" to "visit the Joplin district." There, he declared, "he will see what the white laborer, working on contract, can still accomplish.... Fancy a white man shoveling 50 tons per day in an eight-hour shift!" The *Globe* agreed: the shoveler was "a God-fearing, duty-serving, fun-pursuing, hard-working, whole-souled, mother-revering, patriotic, honest-hearted, good-blooded, devil-defying boy, of which, let us be devoutly thankful, there are thousands and millions in this favored land." As with the district's strikebreakers, the shovelers seemed like perfect workers, perfect American men: productive and loyal, trusting in capitalism, native born, and white.[26]

Not content to reap only honor and respect, shovelers pushed for more money and, in doing so, led the way for other miners to demand more from district companies. Like Joplin strikebreakers had done, the best shovelers sought the highest bids for their services. Many shovelers, meanwhile, continued to seek higher-paying jobs as strikebreakers elsewhere. Their mobility created constant labor shortages that gave all shovelers, no matter the skill, greater leverage with employers and helped reinforce an opportunistic working-class culture in the district. Their leverage, gained amid steadily rising zinc ore prices from 1902 to 1907, worked as long as prices remained high. When ore prices fell during and after the Panic of 1907, however, companies looked for ways to curb the power of the shovelers. By then, some shovelers were determined to defend their position with overtly oppositional strategies, some old, some new, that would embolden many other miners and again bring the WFM back into the district. Once paragons of white American manhood, Joplin's shovelers soon sparked a local working-class rebellion that would unsettle the prevailing order of risk and reward in district mines.

As demand for their labor grew, shovelers moved from job to job, often with no notice, to take higher piece rates. In April 1907, the state mine inspector reported that companies across the district, particularly those working the sheet ground, needed more shovelers. That spring, Charles Landrum, American Zinc's mine manager in southwest Missouri, informed Harry S. Kimball, the company's president, that the "mines are all running in fine shape, excepting short handed as to shovelers." "The situation is serious," he warned and predicted that production would suffer. With ore prices above fifty-three dollars per ton, companies risked losing easy profits with ore stuck in the ground. Landrum was offering piece rates that allowed shovelers to earn an average of $3.50 to $4.00 a day but still predicted that "the shoveler situation is going to be bad all summer," given their new footloose behavior. He also worried that "machine men and other classes of ground labor are likely to go to grumbling" with demands for higher wages to match what the shovelers could earn. American Zinc again raised their pay.[27]

At the same time, Joplin miners continued to receive offers of employment from big companies elsewhere. Local papers reported the presence of recruiters from western districts that spring, particularly Coeur d'Alene. Idaho mining companies were not looking for strikebreakers, since they had crushed the WFM after 1899. They expected trouble, however, from union remnants as the state prepared to try WFM leaders Charles Moyer, Big Bill Haywood, and George Pettibone for orchestrating the 1905 murder of Governor Frank Steunenberg in retaliation for his actions during the 1899 strike. The state's case outraged unionists and reformers across the political spectrum. By hiring nonunion Joplin miners, the mining companies sought insurance against likely protests. Other western operators also sought strikebreakers that summer. For the first time, copper-mining companies from Bisbee, Arizona, sent agents seeking miners to defy the WFM. According to Landrum, shovelers left for these camps by the "car load." The state mine inspector believed they went mainly to take advantage of high wages but also "for the novelty of travelling free of expense and obtaining experience in other districts." Joplin's shovelers might be among the hardest-working white men in America, but they used their leverage for their own ends, whether seeking higher wages or looking for adventure. Their growing power gave them new freedom, in other words, that was rapidly becoming a problem for district companies.[28]

The managers of American Zinc, the largest and most modern company in the district, led the search for alternatives to the shovelers. Kimball encouraged Landrum to try "importing foreign labor," like big lead companies in eastern Missouri did, although he worried that it would incite violence. "There is that serious objection to foreign labor which now exists, and which has always existed" in Joplin, he explained. Both men believed that machine loaders offered the ultimate solution. Landrum was in discussions with the Thew Automatic Steam Shovel Company about its equipment but had not reached an agreement. "I am very much pleased at the interest you take in the shovel matter with me," he informed Kimball, "and I know that it is our only salvation" in the sheet ground mines.[29]

Other mining companies soon joined American Zinc in challenging the shovelers. As ore prices weakened that summer, from a high of $53.50 per ton in April to $46.00 per ton in August, the companies' concern shifted from a shortage of shovelers to a growing need to cut shovelers' wages. They had adopted the piece rate to encourage cheap, efficient production and had not anticipated how enthusiastically shovelers would respond. "The mine owners of the district are beginning to understand that there is something radically wrong with the wage scale in the Missouri Kansas mining district," the *Webb City Daily Register* reported in August, possibly with direction from those same owners. The report recalled that shovelers had earned two dollars a day only a few years before. Now, "many miners are able to make more than $5 per day," and some up to eight dollars per day, on the current piece-rate scale. With such high wages, the paper charged, the shoveler "is making more money than" mine superintendents who earned $150 per month. Officials argued that shovelers should not be paid so highly because their "work is termed unskilled labor." The shovelers had gained power and status so quickly that managers were "at a loss to know how to reduce them without forcing a strike on the part of the miners." If companies did not find a way, the report concluded, many would be driven out of business. This critique discounted the physical cost of the work by assuming that men could perform the brutal labor required to achieve high pay every day. More pointedly, the attack on skill hit at the respect and veneration that shovelers had accrued for making their task so productive. In doing so, it threatened to undermine the racial and gendered arguments that explained the achievements and justified the privileges of Joplin shovelers. If seen simply as overpaid unskilled workers run amok, as this report claimed, they presented a menace that should be stopped. Harry Kimball soon reached this conclusion himself. As he declared privately to a company manager in early 1908, "The shovelers are the nearest to beasts of any men that I have had anything to do with."[30]

In a spirited response published in the same newspaper, a writer called "Shoveler" fought back by asserting that men like him deserved high wages because they were just as invested in the business of the district as company owners and should be treated as equals. "Shoveler" rejected the notion that they were simple, unskilled workers. "If you will stop most any of the shovelers and enter into a conversation you will find that many of them would be an honor to the position of Superintendent because they see and know all the workings of successful mining," the writer declared. Admittedly, "many of them are rough and use language that would not do at a Sunday School Convention," "Shoveler" continued, "but beneath the rough exterior you will find a brain that thinks and an honest heart." They earned high wages for hard work performed as well as for the unseen expenses they assumed to do it. Their work was so demanding, he explained, that shovelers had to replace worn-out overalls and shoes every week and needed new gloves daily. Shovelers also invested their bodies in the job. "Shoveler" claimed that men like him worked harder than anyone else, to the point "of being tired and worn out so he wished for Sunday to come." If one was too tired to work, however, "he is the one who loses, not the company." He reminded readers that companies had chosen the piece-rate system because it increased productivity. High shoveling wages thus reflected commensurate high profits.

"If a shoveler makes more money by contract its because he gets out more dirt and the more dirt goes through the mill the more ore goes into the bin and the bigger the dividend to investors," he explained.[31]

"Shoveler" also made racist and nativist claims on this system of risk and reward. What made the Joplin piece rate so effective, he asserted, was that the shovelers were all native born. "As Americans we have made Webb City the greatest lead and zinc center in the world," he claimed. Shovelers read books, had families, and "prided ourselves on the conditions existing here." He warned mining companies against bringing in "Dago labor" to replace them. Although they might work for less, "Shoveler" acknowledged, foreigners would soon join unions and be on strike. It was easy, he noted, to incite "rebellion among such trash who eat any kind of refuse, sleep any place and send all they earn back to Italy." These claims revealed that some shovelers had come to believe that they deserved high wages because they were hard-working, native-born white men who understood capitalism and would not join unions. Ideas about race, nativity, family life, and unions had come full circle in the district, as "Shoveler" deployed logic once used by mine owners to entice Joplin miners into work as strikebreakers against foreign-born union men in order to now keep foreign-born miners out of the Joplin district.[32]

Above all, "Shoveler" argued that he and his counterparts were themselves entrepreneurs with faith in the market who should be treated as partners. He pledged continued cooperation with companies if they worked with, not against, the shovelers. "When ore is low and expense is high we will take" a pay cut, he promised, "but when it comes up we want our advance in wages"—in other words, a sliding scale that pegged wage rates to market prices. If a superintendent would bother to talk with the shovelers, this writer declared, "he will find the right hand extended to assist rather than retard the progress of this great industry." If mining companies tried to cut the piece rate out of step with market prices, or if they tried to hire foreign labor, "Shoveler" threatened trouble. "No Sir," he concluded, "we have helped to build this mining center and we are proud of it and we won't see any Italian trash come here and reap the benefits."[33]

"Shoveler" made these arguments in the presence of alternatives that could have directed the growing oppositional impulses of district miners toward the labor movement and its political allies. Local AFL organizers, led by the Central Labor Union's Thomas Sheridan, who was elected president of the Missouri State Federation of Labor in 1905, tried to mobilize them around the eight-hour law. The AFL chartered small, directly-affiliated local unions, now called federal labor unions, among miners in Joplin and Webb

City in February 1906, although both collapsed within months. Meanwhile, more radical groups in the area continued to reach out. In the spring of 1906, the small Joplin local of the Industrial Workers of the World (IWW) and the city's Socialist Party local held a series of public rallies to raise money for the defense of Moyer, Haywood, and Pettibone. They received only $2.50 from "a large crowd of the unorganized miners." Area Socialists advocated for the WFM leaders until their acquittal in 1907, in line with the *Appeal to Reason*. The *Appeal* emerged as the mouthpiece of American socialism between 1906 and 1908, when its national subscription base expanded to 325,000. Eugene Debs, Kate Richards O'Hare, and other Socialist luminaries regularly passed through Joplin on their way to visit the paper's office in Girard; Socialist candidates regularly campaigned throughout the district. Most miners either rejected or ignored these voices. The WFM tried again in February 1907 after organizing several locals in the eastern Missouri lead field. "Very few miners" attended organizer Frank Schmelzer's meeting. The problem, he reported, was the piece rate, which kept "the men contented, for they are always in hopes that they will do better the next month." Despite growing tensions with the companies, miners like "Shoveler" were confident that their entrepreneurial labors would command a share of prosperity.[34]

Miners saw that prosperity, and their hold on high wages, challenged by a crisis in financial markets in late 1907 that brought the national economy to a crawl. In the spring a "general depression" in industrial demand weakened the price of all ores but especially copper. When speculators tried and failed to corner the market in United Copper Company shares, they triggered a run on several New York banks and panicked selling on the New York Stock Exchange that soon spread to financial institutions nationwide. Zinc ore prices promptly fell, down ten dollars per ton between August and December to thirty-six dollars. The Panic of 1907, as it became known, lasted into the summer of 1908. During that period, zinc ore sold for an average of thirty-three to thirty-seven dollars per ton, about a third lower than 1907 highs. Mining companies curtailed operations for an extended period. District zinc ore production fell 10 percent and lead ore production fell 8 percent during the panic, while the market value of all mined ores fell 26 percent to $10.4 million, making 1908 the worst year since 1903. American Zinc did not shut down its operations but regularly threatened to do so. Its managers claimed they could not mine the sheet ground for a profit as long as zinc ore prices remained below forty dollars per ton.[35]

American Zinc and other companies responded by slashing wages. The cut was severe: from $3.00 to $2.50 per day for machine men; from $2.25 to $1.75 per day for general labor; and from six cents a can to four cents a can for shovelers. As in the past, miners seemed to accept the move as a legitimate business decision. According to an industry observer in late 1907, "these reductions were received by the men in good humor, and, indeed, in many cases the men themselves proposed that the reductions should be made." While no doubt overstating their equanimity, this writer echoed "Shoveler," who had assured readers that miners understood the relationship between market prices and wages. They also understood, from recent harsh experience, that wage cuts were preferable to mine closures. Union activists, meanwhile, pointed out that they had no choice. "The unorganized mine workers have suffered reduction of from 25 to 40 percent in wages," reported Charles Fear, who had recently started an AFL-affiliated "conservative labor paper" in Joplin. Although Fear heard many complaints, "they can do nothing but to accept the reduction."[36]

If acquiescent while prices were low, district miners made new, more forceful demands for higher wages when the market recovered. In late August 1908, miners at two smaller mines north of Webb City went on strike and received modest raises. To prevent further strikes, some companies agreed to raise shoveler wages 10 percent once zinc ore went above forty dollars per ton, which it did in November. Initially, American Zinc refused to give in. Landrum threatened to "shut down the mines indefinitely" if shovelers at American Zinc went on strike. Internally, however, Landrum worried that the company would not be able to recruit enough shovelers if other companies raised wages. In the meantime, American Zinc continued to test a steam shovel that it hoped would reduce its dependence on troublesome human hand loaders. "We will shake them to action with the automatic shovel," Kimball explained to another manager. Rather than test the company's resolve directly, American Zinc's miners protested in subtler ways. After a rock fall killed a shoveler in mid-September, miners in two sheet ground operations refused to work the following day. Although such acts of respect for the dead were common among coal miners, Joplin miners left little evidence of following the practice until now. By stopping production to mark death in the midst of a tense debate over wages, they claimed a relationship between physical risk and financial reward. Union miners elsewhere had galvanized collective action around common dangers for decades, of course. Joplin miners were learning their tactics, albeit without union affiliation. A week later, shovelers began sending up near-empty cans. Landrum's deputies interpreted the slowdown as a wage demand. He gave in with a half-cent raise per can. "Everybody is busy and happy," a deputy reported, "and are getting fine results." Others were not convinced the peace would last.[37]

WFM president Charles Moyer, for one, launched a new organizing campaign to take advantage of these conflicts after his charges were dropped and he was released from prison in early 1908. Moyer dispatched William Burns to Joplin as part of a renewed drive to strengthen the union's presence in eastern districts. Moyer hoped to make the WFM more stable, a process that began when the union left the tumultuous IWW the previous year. Burns spent three weeks in the district "but was not successful." Undeterred, Moyer sent William Jinkerson in October. Local Socialists helped him set up meetings in Webb City, where miners had just won small concessions from American Zinc. "But they showed no interest in organizing," Jinkerson reported. Most of them, he explained, banked instead on the victory of William H. Taft in the upcoming presidential election because he advocated a higher tariff that promised to raise zinc prices. "They were perfectly confident," he explained, "that the election of 'Taft and Tariff' on zinc would save them." In Joplin, Jinkerson "was unable to find a man who was willing to lend his assistance towards organizing." Instead, he reported, "they were rather inclined to snarl and insult you when you approached them." In November, Taft outpolled William Jennings Bryan in Jasper County; he won Oronogo, Neck City, Carterville, parts of Webb City, and Prosperity, all majority miner precincts. Socialist Eugene Debs, meanwhile, won 5 percent of the county vote. Jinkerson's experience showed that although miners had become more rebellious in their demands, they also remained hostile to the WFM and anticapitalist politics and continued to believe that they could leverage higher wages from market rises.[38]

As the recovery turned into a boom, miners kept pushing to make this true. Ore prices continued to rise in 1909. "Some of our men are grumbling," Landrum reported; he expected trouble "if we do not give them a raise." Landrum gave in. He had little choice, since the steam shovel had turned out to be less efficient and more expensive than human shovelers. Still, the company suffered a shortage of shovelers all summer. While shovelers resumed old tactics of moving around in search for the best piece rate, machine men and common mine laborers also demanded higher pay. In November, miners at American Zinc's Davey mine threatened to strike if they did not receive an additional twenty-five cents per day. Landrum again granted the raise.[39]

For many miners, these skirmishes validated confrontation as a negotiating tactic, especially when companies tried to cut wages as ore prices fell back to forty dollars per ton in 1910. In neither boom nor bust, they struggled to trust company claims that new wage limits were fair in relation to the ore market. Miners responded with three oppositional strategies, some old, some new: strikebreaking elsewhere, making legal claims for workplace injuries, and staging bigger wildcat strikes.[40]

Some went as strikebreakers to South Dakota's Black Hills, where the WFM clashed with the Homestake Mining Company. In the face of union efforts to compel a closed shop in late 1909, the company locked out its miners and resumed operations in early 1910 on an open-shop basis. To accomplish this, they dispatched labor agents to the Joplin district to recruit nonunion miners to Lead and Deadwood, South Dakota, with a promise of free transportation and wages of four dollars per day. While only a few dozen miners accepted the offer in early January, thirty or forty were leaving each day by the first week of February, just as American Zinc and other companies announced a 10 percent wage cut. "The serious part of the business," Landrum reported in March, "is that they are taking the young men and especially the shovelers." Other mine workers also left, although Landrum believed that most of those with families would stay. By April, Joplin newspapers reported that over 1,000 miners had gone to South Dakota. "This system of taking the labor out of the district has resulted in taking away so many of the men from all the camps in the field that mines and mills are unable to secure full crews," the *Joplin Daily Globe* declared. Despite weak prices, the paper claimed, local companies feared that further wage cuts would encourage more miners to leave for the Black Hills. The WFM's Moyer did too. "The unorganized of Missouri are always with us," he lamented.[41]

Other miners pressed for a bigger share of district profits by taking advantage of new laws that expanded their power to sue employers for workplace injuries. In 1907, the state of Missouri had finally answered the demands of recent mine inspectors with three new regulatory laws: one that created a second zinc and lead mine inspector position; one that gave inspectors more scope to scrutinize the health and safety of employees, particularly regarding air quality; and one that made mining companies liable to compensate workers injured, or the families of those killed, in their employment. With the last of these, workers gained new power to seek compensation for injuries, whether caused by company negligence, a coworker, or unavoidable accident. The new compensation law also annulled limited liability contracts between companies and employees and restricted employer liability only in cases where the worker's own negligence caused his injury. Miners started taking advantage of it in late 1908, albeit initially in small numbers. Local lawyers such as Sylvan Bruner helped them file the suits in county court. As tensions over wages grew in 1909, however, more and more miners sued their employers. Miners might reject many government regulations, but they trusted local juries, full of people like them, to rule against companies often owned by outsiders, just as the squatters who had resisted Blow & Kennett in 1857 had done. "The subject of accidents and liabilities is becoming a serious one," Landrum complained in January 1910. It got worse that winter. In a single March week, Landrum reported that miners had filed suits for $12,000 in total damages. Given the broad language of the law, local juries usually sided with injured miners. Some companies started settling out of court. Some, such as American Zinc, took out liability insurance to cover their exposure.[42]

By suing for compensation, miners found another way to claim the financial premium they associated with physical risk taking. This was especially the case for shovelers, who earned the highest wages and suffered the most injuries. In 1910, for example, shovelers at American Zinc suffered injuries at fourteen times the overall rate in district mines and mills; they were four times more likely than those in the next most dangerous job, machine man, to get injured. Shovelers also led the way in compensation suits. In one year at American Zinc's Davey mine, shovelers claimed 43 percent of all compensation awards. Their suits made the financial value of their bodies explicit. Herschel Stringer sued for $300 after cutting his foot. George Young sued for $300 after an ore can fell on his leg, "skinning his shin just below the knee." The company settled with other miners for smaller sums. American Zinc paid Jesse Slater $8.75 for a "mashed" finger. The compensation law provided some relief but also bolstered the entrepreneurial thinking that framed bodily power as investment capital. Miners' aggressive use of the law also made risk-taking behavior costlier, in terms of both wages and potential liability, for employers already keen to rein in labor costs.[43]

In the midst of these upheavals, miners used wildcat strikes to make wage demands only a few weeks after others went to South Dakota as strikebreakers. "There were a great many strikes throughout the district and I presume that as many as 1,000 men struck for the advance," Landrum informed his boss in March 1910. These strikes lasted only a day or two and lacked any organizational coordination but unsettled company officials nonetheless. American Zinc temporarily relented by restoring wages. Other companies continued to try to cut wages, however, in response to stagnant ore prices. Miners responded with more wildcat strikes. Hundreds of miners walked out of sheet ground mines in May and again in July. They attracted unlikely support. The normally antilabor *Engineering and Mining Journal* declared in an editorial that Joplin miners deserved to "get all they can." The journal hoped that the strike would draw attention to the "highly dangerous and unsanitary" conditions in the district and compel the "enforcement of a drastic, sanitary law." In addition to deaths from rock falls and other accidents, the editor explained, many more Joplin miners died from rampant lung ailments. In the sheet ground, he explained, miners inhaled dust composed of small "sharp, angular fragments" of rock that "makes them peculiarly subject to pulmonary troubles, from which a large number die at comparatively early age." This was notable because no major publication had identified the prevalence of miner's consumption in southwest Missouri before. The editor blamed the miners for not taking more active steps to reduce dust, such as wetting the mine faces and shafts, but admitted that ineffective mine regulation also contributed. "Until the miners of the district awake to the dangers under which they work and demand such a law," the editor lamented, "they will continue to court the chance of early demise" by laboring in conditions "as dangerous as work in an arsenic factory." The miners who went on strike in 1910 demanded higher pay, not health and safety regulations. For many shovelers, in fact, time taken to soak piles of ore with water, as the editor suggested, was time lost putting that ore into cans, and cans were what paid. And yet their parallel strategies of wildcat strikes and damage suits suggested, as the editor hoped, that Joplin miners might be on their way to making more coherent demands. In the meantime, the 1910 wildcat strike wave fizzled. A few groups won small pay increases, but most lost. The last strike at American Zinc ended in late July when Kimball declared he would shut down the mines rather than grant the raise.[44]

Although this pattern of rebellion occurred without union direction, the WFM sought to take advantage of the new dynamic. At first, Moyer was skeptical. He admitted to delegates at the 1909 convention that "it is strange" to "see the unorganized workers striking to enforce better conditions, as they have lately in the state of Missouri." As their obstreperousness continued, however, he became more optimistic. "While our past efforts have apparently failed to arouse these people to the necessity of united action," he told convention delegates in 1910, "it is encouraging to know that in late years they at times rebel, and in their unorganized condition strike in protest against the attempt of the employer to further reduce wages." The union's executive board sent Jinkerson to the district again that spring. Echoing the *Engineering and Mining Journal*'s recent observations, the union hoped to build a campaign based on wages and safety. "Owing to the condition of working these mines, many men are killed or injured," Moyer explained. Among those casualties, union leaders noted that "a large per cent of the miners are afflicted with miner's consumption." They admitted that Jinkerson's task would be difficult. The only Joplin miners "who understand organization," the board declared, "are men who have worked in the various camps of the West during strikes and are full of prejudices which makes it very difficult to organize."[45]

Despite these doubts, Jinkerson established a beachhead. In March, he organized a new local in Joplin, 217, with about twenty members, and made good headway in Webb City and Carterville. "Joplin now seems sure of having the miners organized as never before," local AFL leader Charles Fear announced. Even Kimball, American Zinc's president, admitted that the WFM "may be able to form a local union." Local 217 was still small, however. At the convention that summer, C. L. Bailey, its president, called for more organizers. "We must state that as a local union, we are very weak and have but a very small membership" and "are not growing fast," his resolution admitted. "The miner of this district must be educated and his attitude towards organization changed," Bailey declared. "Our members are new to the work, we lack experience, we need organization, and we need help," he concluded. "We urge you to consider our appeal" for the miners of Joplin and "in other districts, which are so affected by the strike-breaker and scab from this district."[46]

Yet the ongoing battle in the Black Hills, the latest in this chain of destructive conflicts, loomed over the union's organizing attempt that summer. In South Dakota, Joplin strikebreakers and union miners revived and even amplified animosities born in earlier clashes. The strikers sang the "Scab's Lament," a version of the tune first composed by their predecessors in Coeur d'Alene. They created new ones, too, such as "The Song of Missourian," which likened Joplin miners to oxen. Strikebreakers also wrote their own songs, such as "The Man from Missouri," which denigrated the WFM and, more important, asserted the independent privileges of white, native-born American men:

> Here's to the Joplin miner, the man from old Missou,
> If the Homestake company sticks by him, he'll surely pull her through;
> There'll be no strikes nor walkouts, no grievances to tell
> For being white and human, he knows when he's treated well.

Similarly, in "Why We Are Here," strikebreakers sang that "the mines were full of Slavs and such, who were really in Americans' places." Now, "the miners are Americans, English-speaking lads, / The pride of Uncle Sam, and not to be called 'scabs.'" They claimed priority by reason of race and nation. Union leaders struggled to imagine how Joplin miners would ever become union men. The WFM's executive board, at the same time it dispatched Jinkerson to Joplin, called the strikebreakers who went to South Dakota "degenerates" and "the 'scab' workingman" a "miserable, cowardly renegade." Many of them believed that most, if not all, Joplin men suffered moral and mental deficiencies. Among the causes of the "deplorable" conditions in Joplin, the board claimed, was the "lack of intelligence and unity among the workers." By using terms that invoked the racist theories of contemporary eugenicists and xenophobes, WFM leaders revealed, at best, a weak understanding of what motivated Joplin miners. Even Moyer still struggled to form alternative explanations. He admitted that "it has been difficult to understand why these workers surrounded as they are by organized labor, seeing and realizing its benefits, should stand aloof, ever ready to sell themselves to the employer for the purpose of assisting him in wresting from their fellow miners conditions which have cost them years of determined effort to secure."[47]

Jinkerson attacked the racist and nativist logic of Joplin miners head-on, looking for total conversion, not pragmatic compromise. The WFM was, after all, a diverse union, with many foreign-born members from southern and eastern Europe, including in the eastern Missouri lead field where Slavic and Italian miners had rallied to the union in recent years. According to Landrum, WFM organizers were making progress until "one of them the other night made the declaration that all laborers were brothers, and that he had equal rights with all, under the union, no matter from whence he came." In the wake of the South Dakota clash, the approach failed. "This started trouble right away among the Webb City shovelers," Landrum continued, "as they did not propose to recognize the 'hunkies' or the 'niggers' as brothers of theirs."[48]

Ironically, the shovelers' animosity to foreign-born workers forced American Zinc to stop an attempt to recruit Russians to replace them. "It is not true that there has been any talk or thought of employing foreign labor," Landrum lied to local reporters in July. "We have the best set of men in the world and have no thought of making any change." According to Otto Ruhl, a local mining engineer, managers such as Landrum feared that "an attempt to import foreign labor would result in solidifying this union sentiment." As these glimpses suggest, Joplin miners gained leverage by strenuously asserting their privileges as native-born white men. Those assertions, however, left them in limbo: in opposition to their employers but unwilling to support the only organization that represented those who did the same work as them. As a solution, some disgruntled shovelers hoped to establish an independent union under local control for native-born whites only.[49]

The WFM altered that calculus when it voted to reaffiliate with the AFL in 1910. The union needed help after its defeat in South Dakota that spring. By allying again with AFL president Samuel Gompers, WFM leaders continued a moderate course that emphasized organization building and collective bargaining over its previous commitments to revolutionary economic change. In Joplin, WFM organizers placed these new goals at the center of their appeals. Charles Mahoney, who led the divorce from the IWW, delivered the news himself in a series of rallies. The organizers had help from Joplin's Central Labor Union, which claimed more than 900 members in various trades. Gompers himself instructed Charles Fear, who was elected to the state legislature in 1910 as a Republican, to give "every assistance" to the WFM. Rechartered in 1911 under the banner of AFL pragmatism, WFM organizers redoubled their emphasis on higher wages and safer working conditions as the main goals of union membership. In September, Guy Miller, the latest organizer to direct union efforts in the district, distributed a circular with this core message: "Unorganized men can make no effective demand for increased wages or better working conditions." Joplin miners had learned this the hard way, he suggested, as strikebreakers and wildcat strikers whose minor rebellions had done nothing to alleviate their long-term problems. Only union men in stable organizations, he asserted, had the power and the allies to win. He challenged them by appealing to their claims of white male supremacy. "You must sink to the level of brutes," Miller concluded, "or rise to the level of men."[50]

By 1911, Joplin miners presented unresolved problems for local mining companies and labor unions alike. They staged no major wildcat strikes after July 1910 because they had little leverage to make direct wage demands, with ore prices stuck around forty dollars per ton. Many miners continued to assert their autonomy in other ways, however. They still moved frequently between employers in search of higher piece rates or daily wages. This aggressive mobility, especially among the shovelers, continued to bedevil companies with labor shortages. "The system thus seems almost to have outgrown its usefulness," Ruhl observed, as the costs of shoveling culture counteracted the productivity gains it yielded. Meanwhile, Local 217 had reestablished the WFM in the district, now with AFL support. Whether the union's new affiliation would overcome the prejudices of local miners remained to be seen. At the same time, miners continued to sue employers for injury compensation, which caused employer liability insurance rates to skyrocket. Mining companies, led by American Zinc, looked to lower these costs. Landrum proposed that the largest operators agree to fire "careless men" and those who were "unreasonable and insisted on going into Court." Landrum's plan did not take, but it revealed that the Joplin miner's great advantage, a willingness to court danger for financial incentives and social status, was fast becoming a central issue of concern for union organizers and mining companies.[51]

Against all expectations, the WFM, backed by the AFL, made a strong and nearly successful campaign to organize the district in the years leading up to the Great War. With ore prices stagnant, organizers interested more miners in union membership as a way to make their work sustainable over the long term. Their message reflected the union's moderate turn since 1910: to deliver steadier, higher daily wages through collective bargaining and to deliver safer workplaces through demands for greater state regulation of the mines. Union organizers focused in particular on the growing crisis of "miner's consumption," the prevalence of which raised new questions about the sustainability of the shovelers' culture. They talked less about the racial implications of union solidarity. Just as the WFM seemed to convince miners across the district to embrace collective security, war in Europe revived ore markets and exacerbated the politics of nationalism. Miners rallied in large numbers to defend their economic and racial interests in an independent movement that would challenge their newfound loyalty to the WFM. They wanted it all: higher pay, stronger health and safety measures, and the privileges they expected as white, native-born Americans.

By the end of 1911, Guy Miller's clear focus on wages and safety, and AFL backing, established the WFM's strongest-ever presence in the district. He organized four new locals to join Joplin 217: Neck City 219, Carterville 221, Webb City 226, and Prosperity 232. The task was not easy. Miller faced ongoing hostility to the WFM—"It nearly always comes from strikebreakers," he explained—and "classic objections" based on nativist racism, such as "I don't want to be a brother to a Dago." But although the locals were small, he believed that the union's more conservative appeals were "changing the sentiment" toward unionism. Miller understood that meant convincing miners to abandon thinking inherited from the bygone days of the poor man's camp. He believed that the continued concentration of mine ownership by firms such as American Zinc would "teach men the necessity of organization more effectively than any organizer can." The locals aimed to achieve a daily wage of $2.50 for all of the district's 5,466 miners, a modest increase when they received an average of $2.32 per day in 1911. Miller still had work to do to convince the shovelers. "They are the only ones who ever make big money," he explained, "but it is at the cost of life and health." Observers who claimed no friendship with unions thought he might succeed. According to Otto Ruhl in late 1911, the miners "have shown a growing tendency toward some cooperation among" themselves "and the present year has seen a respectable number of miners organized into a union." That year the Missouri Bureau of Labor Statistics counted 70 members in the Joplin local, 24 in Neck City, 103 in Prosperity, and an impressive 600 members in Webb City.[52]

While seeking standard safety regulations, the WFM made a bold new demand for the state mine inspector to confront the rising rates of lung disease among district miners. The problem was worsening rapidly. In 1911, the Missouri Board of Health recorded a tuberculosis mortality rate in Joplin of 220 per 100,000 people and in Webb City of 336 per 100,000 people, well above the state's overall rate of 155 deaths per 100,000. Because doctors did not yet acknowledge miner's consumption as a condition separate from tuberculosis, these figures included both causes of death. In contrast to the confusion among doctors, the WFM insisted that miner's consumption was an occupational disease caused by breathing dust-laden air. "In the sheet ground of the Webb City district the men work in a cloud of dust," Miller reported in 1911, with the result that "consumption claims them in a few years." Later that year local union miners were themselves calling for state officials to do something. "We need good mine inspectors," a member of Neck City 219 said, "as bad air is killing more men than anything else." Jack Fox, the former clerk who became a standout shoveler in 1907, died from lung disease in 1911.[53]

Although these were small steps, the locals survived, even amid fluctuating ore prices. New union members received their strongest test in late 1912 when ore prices suddenly soared above fifty-five dollars per ton, a record high. Rising wages convinced many to leave the union. Over the course of the year, the Webb City local lost two-thirds of its dues-paying members. Just as soon as prices had spiked, however, they fell again to just over forty dollars per ton in early 1913. Companies shut down production and cut wages. Shovelers again staged wildcat strikes in protest. Now, however, union locals were in place with answers for miners once again whipsawed by market swings. In the 1912 election, Socialist Eugene Debs won 10 percent of the district vote, while outpolling either Taft or Roosevelt, who split the Republican vote, in several precincts; Debs won a plurality in Prosperity and Duenweg. Organizer Marion Cope, who had replaced Miller in Joplin, held weekly meetings at each local in early 1913. In Webb City, he told a crowd of 150 miners that unless they "organize themselves they must expect to accept whatever rate the operators feel able or willing to offer them." More than 300 sheet ground miners joined the union that year. W. J. Edens, a local miner and member of the Joplin local, believed that his colleagues were finally abandoning their faith in the poor man's camp. "[The miner] knows that the prosperity of the operator of today is not shared by him," Edens informed readers of *Miners Magazine* in April. "He knows that the price the operator gets for his ore has nothing to do with determining the wage he is paid for his labor." Finally, Edens believed, "the miners of the Joplin district are showing signs of an awakening."[54]

The WFM locals withstood another challenge in the winter of 1913–14 when labor recruiters came to Joplin looking for strikebreakers to go to

Colorado's southern coalfields and Michigan's copper range. The union and its allies worked hard to counter their promises. Reuben T. Wood, president of the Missouri State Federation of Labor, and organizers from the United Mine Workers of America (UMW) rallied support. Cope used two strategies: accusing the agents of misleading potential recruits about pay and conditions and appealing to the growing sense of union solidarity in the district. Union activists took care to tailor their arguments to the nativist limits of that solidarity, however. Rube Ferns, a UMW organizer from Scammon, Kansas, and former world welterweight boxing champion, claimed that many of the strikers were not foreigners. He explained that "40% of them are Americans, Irish, English, Scotch and other English-speaking people" on "strike for a living wage and conditions to work where their lives will not be impaired" and thus worthy of respect. Before leaving in January 1914, Robert Copeland, the agent for Rockefeller's Colorado Fuel and Iron Company, recruited only 100 strikebreakers in the district, a tenth of the number that went to South Dakota. In April, union leaders deterred recruiters from Michigan in similar fashion. Regardless, the UMW and WFM lost both strikes, most notably in Colorado, where, beginning with the infamous Ludlow Massacre in April 1914, strikers and the state National Guard went to war. For the first time, Joplin miners had generally stood aside. According to Wood, "the Joplin district purged itself of the name of disgrace." Both he and Moyer believed they would now complete their organization of the district's miners.[55]

Indeed, the union movement showed life in 1914. In January, Cope and Ferns organized a new local, number 138, in the new camp of Commerce, near Miami, Oklahoma, thirty miles southwest of Joplin. The mines in Oklahoma had been marginal until 1913, when a series of new zinc ore discoveries spurred development and production, which exceeded 32,000 tons in 1914, a threefold increase in two years. Meanwhile, in Joplin and Webb City, miners continued to join the WFM. Local 226 remained the largest, now with over 800 members.[56]

Union miners could also take heart because the problem of miner's consumption finally attracted serious attention. In 1912, the state mine inspectors began urging companies to reduce dust in the mines as a means of preventing the disease. They were joined by a group of middle-class women, led by Bess Hackett, a society editor for the *Joplin News Herald*, who formed the Jasper County Anti-Tuberculosis Society. In 1913, the Missouri Board of Health reported that deaths from tuberculosis in Joplin and Webb City had jumped by 40 percent over the previous year's already very high number. Although the state figure included deaths from miner's consumption, Hackett feared that it still undercounted the true total. To get a fuller picture, the society surveyed the people most acquainted with the disease: the wives, mothers, and daughters of the district's miners. By communicating directly with these women, the society identified many more miners who died of lung disease but did not get counted: those who sickened and died without ever visiting a doctor and those who went to die at homeplaces in surrounding counties. Their findings were shocking. The society reported that more than 1,100 miners from Jasper County died of tuberculosis "or miners' con" during the eighteen months of their survey. In one mining town, the society found 132 widows of miners who had died from lung disease. They believed that as many as 60 percent of miners currently working in the sheet ground had some form of lung trouble.[57]

These findings emerged at a time when the district had never been busier or richer. Between 1912 and 1914, ore and mineral sales exceeded $46 million. Joplin had become a prosperous city of over 32,000 residents with a sizeable middle class of professionals, merchants, and skilled workers. They enjoyed clubs and fraternal organizations, public parks, paved sidewalks and streets, and an interurban train system that provided fast, relatively cheap travel to neighboring towns that wanted to emulate Joplin, such as Webb City and Carterville, home to over 11,000 and 4,500 residents, respectively. Prosperity, however, created and widened class divisions that strained the district's older sense of social equality. The Anti-Tuberculosis Society was sympathetic toward the sick but also afraid of them.[58]

Many middle-class residents increasingly worried that the costs of the mining industry would ruin their good fortune. In late 1912, Dr. Alice Hamilton surveyed the people who worked at and lived around the Picher Lead and Zinc Company's Joplin smelter for evidence of lead poisoning in a study commissioned by the U.S. Department of Labor. Her report found a few people "with lead convulsions or insanity" in the immediate vicinity of the smelter. The crisis of miner's consumption, meanwhile, threatened to taint the district with "one of the worst reputations," that of being plagued by tuberculosis, a disease most associated with the poverty of African American and immigrant communities, even though most of the deaths were caused by mine dust, not bacteria. In 1914, the Anti-Tuberculosis Society hired a nurse to visit sick miners and set up an open-air tent to care for the worst cases.[59]

Middle-class antituberculosis reformers blamed the miners themselves for the epidemic. Although still unsure how miner's consumption related to tuberculosis, most local observers now agreed that "the dust evil" in the dry sheet ground mines was the main physical cause. They also agreed that the best way to stop the dust was to add waterlines to all machine drills and to wet down blasted rock before it was shoveled. Civic leaders were doubtful that the miners would cooperate, however. "The miners themselves seemed to be the least interested" in protective measures, mine owners declared, because they "give little consideration to their personal health." The *Webb City Daily Register*, meanwhile, believed that "the miner's own carelessness" regarding personal health exacerbated lung injuries caused by dust. "In many instances the miner's system is undermined by bad habits, and either through ignorance or carelessness [miners] fail to protect their health either in their habits or in their manner of living," the paper declared, with heavy class-based condescension. Men considered American heroes ten years before were now a threat. Despite the growing consensus about dust, middle-class commentators concluded that the miner needed to be educated and, if that failed, actively prevented from "infecting his fellow toilers." The Anti-Tuberculosis Society agreed to pay for "government experts" to come to the district "to remain among the miners and educate them." In October 1914, Hackett sent a request for assistance to Rupert Blue, the U.S. surgeon general. Sensing "a good opportunity for a scientific study of 'miner's consumption,'" Blue dispatched A. J. Lanza, a surgeon with the U.S. Public Health Service, and Edwin Higgins, an engineer with the U.S. Bureau of Mines, to Jasper and Cherokee Counties to conduct a month-long field study.[60]

In preliminary findings, Lanza and Higgins concluded that "miner's consumption" was actually silicosis, an occupational disease caused by inhaling silica-rich dust, echoing what the union and miners themselves had been saying for years. The pair examined ninety-three miners from Webb City and Carterville who volunteered for the study, sixty-four of whom had diseased lungs. Although they had tuberculosis-like symptoms, very few tested positive for tubercle bacillus. Their disease was silicosis, sometimes with a secondary tuberculosis infection. With lungs lacerated by minute, razor-sharp silica, these miners found it harder and harder to breathe and grew weaker until they could no longer work. Most were reduced to permanent bed rest, where they slowly suffocated. Those in most danger were the machine men and the shovelers. Lanza and Higgins concluded that piece-rate incentives, "when unrestrained except by the individual strength and willingness of the miner," encouraged the most detrimental behavior. They found shovelers "already on the down grade" after five years on the job. "Hard, constant work had broken these men down," they reported. "The whole picture furnishes an example of burning the physical candle at both ends." Although Lanza and Higgins believed that miners should be encouraged to improve their living conditions above ground, their recommendations focused on health and safety measures in the mines. They called for installing waterlines on all drills and mine faces, prohibiting the employment of anyone under twenty years old in shoveling, enforcing "maximum daily tonnage for shovelers, so that they can not injure their health through overwork," and implementing an extensive educational campaign about the causes and remedies of the disease. Rather than seek out the union, however, Lanza and Higgins worked most closely with district mine owners, whom they convinced to form an organization, the Southwest Missouri Mine Safety and Sanitation Association, to enact their recommendations. American Zinc's Howard Young was its first president.[61]

While the new association considered workplace reform, company officials emphasized the need to impose greater control over the miners if the reforms were to stick. In late 1914, the association announced an ambitious aim to establish "rules which may be introduced over the entire field, looking to the reduction of rock dust in the mines, the amelioration of conditions now pronouncedly unsanitary, and other rules for the better safe-guarding against ordinary accidents." Companies would enforce these rules on a compulsory basis, "as will tend to educate the miners." The association wanted Lanza and Higgins around to help install the new safety regime. Young, for example, praised Lanza and Higgins for getting "the miners in a frame of mind to cooperate with the operator and helping better working conditions under ground." The pair had left the district after completing their study, but the mine owners soon invited them back. Lanza and Higgins returned in February 1915 with plans to stay for six months to write a full report. They spoke to large crowds, some over 1,000 strong, of miners and their wives about progress on reforms. The presence of so many women at meetings about the inherent vulnerability of hypermasculine men reflected the immense psychological and economic burdens foisted upon women in the families of sick miners and suggested that they were a major force for change. Lanza and Higgins reported receptive audiences. In March, the Southwest Missouri Mine Safety and Sanitation Association, with the help of the state mine inspectors, secured three new mine safety laws that fell well short of what Lanza and Higgins had recommended: two focusing on sanitation and one mandating the installation of waterlines.[62]

District companies were particularly keen to exert greater authority over their workers as war in Europe sparked a new metal market boom. In addition to bringing Lanza and Higgins to "educate" the miners, leading companies also punished union members. Cope informed the WFM executive board in late 1914 that "employing companies were discriminating against all who joined the union" with blacklists. At the same time, according to

American Zinc's internal records, companies tempted miners with market-based raises as ore prices skyrocketed, from an average of forty-three dollars per ton in November 1914 to sixty-seven dollars per ton in February 1915. American Zinc offered loyal workers a sliding scale for raises that followed the price of ore. In February, for example, Young announced that American Zinc would pay a bonus of one cent per can for shovelers and twenty-five cents per shift for all other miners when ore went above sixty-five dollars per ton, so long as shovelers worked twenty days and other miners worked eighteen days a month. To prevent the recurrence of labor shortages, meanwhile, Young decided to hire "foreigners from the Kansas coal fields" as shovelers. The move enraged the company's native-born miners. "The American shovelers started to running these fellows out and it has cut our night shift force down very much," Young informed Kimball in March. But the company continued to hire more. Taken together, these moves reflected a concerted effort by the district's largest company to take full control of its miners once and for all.[63]

At this crucial moment, the WFM faltered. Crushing strike defeats in South Dakota in 1910, Utah in 1912, and Michigan in 1914 depleted the union treasury. In the summer of 1914, the WFM's largest, richest, and oldest local, the Butte Miners' Union, collapsed amid violent infighting between conservative leaders and radical insurgents backed by the IWW. The union's locals in eastern Missouri also shattered along political and national lines. In the meantime, Moyer and the WFM leadership tried to merge with the UMW, itself reeling after defeat in the 1914 Colorado coal strike. The coal miners rejected the proposal because of the dire state of the WFM. Without dues from Butte, the union nearly dissolved. In August, *Miners Magazine* cut its format from eighteen pages to four as "a matter of economy." That same month, rather than sending reinforcements to help Cope, the executive board temporarily laid off all organizers in order to remain solvent. Cope returned to Colorado. The locals quickly weakened; Commerce 138 folded. After four years of union presence, Joplin miners now faced American Zinc with little support from the WFM.[64]

Without WFM direction, miners across the district relaunched an independent drive to get higher wages and to protect their nativist privileges. In May 1915, miners around Webb City started a series of small wildcat strikes for more pay. They also moved against foreign-born miners. At American Zinc, Young informed Kimball, "American day-shift shovelers ordered all foreigners to leave the district." The implosion of the WFM exacerbated the situation. Several dissident union organizers from other districts, including former WFM and UMW members, led by a man named George Wallace, came to take advantage of the organizational disarray. They convinced many miners to start their own local union. They called it the American Metal Miners Union (AMMU). S. E. Graves, a "renegade" from a WFM local in eastern Missouri, was elected president. Charles Fear, the local publisher, politician, and AFL organizer, supported the move. By the end of May, the AMMU claimed 600 members, most of them in the sheet ground mines.[65]

The AMMU promised to give miners what the WFM seemingly could not: the benefits of stable and safe terms of work and the power of racial and nativist exclusion, free of outside interference. The union stated aims that corresponded to the "pure and simple" collective bargaining doctrines of the AFL: employer recognition of the union, a district-wide wage scale agreed through contracts, the arbitration of disputes, and "safer and more sanitary conditions." Most important, the AMMU proposed a daily wage for all classes of workers, including shovelers, with a sliding mechanism that would go up or down twenty-five cents per day for every ten-dollar move in the market price of ore. For shovelers, that scale would start at $3.00 per day when ore was $40 per ton and rise to $4.75 per day when it reached $110 per ton. While the sliding scale demonstrated how miners remained committed to market thinking, the union's abandonment of the piece rate revealed a willingness to moderate the most flagrant entrepreneurial commitments, in line with recommendations from Lanza and Higgins. The AMMU explicitly limited its membership and benefits to white, native-born Americans. "This is a white man's camp; down with the foreigner," the AMMU blared. Union leaders publicly refused to affiliate with the WFM after Cope returned to try to regain control. They attacked the WFM as a friend of foreign-born miners, "an aggregation of 'Bohunks.'" At one rally, Cope tried to speak but was pulled down from the platform. "Take him down," members of the crowd yelled. "He wants the foreigners in." "We are forming our own independent union," another union leader told the press. "It is the purpose of the men who are responsible for the organization to keep it Americanized just as it has been in the past. Foreign labor will not be admitted and we do not want it understood that we are affiliating with any other organization." The AMMU, Graves explained, "is to be a white man's organization." With ore prices nearing $100 per ton and nationalist fervor rising after the *Lusitania* sinking, the AMMU was popular. Within a month, it claimed over 2,000 members.[66]

Confident and aggressive, the AMMU launched an all-out strike on June 29 to reshape labor relations in the district. The strike began in the sheet ground, where over 1,000 miners quit work, at first to reverse a recent wage cut. The rebellion grew as they demanded "a fair share" of wartime profits. Miners deserved these rewards, AMMU leaders argued, because they faced "the hazard of underground work" to produce the ore. According to the AMMU, anything less than a wage that matched surging market prices could not be "fair, honorable or patriotic"—a betrayal of their white, nativist prerogatives. The striking miners made brazen shows of force in a series of parades and rallies in Webb City, Carterville, and Joplin. In Joplin, 2,500 union miners and supporters paraded down Main Street. They carried American flags, large and small. The union continued to grow. By July 1, nearly 3,000 miners were on strike. By July 3, the union had shut down every mine in Jasper County and planned to extend into Oklahoma. Some observers estimated that as many as 8,000 miners were on strike by July 4.[67]

Although at first dismissive of the strike as a "good natured holiday," district mining companies soon took a publicly respectful but firm stance. They hoped to contain, not antagonize, the miners and did so by appealing to the entrepreneurial traditions of the poor man's camp and their sense of racial prerogative. A contingent of operators, led by Young, met with union representatives to refuse their proposed scale but reported that "the conference broke up with good feeling." He believed that most men wanted to go back to work and would do so with some pressure and inducements. The same day, the Joplin city council passed an ordinance requiring a permit to parade. Meanwhile, American Zinc directed the sheriff and police to protect mine property, particularly the water pumps. On July 6, ninety mine operators announced a compromise to end the strike. They offered to pay men who went back to work the prestrike wage of four dollars per day or ten cents per can and would institute a temporary sliding scale for all workers "based on the price of ore, so that the miner, as in the past, will share in the prosperity of the district." They would not attempt to hire foreign-born workers. Neither, however, would the companies "recognize the so-called union nor stand for union domination or control." Joplin mayor Hugh McIndoe urged the men to remain true to the district's poor man's traditions, which he believed were still attainable. "There is but a small space that separates the miner from the operator," he said. "I hope that within the next year you men will make enough that you will be operators and dealing with employees who are asking for more wages." If compromise did not work, American Zinc was prepared to compel them. Young hired fifty professional strikebreakers from the Bergoff and Waddell firm, which had broken the 1914 WFM strike in Michigan, in case these appeals failed.[68]

The AMMU capitulated immediately. Two days after the company announced its compromise, miners in Neck City voted to go back to work. At Chitwood, miners were "anxious to return to work at the former wage scale" with the new sliding mechanism. Miners in Webb City and Carterville abandoned the strike on July 10. The Joplin local, the last to hold out, decided to go back on July 11. The "roar of assent was so decisive that a vote was not taken," the press reported. The local president, Adam Cullifer, a fifty-five-year-old mine engineer who had worked in the district since the 1870s, encouraged his colleagues to take advantage of the high wages on offer, more than four dollars per day in most mines, and the promise that pay would rise further along with the ore markets. "I desire to get all I can for my work and to work under the best conditions possible," Cullifer explained. "If we can't be the operators," he declared, "let's work for the operators, if we can, and be satisfied when we obtain good wages." He and others proclaimed that they would never go on strike again. The strike might have been a mistake, but viewed in terms of the entrepreneurial and white nationalist commitments of many district miners, it had not been a complete failure. They had forced the companies to commit, albeit temporarily, to a wage scale based on market prices and to the exclusion of foreign-born miners from the district. Without a written agreement, they trusted the verbal pact between white men. Young reported that American Zinc's miners "are apparently in good spirits." The Neck City AMMU local celebrated with a barbeque and baseball game.[69]

Surprisingly, the AMMU and the district's union movement survived the strike. Many members, including Cullifer, wanted to stay organized to make the sliding scale permanent and defend their nativist privileges. They believed that the best way to do that was by affiliating directly with the AFL, without the WFM. Union leaders approached Gompers about doing so, but he reaffirmed the WFM's jurisdiction over all metal miners. He urged them to "become part of the Western Federation of Miners and make common cause with the organized workers of America." Moyer visited the district himself in the weeks after the strike. WFM leaders had been in denial over the AMMU; Cope explained its bellicose popularity as the result of foreign interference, the dirty work of clandestine German agents who wanted to disrupt Allied metal production. Moyer, however, tried to harness the nationalism of the AMMU by portraying the WFM as little more than a gateway to the AFL. He promised to make the AMMU a district council within the WFM, a designation that provided maximum local autonomy and tacitly approved its racist and nativist aims. Gompers approved the strategy. Other union leaders followed, including AFL mining representative James Lord, former Butte Miners' Union leader James Lowney, Reuben Wood, and mining union luminary Mother Jones. They all reiterated the pledge that the AMMU would retain near-total control over its affairs. It worked. In early

August, the AMMU locals in Joplin, Webb City, Chitwood, Oronogo, Prosperity, Duenweg, and Carterville affiliated with the WFM. Miners formed new locals in Zincite, Galena, Sarcoxie, and Commerce in September and October. "I do not know just how this is going to work," Young informed Kimball, "but it looks as if the Western Federation are getting a stronger hold on the Joplin miners than they have ever had before." By the end of 1915, the WFM had more than 5,000 members in the district, although most identified their membership primarily with the AFL.[70]

In a significant turnaround, Joplin miners now represented one of the strongest contingents within the weakened WFM, a dire sign for the union. Of seventy-seven locals in good standing, eleven were in the Joplin district. In late 1915, Charles Moyer informed Guy Miller, who was organizing in Arizona, that the WFM considered Missouri "the most important field in its jurisdiction." After almost twenty years of ruinous conflict between Joplin miners and the WFM, he believed the union was on the verge of having them completely organized, albeit on a promise that the WFM would leave them alone. Despite the racial and ethnic enmity at the heart of that conflict, Moyer had even come to value the district's staunch white nativism. "They are all practically one nationality," he told Miller, which he now acknowledged as a source of strength and unity that benefited organization. If anyone doubted that Joplin would soon be "the best unionized district in the United States," WFM organizer Joseph Cannon declared in September 1915 with a highly ironic flourish, considering past events, "just watch them. They will 'show you.'"[71]

After a decade of chaotic wildcat opposition to company shutdowns, pay cuts, and threats of mechanization and cheap labor, Joplin miners had by 1915 forced area mining companies to honor two key demands: a market-based wage scale and the exclusive hiring of native-born white men. To do so, they had wielded collective force through an independent union that reaffirmed their faith in the power of physically strong white American men to command high wages and social respect. Despite defeat in the 1915 strike, a majority of miners for the first time seemed ready to join the WFM, if only as a means to AFL affiliation, as a way to defend that social and economic bargain. Their swing toward the WFM and the AFL was not a recantation but rather a reconfirmation of their commitments to local control, white nationalism, and the entrepreneurial incentives of the sliding scale on the eve of American entry into the Great War.

CHAPTER 6
RED-BLOODED, RUGGED
INDIVIDUALS

While the first year of the Great War revived the fortunes of Joplin miners, American belligerence in 1916 and after gave them new confidence in old ideas about the chances of white working-class men to claim the nation's prosperity. In 1915 they had organized, first in the American Metal Miners Union and then in the Western Federation of Miners (WFM) and American Federation of Labor (AFL), as a last resort to demand better pay and preferential hiring from recalcitrant employers as wartime demand pushed zinc and lead prices to record highs. Just as a movement started to build, however, the Wilson administration's war stance created dynamic new conditions that challenged organized labor and once again encouraged white working-class entrepreneurialism. In 1916, despite an official policy of neutrality, the U.S. government increased military spending, which in turn sent zinc and lead prices even higher. Joplin miners realized the gains through informal sliding-scale contract provisions as a surging industrial economy tightened regional labor markets. With the Wilson administration moving toward war, its simultaneous calls for patriotic loyalty spurred a newly aggressive nationalist sentiment among native-born whites who ramped up their intimidation and persecution of foreigners and suspected radicals, particularly the Industrial Workers of the World (IWW) and associated groups, such as the WFM. To Joplin miners, the political economy of American belligerence from 1916 to 1918 delivered in full what unions had only promised and seemed to show that hard-working white men like them could still prosper and do so on their own.

Nowhere did Joplin miners see the lucrative benefits of war more than in Ottawa County, Oklahoma. Beginning in 1915, as ore prices soared, mining companies accelerated the development of metal discoveries thirty miles west of Joplin—the district's first major expansion since the 1880s. By the following summer, a series of rough boom camps extended north along Tar Creek from Commerce to just across the Kansas state line: places called Cardin, Treece, Douthat, Hockerville, and, soon the largest of all, Picher. Observers predicted that the new field would "prove the richest in the world." At a time when coal and metal mines elsewhere were fully mechanizing, these firms continued to rely on shovelers, the largest and most important group of miners. Pay was high. "The most lucrative of our laborious professions is that of plying the shovel," a company official declared. "Many of our shovelers make as much as $6 and $7 a day." Just as important, companies in Ottawa County hired only native-born white men; the new field was "purely American in its men and methods."[1]

The timing of the Oklahoma discoveries was indeed fortuitous. The wartime boom exhausted the old sheet ground mines around Webb City and Carterville. As companies shifted production to Oklahoma, the heart of what now became known as the Tri-State district, miners followed the work. By 1920, nearly 16,000 people, most of them miners and their families, lived in Picher, Cardin, Treece, and Commerce—10,000 of them in Picher alone. None of those places had existed ten years earlier. Most came from the old Missouri boomtowns; Joplin, Webb City, Carterville, and Neck City lost almost 9,000 residents combined in the 1910s. Their labor drove zinc ore production in the new field: from 55,285 tons in 1916 and 557,066 tons in 1920 to an all-time high of 749,254 tons in 1925. Meanwhile, production in Missouri plummeted from a high of 322,123 tons in 1916 to 49,786 tons in 1920 and 28,865 tons in 1925. The collapse was almost total in the sheet ground mines: from 113,835 tons in 1916 to 157 tons in 1925.[2]

For Tri-State miners, the transformations of the war years reaffirmed their faith that capitalism and nationalism would reward working-class white men. They believed, not incorrectly, that their brute strength and reckless pursuit of individual incentives powered the district's rebirth in Oklahoma, which profited companies and supplied the American and Allied war efforts. After flirting with union solidarity and security in 1915, Tri-State miners, led by the shovelers, once again indulged ideas about work that linked remuneration with masculine power and risk. They claimed special status as loyal, native-born white men who courted danger on the job, a status further valorized by frenzied wartime campaigns for loyal, 100 percent Americanism, the suppression of anticapitalist radicals, and immigration restriction. Despite the conservative appeals of the American Metal Miners Union and the AFL, the latter an enthusiastic war supporter, Tri-State miners turned against unionism once again. By the early 1920s, they expected that hard work, defense of race and nation, and fidelity to the sliding scale would assure them, individually, a fair share of the district's rising profits.

Like many other American workers in the war's aftermath, Tri-State miners and their families also sought improved living conditions. Life in the

Oklahoma boom camps was rough, as social disruption, unsanitary housing, hazardous working conditions, and inadequate medical care kept disease and death close. They expected better. Tri-State mining families wanted, in short, the ideal American standard of living presented in postwar consumer culture: homes with modern conveniences, good health, access to public schools, and the money and time to enjoy leisure activities. Tri-State families claimed that ideal in racial and national terms, as white Americans. Yet these deepened commitments—to white nationalism, dangerous masculinity, and working-class entrepreneurialism—would restrict and bedevil their attempts to find solutions and give new license for mining companies to subdue and control them with unprecedented powers born of wartime government support.

The Oklahoma boom was different from any the district had experienced before. In the past, mining firms had followed the discoveries of individual or small-scale prospectors who opened new fields. Now, large companies controlled the exploratory process by methodically drilling test holes on large tracts of leased land. Much of that land belonged to members of the tiny Quapaw Nation, whose property transactions were subject to oversight by the U.S. secretary of the Interior, a power imbalance that favored big mining companies with political pull. These companies exploited new discoveries with sprawling operations that required thousands of miners working for daily wages or piece rates. Although the new field resembled mining districts elsewhere, its settlements looked a lot like the area's old boom camps. Companies invested nothing in housing, schools, or basic services. Wages, however, were high. As the old field around Joplin collapsed, miners abandoned recent commitments to safety, stability, and well-being to take lucrative work in a place that promised none of those things. As they did so, the war logic that made their labor so valuable also encouraged ideas of white nationalism and rough masculinity that reassured many men and women that a prosperous future awaited them in the new Oklahoma field.

Although small companies had mined in Indian Territory since the 1890s, the Miami Royalty Company developed the first significant deposits of zinc ore north of Miami, Oklahoma, in 1908. Its operations were modern but small compared to those of companies in Webb City and Joplin. By 1912, however, Miami Royalty's expanding operations attracted the interests of Picher Lead and Zinc Company engineers, who convinced their boss, Oliver S. Picher, to lease mineral rights nearby. Picher, the son and nephew of the firm's founders, had set an ambitious course since assuming a senior leadership position in 1906. With an $800,000 investment from the Cincinnati-based Eagle White Lead Company, he purchased a lead smelter in Galena and diversified the firm's range of products. Picher's decision to produce more zinc had the company looking for untapped ore deposits. In 1912, Picher Lead leased extensive mineral rights on 2,700 acres north of Miami; Miami Royalty also acquired new holdings and renamed itself the Commerce Mining and Royalty Company. Members of the Quapaw Nation owned much of this land in sizeable allotments that had only recently been made available for mining. In 1897, the U.S. Congress gave the Quapaw, of whom only 234 members remained, the right to lease their allotted lands for mining purposes for ten years unless declared incompetent because of age or disability, in which case the secretary of the Interior would make agreements on their behalf. In 1907, the commissioner of Indian affairs issued new rules for declaring incompetence that gave the federal government increased power over leasing arrangements. When securing their first leases in 1912, Picher Lead and Commerce Mining made some agreements directly with Quapaw allottees and the rest with the Department of the Interior. Soon after, both firms deployed drill crews to explore the new holdings. In August 1914, drillers for Picher Lead hit a thick vein of ore 300 feet deep on Harry Crawfish's allotment just east of Tar Creek about a mile south of the Kansas state line. Picher Lead sent nine additional drill crews to map the surrounding territory. Their drilling hit dozens of other ore deposits that assayed from 6 to 30 percent metal.[3]

Picher Lead exemplified the new scale of production in Ottawa County. The company kept the drill results secret until early 1915, just as the outbreak of war in Europe sent zinc prices soaring. Picher Lead decided to operate the mines directly, as the American Zinc, Lead, and Smelting Company had done in the sheet ground, and not sublease the work. "The findings at certain places have been so rich," the *Joplin Globe* reported, "that the company decided it could not afford to turn the land over to someone else, but should mine for itself." To oversee operations, Picher Lead hired A. E. Bendelari, an experienced engineer who understood the district's history. Bendelari directed construction of the first shafts and mills in March 1915. Flush with confidence, the company increased its capital stock to $5 million, leased 3,000 more acres of Quapaw land, bought a zinc smelter in Illinois, and built a new zinc smelter in Henryetta, Oklahoma. By the spring of 1916, Picher Lead was running six mines and four mills, including the Netta, the district's largest mill, which could process seventy-five tons of crude ore an hour. That summer the company merged with Eagle White to form the Eagle-Picher Lead Company, one of the largest zinc and lead firms in the world.[4]

Most miners in the Joplin field were initially slow to comprehend what was happening over the state line. With the American Metal Miners Union and the WFM gaining a foothold in 1915, many looked to a better future at the district's historic center. In January 1916, the WFM claimed over 1,000 members in eleven locals. Working together in the new district council, these locals pressed demands for collective bargaining and improved health and safety regulations. After the most profitable year in district history, when companies sold over $19 million in ore and mineral, union miners called for a district-wide contract with a minimum wage and bonuses determined by a permanent "sliding scale that would give the miner a measure of prosperity." For shovelers, the union now wanted a modified piece-rate system: a base wage of $2.75 per day plus seven cents per can, with a half-cent increase per can for every ten-dollar rise in the price of ore. The union also wanted more done to counter silicosis, another plausible demand given the high level of civic attention to the crisis. According to a U.S. Public Health Service (PHS) doctor, there was "now a well grounded conviction among miners and opera-tors that rock dust is dangerous to life and health" and there was "the great-est enthusiasm" for prevention. "Join the union," the district council urged in June 1916, "thereby helping to increase wages, improve the conditions, protect health and lengthen life." This campaign reflected five years of tough organizing. At best, the WFM sought a moderation of the district's mining culture by calling for a sliding scale with the security of a minimum base wage. That summer union miners could claim partial success on all of these issues, all while companies excluded foreigners and African Americans from work. Union miners thought more was possible. They asked nonunion men, "Will you join with us in making the Joplin district a fit place to live in and raise our families?"[5]

By early 1916, however, these same miners faced an increasingly uncer-tain economic future. After thirty years of production, capped by a furious wartime surge, the sheet ground mines were thinning. American Zinc and other companies were mining dirt that yielded only 2 percent of its weight in concentrated ore for sale, half the yield of a decade earlier. With prices over $100 per ton, these firms could make money from such low-grade ore by processing ever-larger amounts. In 1916, for example, companies in the Mis-souri section of the field produced over 302,000 tons of concentrated zinc ore worth a record $24 million but only by extracting more than 13 million tons of crude ore, 40 percent more than the previous high in 1912. Once ore prices began falling in March 1916, most companies could not afford to mine and mill sheet ground ore. Smaller firms shut down, some for good. Ameri-can Zinc cut wages several times, prompting union discussions of a strike, but continued operating. No one had given up yet. American Zinc leaders even bet on a revival by buying the entire holdings of the Granby Mining and Smelting Company in June 1916.[6]

Declining profitability in the sheet ground tested the WFM district council just as the national union itself succumbed to internecine fighting amid growing hostility to radicals and foreigners. The WFM was riven by sectarianism in 1916 as a radical IWW faction tried to wrest control from WFM president Charles Moyer and his more conservative allies. This fight reached Joplin in the spring as local union leaders acquiesced to recent wage cuts rather than strike again. In June, Frank Little and several other IWW, also known as Wobbly, organizers arrived in the district with hopes of capturing the WFM's unlikely stronghold. At a meeting of Joplin 217, they took the floor to rail against the union's impotent leadership, both nationally and locally. They demanded a strike. Most miners responded with hostility to the Wobbly appeal, in part due to the lessons of the 1915 strike but also out of revived animosity. The week before Little's group arrived, the Missouri National Guard had launched a recruiting drive in Webb City and Carterville ahead of an expected deployment to the border with Mexico as Pancho Villa's raids escalated hostilities. The military's appeals to loyal American manhood fueled a violent reaction to the radical organizers. At a mine near Carterville, a confrontation between the Wobblies and a group of shovelers and machine men ended in "a free-for-all fight." According to American Zinc managers, "the men who wanted to work cleaned up the 'I. W. W.'s' in grand shape." The clash also pleased WFM organizer Marion Cope, who reported that the IWW men "were given a dose of their own medicine, 'direct action,' with the result that several of them returned with broken heads." Little and his colleagues soon left the district. The miners accepted the wage cut. The whole scrape lasted less than two weeks, but it conjured up old demons that boded ill for the WFM, despite Cope's gloating: yet again, Joplin miners fought with unionists to stop a strike.[7]

The WFM rapidly lost support in the Joplin district and elsewhere as the nation went to war. Less than a year old, the district council was hit from left and right, dogged by the union's radical past yet unable to resist wage cuts in the sheet ground. Government jingoism exacerbated its agony. Jasper County's first contingent of National Guard soldiers mustered for training in late June 1916, two weeks after the Wobblies left the area. Others continued to sign up for service amid flag-waving rallies. Although Moyer prevailed over IWW challengers at the WFM convention in July 1916, and the union signaled a new start by adopting a new name, the International Union of Mine, Mill and Smelter Workers (hereafter referred to as Mine

Mill), it was not enough. By January 1917, every local in the district council, save for Joplin 217, was defunct. Mine Mill could not shake its association with the IWW as the blunt logic of belligerence turned all dissenters into enemies. When President Wilson called for a declaration of war in April 1917, he pledged to counter disloyalty with "a firm hand of stern repression." Soon after, he signed the Espionage Act, which gave the federal government broad power to punish anyone considered a threat to national security. That summer, IWW organizers led former Mine Mill locals in Arizona and Montana into costly strikes that ended with the deployment of U.S. soldiers and further union turmoil. Vigilantes did much of the work of repression. In Bisbee, Arizona, a posse expelled, or "deported," as they called it, more than 1,000 miners into the desert. In Butte, Montana, a lynch mob tortured Frank Little and hanged him from a railroad trestle. In the eastern Missouri lead field, vigilantes beat, robbed, and forcibly expelled more than 1,000 foreign-born men, women, and children, many of whom had recently belonged to the WFM, in what they called a "hunky riot." The following month, the nascent Bureau of Investigation used its powers under the Espionage Act to raid every IWW office nationwide. With little political room to operate and weaker than ever, Mine Mill languished. The Joplin local collapsed in early 1918, an organizer later explained, "owing to the traditions, superstitions and prejudices existing among the miners of this field."[8]

As the union floundered, the mines in the Joplin district finally stopped producing. Although ore prices rebounded to above eighty dollars per ton in early 1917, American Zinc could not make the sheet ground deposits pay. Its exploration of the Granby Mining holdings yielded nothing. The value of production in Jasper and Newton Counties plummeted from $21.6 million in 1917 to $6.9 million in 1918. It was a "disastrous" year for "nearly all the operators of zinc mines in southwestern Missouri," including American Zinc, which began to liquidate its holdings.[9]

Miners were streaming into Ottawa County from Kansas and Missouri by then. Some men went soon after hearing about the new discoveries, before the sheet ground companies began to falter. "A substantial boom is on," the *Missouri Trades Unionist* reported in the summer of 1915, "and men are in great demand." The paper reported that 2,300 miners were already there, mostly "former residents of Joplin, Webb City and Carterville." Over the next five years, thousands more followed. By 1917, wageworkers could make "30% above normal" in Oklahoma. Shovelers could earn much more, up to eight dollars per day on prevailing piece rates. "The activity in the Oklahoma field is so great that miners are being drawn from the older camps," a correspondent reported that March, "and there is a dearth of men, particularly shovelers, in Joplin and Webb City." Harry Hood's father, for example, abruptly quit his job in Carterville in 1916, assured that he would find something better in Oklahoma. "There were jobs," Hood recalled his father saying, "plenty of them." Two years later, eighteen-year-old Joe Nolan left Webb City, where his shoveler father was dying of silicosis, to take a plum hoisterman job. Older men were also drawn to Oklahoma. By 1920, Scott McCollum, the former Leadville strikebreaker, now forty-eight years old, worked as a miner again in Picher. No longer an electrician, McCollum left Joplin after his oldest son, Otto, a twenty-five-year-old laborer in a lead smelter, died of pneumonia in 1918.[10]

Missouri miners did not take the WFM with them. The small Commerce local had collapsed without WFM support in 1916, just as the rush got under way. No one tried to transplant the other locals to the new field, where, without a union, miners could make over one dollar per day more than the union was trying to secure in Missouri. The WFM, later Mine Mill, made no noticeable effort to bridge the transition either. Instead, the union focused its meager resources on organizing zinc smelters in the surrounding coalfields. Starting in 1916, Emma Langdon joined Cope to lead drives that built new locals in places like Fort Smith, Bartlesville, and Henryetta, but the union registered no presence among Oklahoma's zinc miners. If any diehard union sympathizers remained, they stayed quiet.[11]

Many mining companies also relocated operations to Ottawa County. Eagle-Picher and Commerce Mining created favorable conditions for others to enter the new field. As early as 1917, dozens of firms leased land near proven discoveries as federal administration of Quapaw lands and improved infrastructure eased new development. In 1917, the St. Louis–San Francisco Railway, which ran the old Atlantic & Pacific mainline and was known as the Frisco, completed a spur line that connected Picher, Cardin, and Douthat to its existing network, which offered direct routes to Joplin, Galena, and other smelter towns in Kansas. In 1918, the Southwest Missouri Railroad extended the district's interurban trolley line to Picher. By 1920, many small companies with fewer than fifty workers, such as Mike Evans's Keltner Mining Company, operated in the Picher field. Large firms dominated employment and production, however. Federal Mining and Smelting Company, a subsidiary of the American Smelting and Refining Company, employed 150 men. Golden Rod Mining and Smelting Corporation employed 500 men in thirteen mines. Admiralty Zinc Company ran four mines that employed 250 workers. Commerce Mining employed 400 men in eight mines. Eagle-Picher was the biggest of all, with over 1,000 employees.[12]

Miners who came to Picher, Cardin, and Treece found work and high wages but also crude, barely governed camps. Unlike in many other single-industry areas, these were not company towns with employer-provided services. Mining companies built offices, a church, and initially a few bunkhouses to accommodate the newcomers. For medical care, Eagle-Picher employed one doctor, Lee Connell, who oversaw the construction of a thirty-bed hospital in 1917. Otherwise, companies rented out unused plots of leased land for all private construction, business or residential. They retained the right to demolish any structures with thirty days' notice if the rented land was needed for mining work. The companies built no housing themselves and provided no infrastructure such as water or sewer systems, which, according to one report, "produced a typical frontier mining camp." People who moved to the field built houses on these plots, which were atop and around active mines. Most houses had no foundation and only a shallow drop toilet in the yard. People bought drinking water from delivery trucks that came once a week. They stored it outdoors in large barrels. In 1917, an Oklahoma City reporter found almost 8,000 people living in Picher in houses like these, "of the shack character." The following year, an American Zinc official said Picher had "the worst living conditions I have ever seen." These conditions persisted. "The town is unlike other towns," a 1919 survey declared, "in that there are only the poor or tenement sections." In 1920, another observer found homes "very much below standards found in similar mining communities." People had no incentive to build for permanence. Eagle-Picher evicted tenants as its operations grew. "Of course I can only want a shack," one miner explained in 1920. "If the mining company wants to put a shaft under my front door, or a tailing pile on the kitchen stoop, then I've got to move. Even if there were sewers and a water system I wouldn't want to connect with them," he concluded, because the company could evict him any time. "If I owned hell and Picher," one resident joked in 1918, "I would rent out Picher."[13]

Some miners resisted permanent relocation. In 1923, about half of the men who worked in the Picher field returned on weekends to family homes in Webb City, Joplin, and Galena. After the extension of the interurban line, miners could make the trip in an hour and fifteen minutes for ninety cents each way. They lived in boardinghouses or rented a house with other miners during the week. These commuters wanted to hold on to the better living conditions they had in communities that offered amenities such as schools, churches, leisure activities, and municipal water and sewer connections as well as affordable, frequent trolley service to Joplin, still the fourth-largest city in Missouri, with over 29,000 residents. These communities were also awake to the silicosis crisis. Webb City opened a public hospital for lung disease patients in 1918, a direct result of union and middle-class campaigns earlier in the decade. By then, however, most of the mining jobs were in the Picher field.[14]

In Oklahoma, miners sought individual incentives on terms that ignored health and safety considerations. Eagle-Picher and other companies tied all wages, whether hourly or piece rate, to the market-based sliding scale. With ore prices at historic highs, the sliding scale encouraged heedless work practices. Eagle-Picher, Commerce Mining, and other Oklahoma companies operated with "no safety devices," according to the state factory inspector in 1920, despite investing in state-of-the-art processing mills. Miners descended to the bottom of shafts, often 250 or 300 feet deep, in open, free-swinging tubs with no guards or catches. Only the Anna Beaver Mining Company used cages, not tubs, which a leading engineer sarcastically deemed "an innovation here." Companies did not run water supplies to the machine drills or sink dedicated ventilation shafts, despite the Lanza and Higgins report about the dust danger. Machine men and others were allowed to work with shovelers below them, which made for constant danger from rock falls. No one seemed hurried to operate differently. They did not have to. Oklahoma state mine inspection law did not explicitly cover zinc and lead mines until 1929. The state's three assistant inspectors focused their efforts on the coal district over 100 miles away until 1927, when the state appointed a fourth assistant inspector to cover the Picher field. Meanwhile, accident and death rates soared above the averages in all other American metal mines. Yet according to one resident, "with lush profits and top wages, no one was concerned with health measures, least of all the miners."[15]

Courted with good wages, miners again indulged a reckless ethic that valorized male aggression as the means to high pay and social respect. The shovelers led the way. "Straining every muscle for hours under the earth, working in an atmosphere of rock dust thrown off by the drills," according to a journalist visiting in 1916, they "worked in a veritable fury under the stimulus of extra pay." Like many before them, these "raw boned natives" cast aside concern and caution, abandoning unions and reform, to pursue big pay "with nothing but a pair of arms and a shovel and a strong back." Four years later, an observer noted, shovelers remained "the central cog and upon which the whole underground work rests."[16] Those who came from the Missouri field, particularly the experienced shovelers, needed no introduction to the district's stories about the power of hard work. The new field also attracted young men who had never mined before. Many sons of miners took their first job in the Picher field. Others came from regional farms. Tony McTeer moved to Picher in 1919 at age twenty from near Sparta in Christian County, Missouri. He had struggled to make a living as a farmhand since being orphaned in 1913. The newcomers were all attracted by the immediate prospect of high pay. Men predominated in the camps, making up two-thirds of the population in the early 1920s. According to a local historian, miners "in those early years were little concerned about the future safeguards or future practices."[17]

As they had in the sheet ground, young miners reveled in the physicality of their work. Charles Chesnut got his first job as a "screen ape," breaking boulders with a hammer in a mill. He went to bed exhausted at 4:30 P.M. after his first day but could soon "bust boulders all day and was in excellent shape." Just as in Webb City and Joplin, the shovelers exceeded with entrepreneurial brawn. The local press celebrated their productivity. "It has been common knowledge," the *King Jack* claimed, "that western miners are not efficient in the matter of shoveling compared to those of the Tri-State," whose "tonnage is something enormous compared with that of other fields." The shovelers awed Charles Morris Mills with their "characteristic energy" when he came to Picher in early 1920 to investigate conditions on behalf of the Interchurch World Movement, an ecumenical organization that briefly sought to solve industrial problems with Christian principles. When Mills asked for explanations, a mine boss told him, "Well, they're naturally hard workers, being good Americans, and we pay 'em damn high wages." Mills was scared of them. He considered the miners "uncompromising and almost unapproachable," under the sway of what he called a "feverish unsteadiness" that was "distinctly hideous."[18]

Miners understood their claim on the incentives of the sliding scale as a function of white nationalist privilege, a benefit of Americanism. They defended it by insisting on the exclusion of foreign-born and African American workers, who they feared would work for less. We "want this to be a white man's camp," a miner called Mac bluntly told Mills. They succeeded. Picher "is an All-American camp, and no foreign labor is tolerated by the miners," a visitor recorded in 1920. "Negroes are conspicuous by their absence, and are not wanted in any capacity," another visitor noted. Workers actively policed the "unwritten law" that kept them out. Mac explained that during the 1919 Kansas coal strike "some dirty Austrians and hunkies tried to work here." "Before you could say Jack Robinson," he recalled, "there was a gang ridin' 'em out in box cars right back to Kansas." Their deep-rooted xenophobia and racism had grown stronger as exclusionary Americanism intensified in the years after the war, when leading politicians forcefully advocated immigration restrictions and white vigilantes unleashed staggering levels of violence against African Americans, especially in the 1921 massacre in Tulsa, Oklahoma. Mining companies did not resist their demands for exclusion.

The masthead of the *King Jack*, a newspaper that relied on company favor, plainly announced who was welcome and who was not: "No foreigners; No niggers, but a class of citizens who respect the Flag of our country." Of the 41,108 people who lived in Ottawa County in 1920, there were only 377 born outside of the United States and eighteen African Americans. Ten years later, only 252 foreign-born people and two African Americans remained. None of them lived in Picher.[19]

While not new, this emphasis on racial and national exclusion reinforced a resurgent antipathy to unions in the Picher field. The Wilson administration's campaign against domestic radicals during the war and after, alongside the new threat of Bolshevism, gave these antagonisms new life. "We don't want no Bolos or I. W. W.'s or labor grafters who steal the pot before the draw," Mac told Mills. Some miners had joined the WFM in 1915 because it promised to protect white national dominance in the field. But now, Mac claimed, "we don't have to have no union to keep out the greasers." They could do it themselves with vigilante violence. Mining companies acceded to demands for racist exclusion, despite chronic labor shortages, as a means to forestall potential union appeals. The *King Jack* gave voice to this informal pact when it lambasted foreign coal miners for striking in 1919. "These 'foreign borns' prior to coming to the good old U.S. were willing to work long hours for a miserable pittance," claimed Frank Hills, mining editor of the *King Jack*, "yet when they are cared for in this land of the free, their stomachs filled, their bodies clothed, a few dollars to jingle in their pockets, they become more autocratic than any despot." When the IWW blocked a train of strikebreakers on their way to Pittsburg, Kansas, Hills called them "uncivilized bohunks" and a "bunch of heartless savages." No one "had the right to say that no one should work in their place," he added. "Unionism does not strike to get the work done and it never means better work, because better work demands greater devotion, loyalty." In some accounts, the old alchemy of white nativity, antiunionism, and aspirational common interests crept into descriptions of the new field. According to a 1921 report deeply informed by local leaders, "many of the operators and practically all the superintendents were once just common miners, who raised themselves step by step, as an intelligent native American will do when given the opportunity, freed from the trammels of restrictive unionism and the poison of specious propaganda." The idea remained powerful. Mention unions to a Picher miner, Mills said, and "he will immediately brand you as a 'Red' and mark you for deportation."[20]

Despite living as tenants on leased land, miners increasingly looked to settle in Ottawa County. Picher's residents incorporated the camp as a town in 1918. They elected a mayor and board of trustees. The new government raised money for schools, new sidewalks, and electricity service. As more miners moved their families to Ottawa County, businesses expanded. In 1920, Picher's residents tried to buy some of the leased land the town occupied so that banks would loan more money for further development. To help, Congress had passed a law in late 1919 that allowed the secretary of the Interior to unilaterally sell Quapaw allotments for townsite purposes. The mining companies objected. They argued that any transfer of property would infringe on their leasehold rights and impede mining development.[21]

As Picher's residents pushed back, they voiced new opposition to company authority that used the camp's valorization of dangerous work and assertions of nativist privilege to claim decent living conditions. "What we are striving for is to protect our selves, a great population who has to live here to make up a city, inhabited by the very ones who work for the operators and go into the ground and face death every day to take out the ore," petitioners informed the commissioner of Indian affairs. "The miners like all other people like to live close to the work," they concluded, "they also like the same rights under which to live that other American citizens enjoy." In 1921, they struck a deal with the blessing of the secretary of the Interior, Albert Fall. The mining companies allowed the sale of eighty acres for town development and received in exchange a ten-year extension on all other leases, thousands of acres in total. Companies retained the right to evict anyone from leased land. With that compromise, Picher continued to grow, albeit in an uneven fashion that left the town perpetually disfigured: while a regular street grid took shape in its eastern half, its western half gave way to growing chat piles that pushed everything else aside.[22]

Picher remained a rough town that catered to the unrestrained appetites of young working-class men. Although Oklahoma was dry, local bootleggers supplied several illicit bars, where miners drank alcohol with impunity. Local police did not enforce prohibition law, state or national. Pool halls, gambling houses, and brothels also violated state law to serve local workers. None hid. All of the brothels in Picher were on the same street, known locally as "chippy town." "If a tent were stretched over the entire town of Picher," one resident recalled, "it would constitute a giant bawdy house, not to mention other vices thrown in for good measure."[23]

Many women saw opportunities in the new camps and supplied the emotional and physical labor that created and reproduced social bonds. According to a 1919 report, "the discovery and rapid development of the mining field attracted a large number of women" to Ottawa County. As many as 2,000 of them were single, or "floating women." They looked for work, excitement, and possibly a spouse. Women ran and labored in boardinghouses, and some worked as prostitutes. They cleaned, cooked, and washed clothes for money. Most of the camp's adult women were married, however, especially by the 1920s. Wives and daughters of miners performed domestic labor: mending, cleaning, cooking, raising children, and perhaps keeping chickens for eggs and meat. Some also catered to boarders who rented out spare or partitioned rooms. The wives of miners took great pride in preparing lunch for their husbands. As Iva Simpson recalled, they packed the food in big tin dinner buckets, with coffee on the bottom, a center tray with two cups for "scrambled eggs, chicken and noodles, beans, soup, corn, or whatever she had on hand … with room to stuff a couple of biscuits between," and a shallow tray on top that "held slabs of pie or cake, thick slices of homemade bread and butter and a spoon." Many women sold lunches like these to single men. "The man whose wife packed a good bucket was the envy of the crew," Simpson recalled. He would have "bachelors, tired of boarding fare, beating a path to his door to ask his wife to fix buckets. Many a housewife did a good business packing buckets."[24]

Many women relished the achievements of Picher, despite the struggle. "There was so much work" in the early days, Mrs. B., the wife of a Picher miner, recalled, "that people rushed in here and started to get at it before they even had a roof over 'em.… Everybody that come here then had to work hard but they didn't complain none." She delivered her husband's lunch to the mine herself. "I'd put on my big rubber boots an' with the bucket in one hand an' the baby in the other, off I'd go." She liked the sense of contributing to something dynamic. "Watchin' the new mills go up an' the new mines go down wuz as excitin' as watchin' kids grow," Mrs. B. said, "an' everything wuz like that—growin' over night, an' the people felt like they wuz part of it."[25]

As wartime nationalism emboldened working people's claim on the district's prosperity, the federal government helped mining companies consolidate control of the industry. Eager to smooth production, Pope Yeatman, chair of the nonferrous metals section of the War Industries Board, urged mining and smelting companies in early 1918 to form an industrial association. Such associations became increasingly common during the war, as the board effectively waived antitrust regulations to promote the rationalization of private businesses and discourage ruinous competition. Mining companies in the Tri-State district had barely cooperated with one another since the Missouri and Kansas Zinc Miners' Association collapsed in 1906. American Zinc had dominated the district in the decade that followed. The opening of the Picher field, however, called for more industry organization. With Yeatman's blessing, mining and smelting firms from Kansas, Missouri, and Oklahoma formed the American Zinc Institute (AZI) in late 1918. Eagle-Picher, Commerce Mining, and American Zinc all took part, as did a number of smaller companies. For Tri-State firms, association promised not just a means to industry stability but also a way to impose order in the district.[26]

The AZI drew on company networks established in response to government-backed efforts in 1915 to address the district's tuberculosis crisis. Initially neglected by Oklahoma operators, the Southwest Missouri Mine Safety and Sanitation Association relocated to Picher in 1918 under a new name, the Tri-State Mines Safety and Sanitation Association, and a new leader, Eagle-Picher's A. E. Bendelari. In April 1919, the association reorganized as the Tri-State AZI chapter. Richard Jenkins, who had served as the association's founding secretary, became the secretary-treasurer of its latest iteration. The chapter inherited its predecessor's health initiatives, if not its priorities, including the new Webb City sanitarium. It also took control of the Picher hospital. In line with the new Republican administration that took power in 1921, however, the AZI chapter's main business was business.[27]

Mining companies used the AZI to navigate the severe national economic crisis that hit in 1920. As war industries contracted, the price of zinc ore fell from over fifty-three dollars per ton in January to thirty-six dollars per ton in December and continued to fall through the summer of 1921 to around twenty-one dollars per ton, the lowest price since 1897. AZI chapter members negotiated district-wide shutdowns in June and again in October 1920; all cut wages. The district produced only 313,569 tons of zinc ore in 1921, down from over 569,000 tons in 1920. While the collapse strained family economies and local charities to the breaking point, the AZI chapter stepped in to provide respite through a central relief committee. The committee raised over $5,000 for this "welfare work." The committee bought credit at local grocery stores and issued it as vouchers to unemployed miners who agreed to repair county roads in exchange. Only miners with dependents were eligible, and they could redeem the vouchers only for food. Officials claimed that miners agreed with these measures. Not everyone did. Some men left the district. Meanwhile, AZI representatives in Washington used the example of unemployed Picher miners to lobby for higher zinc tariffs. The Fordney-McCumber Tariff, signed in 1922, more than doubled rates on imported zinc.[28]

There were still limits to AZI unity and authority. As the economy recovered, AZI firms colluded to try to prevent wages from returning to pre-recession highs. They rehired miners in 1922 at relatively modest rates: three dollars per day and eight cents per can. As prices continued to climb, miners demanded raises in accordance with the now customary sliding scale. When companies resisted, a few dozen miners launched a wildcat strike at two small mines. Both companies, Chanute Spelter and Kanok Metal, quickly conceded raises of twenty-five cents per day and the miners went back to work. Unwilling to incite further strikes, most other companies gave similar small raises. Miners had defended the sliding-scale principle as production rebounded in 1922 to 524,265 tons of zinc ore worth over $31.3 million, a sharp rise from the previous year's sales of $11.2 million.[29]

The AZI chapter's concern with labor costs refocused attention on the physical health of miners and their families. Conditions in Ottawa County remained grim. "Many communicable diseases were epidemic, especially smallpox," the county health superintendent reported in 1921, due to "filth and over-crowding." According to his survey, "venereal disease was very prevalent" and "the tuberculosis rate was many times the state average, due largely to the silicosis which is so prevalent in all zinc and lead mines." Charles Morris Mills reported similar findings. His 1921 account described tens of thousands of "Americans, working amid highly dangerous surroundings, living in filth and disease, purely individualistic and lacking the commonest incentives for decency." He blamed the operators, whose failure to provide for the public welfare had let ideas of "individualism and freedom" run rampant among the workers, who in turn "developed an utter irresponsibility in regard to living and housing conditions." Even if mining companies had no social conscience, he admonished, they should at least recognize the business costs of a perpetually sick and injured workforce. "The problem here is to Americanize Americans," Mills quipped. Richard Jenkins agreed. In 1921 he advised the AZI chapter to expand its "welfare work" to include permanent community medical services at the Picher hospital. Like A. J. Lanza and Edwin Higgins, who had conducted a field study on miner's consumption in Missouri, Jenkins believed that companies had to act because the miners would never better themselves. "They apparently give no thought as to the future," Jenkins mused. He assured his AZI colleagues that a "humanitarian" welfare program also made good business sense because it would result in "shortening the time of our sick and injured in getting back on the job and, in a general way, also in increased efficiency." The AZI chapter gave Jenkins money to open a free clinic and nurse service.[30]

Jenkins's initiative drew on the support of federal government agencies that were also concerned with the district's health crisis. In the summer of 1918, a U.S. Army doctor investigated the alarming rates of venereal disease found in soldiers from the Tri-State. That fall, government officials from all three states worked with the PHS to establish a federal sanitation district in the area. Jenkins, who chaired the Jasper County PHS committee, was involved. The PHS began administering these districts in 1916 to create sustainable county-level health services in rural areas with vital war industries. The PHS created the Tri-State Sanitary District in January 1919 and sent Royd Sayers, an army doctor, to make an initial survey. In July, the PHS stationed Thomas Parran, a commissioned assistant surgeon, in Joplin to begin the work. Under Parran's leadership, the project gave immunizations, ran prenatal clinics, treated venereal infections, and built over 3,500 sanitary privies across the three counties. The project also employed a full-time health officer in Ottawa County. Parran hoped that the officer, "in addition to dealing with general health problems, would carry out some intensive work with miners on silicosis." Although Parran was reassigned elsewhere in 1921, the PHS continued to fund the county health units. The Ottawa County service, however, could not handle the scale of the growing silicosis problem.[31]

In 1922 the AZI chapter asked the U.S. Bureau of Mines to commission a study to further explore the causes of the disease in the Picher field. Sayers, now the bureau's chief surgeon, consented. According to Parran, who briefed Sayers on the situation, the mine operators believed "that the deaths were occurring largely among miners who previously had worked in the old (Webb City) field" and were not due to their operating practices. His surveys, however, showed that more than half of those afflicted had only ever worked in the Picher field. Sayers dispatched a team of mining engineers and doctors, led by Daniel Harrington and Richard Ageton, to Picher in February 1923. Eagle-Picher provided them two rooms in its headquarters for the work and lodging at the company's staff boardinghouse. During eight months of research, Harrington and Ageton visited forty-six mines in Ottawa County and examined 309 miners.[32]

Their study concluded that miners were developing new cases of silicosis at the same rate that Lanza and Higgins had observed in Webb City in 1915. One-third of the miners they examined had silicosis, and another third showed early-stage symptoms. The investigators blamed both the companies and the miners. They found miners drilling, blasting, and shoveling in totally dry conditions with no water provided to mitigate dust and no ventilation systems in place to clear the air. What surprised Harrington and Ageton was that miners did not demand safety measures. "Many seem willing to take precautions when their attention is called to them, many do not know how to protect themselves and again a large number, even a large percentage, seem indifferent, even fatalistic, and will take precautions only if compelled to do so," they reported.[33]

Why did so many miners resist measures to improve air quality? They had known that dust caused deadly lung disease since the 1900s. What miners rejected was the means of dust abatement, particularly the water soaking of underground workings. When miners voiced objections, they explained that wet drilling and wet ore piles made the work slower and more difficult. Shovelers disliked wet ore because it clumped together and was heavier, which reduced the number of cans they could load. They also claimed that wet workings gave them other diseases, such as pneumonia or rheumatism, an inflammatory disease that stiffened joints and weakened muscles. Harrington and Ageton reported that miners expressed a preference "to die of miners consumption from the dry drills rather than of rheumatism." Tri-State miners understood the risk of silicosis but resisted precautions, especially those pushed by government doctors and mining companies, that threatened the relationship between their self-conception as physically strong, unrestrained men and their earning power.[34]

The federal team recommended that mining companies take "drastic" steps to address the crisis. These should include "the establishment of up-to-date equipment and practices underground" as well as a thorough education campaign. Once companies took these steps, the team declared, it was crucial that "the proper supervision and discipline be maintained to make the original effort effective." The ultimate success of such a campaign against silicosis, in other words, depended on company willingness to compel obedience from miners and others who resisted. The investigators advised mining companies to coordinate the campaign through the AZI chapter.[35]

The team also noted very high injury rates. Harrington and Ageton again asserted that ultimate responsibility for safe working conditions rested with the companies. "There is not any coordinated and controlled safety work in the district," Ageton reported to George Rice, the chief engineer of the Bureau of Mines, "nor do any of the companies pay a great deal of attention to accident prevention and care of injuries other than the trimming of the roofs in the stope and making the hookers wear trench helmets." Their report recommended a series of safety measures.[36]

Linking safety and health, they suggested that miners with lung diseases, especially those in demanding jobs like shoveling, were more likely to get hurt on the job. The team urged companies to require miners to have a physical examination every six months, a recommendation based on studies of medical screening in South African gold mines. Men with damaged lungs should be encouraged to leave the mines. "Such an examination would also be of advantage to the employee," Ageton concluded in a separate report, "as it would show them their physical defects and most certainly men do not want to work at an occupation for which they are physically unfitted and thereby shorten their life."[37]

The federal investigators showed companies that greater control of workers in the mines would reduce operating costs, most directly in terms of insurance expenses. Concerns over the cost of liability insurance had never been far from company efforts to increase mine safety, especially in Missouri where workers could sue. Oklahoma passed a workmen's compensation law in 1915, however, that created a binding system of compensation for accidents that followed a set schedule of monetary awards decided upon and administered by a state industrial commission. Meant to eliminate costly adversarial court cases, the law required companies in hazardous industries to purchase commercial compensation insurance or show the financial wherewithal to insure themselves. Many smaller companies bought compensation policies; larger companies, such as Eagle-Picher, self-insured. As mining accidents increased along with production, however, insurance premiums increased—by 50 percent between 1921 and 1923. Many insurance companies stopped issuing new policies in 1923. Metropolitan Life Insurance Company raised premiums by another 10 percent and established its own private "mine inspection service and medical supervision" office to monitor working conditions. The federal team believed that its medical exams would help companies screen out men who represented higher insurance liability risks, such as those with silicosis and old injuries, especially to body parts that merited high compensation payments, such as arms, hands, legs, feet, and eyes. In turn, insurance companies would lower premiums.[38]

Mining companies strengthened their cooperation in response to the recommendations. In late 1923, the field's smaller companies established their own association, the Tri-State Zinc and Lead Ore Producers Association (OPA), based in Picher. The OPA acted quickly to address the "very large burden upon the district of compensation charges." It hired Ageton to run a campaign to reduce accidents and improve the health of mining families. Beginning in early 1924, he created and circulated a statistical register of all accidents. He also promoted the use of mining helmets as well as enhanced ventilation and wet drilling to reduce dust and "the silicosis-tuberculosis evil." Although Eagle-Picher, Commerce Mining, and other big companies did not join the OPA, these firms cooperated with and funded Ageton's campaign through the AZI chapter. To convince miners to cooperate, Ageton advised company officials to speak of the safety program in terms of only humanitarianism, not labor costs. "We should do all that is possible to make the men in the mines and their families feel that this is *their* accident prevention campaign," Ageton said, "and that it will be to their physical and financial advantage to cooperate in every way." That summer, the AZI chapter and the OPA asked the Bureau of Mines to send Frederick Flinn, an industrial toxicologist from the 1923 study, to continue silicosis research in the district. Working in AZI facilities, Flinn opened a small clinic for examining miners for lung injuries in late 1924.[39]

Better organized and backed by federal agencies, Tri-State mining companies flourished once more as the economy roared to life. In 1923, district firms produced over 688,000 tons of zinc ore and 111,000 tons of lead mineral worth more than $38.4 million combined, the richest year in Tri-State history. They did even better in 1924: 740,569 tons of zinc ore and 118,770 tons of lead mineral, worth more than $43.3 million combined, all district records. Eagle-Picher reported a profit that year of more than $3 million. These gains relied on the hard labor of the district's 8,000 miners, about 3,000 of whom were shovelers. In 1924, the richest year in district history, these men moved over 12.4 million tons of crude ore by hand with the undimmed expectation that they might reap some of the rewards. They certainly bore the costs. Ageton's first accident survey revealed that shovelers received over 51 percent of all injuries in the mines. Earlier research already confirmed that they, along with machine men, suffered the highest rates of silicosis. Any widespread effort by mining companies to reduce either accidents or lung disease would have to confront these men.[40]

As prosperity returned, miners and their families grew restless for their share. The operators' associations, meanwhile, expanded production cautiously to avoid an ore glut. Some companies put workers on a five-day week, while others closed operations for a week every two or three months. "It was just about the first time on record that producers in this field have indicated intelligence enough to unite on a program," the *Joplin Globe* observed in late 1923. For miners, these strategies created continual uncertainty about available work and kept earnings below the level of a full, six-day workweek. Despite company commitment to the sliding scale, pay stagnated: $3.50 to $4.00 per day for machine men, the highest earners, and 10.0 to 10.5 cents a can for shovelers. Poor living conditions compounded their frustrations. Residents of Ottawa County experienced overall mortality and infant mortality rates that not only were the highest in the state in the mid-1920s but were two or three times the state average. Mining families had come to Picher for high wages and now they wanted a better quality of life. They wanted to live how they thought white Americans deserved to live in the prosperous 1920s. Instead, their lives were getting worse. Faced with increasingly powerful operators, some turned for help to the AFL's state affiliate, the Oklahoma State Federation of Labor (OFL). But as Picher's miners looked to restore the balance of power, most could not deny the promises of their poor man's entrepreneurial tradition, promises that made it hard to imagine unions, even conservative ones, as vehicles of prosperity.[41]

Since the demise of the Mine Mill locals in 1916, district miners had shown little interest in organizations outside of the political mainstream. At the ballot box, they split votes more or less evenly between Democratic and Republican candidates. Democrats won in Ottawa County in 1916 and 1918, although Republicans ran ahead of their vote elsewhere in Oklahoma before sweeping the county in 1920. In the gubernatorial election of 1922, Oklahoma elected a champion of white working-class interests, John C. Walton, under the banner of the Farmer-Labor Reconstruction League, a consortium of Socialists, United Mine Workers of America (UMW) and AFL members, and liberal Democrats. Walton won Ottawa County but only by 93 votes out of over 8,200 cast. He got less support than the Democratic candidate for Congress, who won the county by 420 votes. Two years later, Republican candidates for president and senate won Ottawa County again. Meanwhile, some miners might have joined the Ku Klux Klan, but middle-class people dominated its nearest groups in Miami and Joplin. By contrast, Homer Wear, an organizer for the IWW's Metal Mine Workers, a union formed in Butte in 1917 to rival Mine Mill, canvassed the district in the summer of 1923. He elicited little interest beyond the Picher police, who arrested him on charges of criminal syndicalism. Although presented with a wide range of political ideas and organizations after 1915, the field's mining communities registered little collective enthusiasm for any.[42]

As miners watched company profits rise in 1923 while their wages remained flat, some sought assistance from the OFL's ambitious new leader, Ira Finley. Late that year, he "received letters from lead and zinc miners telling of the conditions in the fields and urgent need for unionism." Finley, a Socialist, took them seriously. Just elected, he promoted an aggressive new organizing drive at a time when AFL unions nationally were in full retreat as employers pressed a new antiunion open-shop strategy called the American Plan. In the missives from Picher, Finley sensed a chance to establish a new union beachhead that could buttress the OFL's strongest contingent, the dozens of UMW locals in the coalfields of southeastern Oklahoma. He referred the requests for assistance to Mine Mill's office in Denver. Although still an AFL affiliate, Mine Mill barely existed in 1924. The union had not held a convention, published *Miners Magazine*, or paid organizers since 1921. It claimed a few thousand members at most, mainly in Montana. Not long after Finley's letter arrived in Denver, however, Charles Moyer, still Mine

Mill president, dispatched John Turney, a longtime board member, to north-eastern Oklahoma. Turney arrived in March 1924, and Finley soon joined him. John J. Beggs, a Picher miner and former UMW member, helped orient them; he likely had written one of the initial letters to Finley. By the end of the month, against all odds, these three had organized five new Mine Mill locals: Treece 130, Picher 134, Hockerville 136, Cardin 138, and Commerce 139. Although conducted under the banner of Mine Mill, their campaign was really led by the OFL, which respected the economic and racial views of Tri-State miners perhaps more than any union ever to enter the district.[43]

Turney and Finley declared that Mine Mill and the OFL would help miners gain the benefits of true Americanism. In speeches to mass meetings of hundreds of miners and their families, they said Mine Mill was "being organized along the conservative American labor movement lines." The organizers emphasized their primary aims to raise wages and improve working conditions. They also criticized the social problems that plagued mining communities, particularly the effects of silicosis and the consequences of poor housing and the constant threat of evictions. Turney and Finley assured those who listened that as key workers in a profitable American industry they deserved better. The organizers reported serious and eager audiences. "Having tasted of the 'glorious benefits' of the so-called 'open shop' 'American plan,'" Finley informed readers of the *Oklahoma Federationist*, "they know that it is the most un-American plan ever devised by the evil mind of an industrial exploiter and they are determined to have no more of it."[44]

Turney and Finley heralded the union as the true champion of white, native-born Americans by showcasing its robust support for the new immigration restriction legislation then up for debate in Congress. Finley reported that Tri-State miners "are enthusiastically supporting the Johnson immigration bill," a law that proposed to set very low immigration quotas based on national origin, with a particular eye to excluding people from southern and eastern Europe deemed racially unfit. According to him, they understood that more arrivals "from the pauperized nations of Europe" would "drop this country into the same bottomless pit into which those countries have fallen." The OFL's nativist assertions reflected broad support for the legislation, which passed the house on April 12, among union affiliates in Oklahoma and the AFL more generally. Finley featured it in most of his stump speeches; the *Oklahoma Federationist* reprinted Samuel Gompers's letter in support on its front page. To Tri-State audiences, however, these invocations had powerful effects. Positioning the union as a nativist bulwark, Turney and Finley affirmed the long-held racist claims of Picher miners on the district and its profits. Perhaps that was necessary if the OFL campaign had any chance in Ottawa County in 1924. In the long term, however, the OFL's anti-immigration campaign fed local working-class hatreds that had always decimated efforts to build solidarity there. Moyer had tacitly accepted this strategy in 1915. While it had failed then, the OFL hoped it might work now. Picher miners "are all American-born white men," Finley explained, and "are being organized in the open, like real American citizens should be organized." Beggs hoped that their commitment to the "ideals of Americanism" would "never weaken."[45]

While union activists reaffirmed the racism and nativism of Tri-State miners, they challenged them to abide union standards of manly responsibility to others. They appealed especially to men with wives and children. Only through collective bargaining, Finley asserted, could "they benefit themselves and their families." Mine Mill and the OFL were unions of breadwinners, they argued. Men who chose solidarity and responsibility would not only gain higher wages but also reorient their families around the American standard of living and its associated gender roles. Through collective bargaining, Beggs contended, miners with families could show employers that earnings of twenty-four dollars per week for skilled workers were insufficient when the cost of living in Picher was that or more. Paid at or below his "cost of production," Beggs explained, a miner had "nothing left for clothing for himself and family unless the stomach pays for it. There is nothing left for medicine or doctor; nothing for needed amusements, such as picture shows and picnics." "We have every confidence," the OFL declared, "that the new organizations will accomplish wonderful things for the miners, their wives and children."[46]

Union organizers had harsh words, however, for men who believed that they could get ahead on their own, a risky criticism that confronted the district's primary male tradition. In public speeches, Finley "placed the blame for the conditions of the miners solely and squarely upon the miners themselves." "They could have solved their own problems" if they had joined the union earlier, he declared. Everywhere Finley touted the costs of their failure: "unjust conditions," "miserable houses," and "a considerable number of men suffering from" silicosis. The "victims" of the disease "stalk around looking at the world out of hollow, sunken eyes from which all hope has fled," he wrote. Beggs also lacerated the miners for their pursuit of self-interest over everything else. He claimed that children "who die for lack of sufficient food and medical care" go to their deaths believing "Daddy's so big and strong, can do just anything." But in the end, Beggs scolded, the children "must tell the God they return to, that Daddy failed them, that he cared more for a soulless corporation than he did for them." "Men of the Tri-State, the remedy is in your hands," he concluded. "May God deal unto you as you deal unto yours, me and mine." Unlike Ageton and the operators, who thought the miners were too ignorant or fatalistic to care for themselves, the OFL organizers understood that their behavior reflected stubborn attachments to the individual rewards of brutal male power. The union called on Picher men to restrain themselves, to admit their vulnerabilities and accept the need for safety, for the benefit of their families. Whatever these men feared losing in terms of individual opportunity and respect, the union would make up for with new collective strength.[47]

As responsible union men, OFL organizers argued, Tri-State miners should serve as partners with the companies to make district production more beneficial for all. "These men are not joining the union with the purpose in view of antagonizing their employers," the *Oklahoma Federationist* explained, "but for the purpose of solving the many problems that confront the zinc mining fields." In public rallies, Finley "dwelt upon the necessity of every group of people being organized," the paper reported, with an appeal to the new 1920s idea of "regulatory unionism" whereby unions offered to help companies make production more rational and profitable. Finley went further to publicly denounce strikes and anyone who advocated them. "In my talks," he wrote, "I am warning the miners against heeding the agitators who will attempt to bring about a strike. We do not want these people to fall into a strike trap that will probably be set by some enemy of their cause." The OFL did not rule out strikes altogether but rather urged members to consider direct action only as "a last resort when all other methods of adjustment have failed." While these warnings seemed to recall the lessons of past conflicts, Finley and Turney also seemed sincere when they instructed new members that successful union "organization is a business proposition."[48]

Many miners joined the union that summer. Mine Mill added six more locals: Douthat 140, Baxter Springs 143, Quapaw 144, Galena 146, Miami 153, and Zincville 155. In June these locals organized a central leadership group called the Tri-State Council. At its height, the council claimed 1,000 or more members. Picher 134 was the largest local, with over 250 members, although the Treece, Douthat, and Cardin locals each boasted over 100 members. According to industry observers, they met regularly, collected dues, and took in new members. The *Engineering and Mining Journal* predicted in June "that in the near future every worker will be asked to show his union card when he appears for work at the mine in the morning."[49]

Women gave Mine Mill conspicuous public support because organizers explicitly addressed the problems facing families. During one of Finley's speeches in Picher, a miner's wife handed him "a large bouquet of roses with her compliments for the work we are doing." After the talk, he reported, "a number of miners' wives, some with babes in their arms and all with smiles on their toil-worn faces, came forward to shake hands with" him and Turney and to thank them. Women also organized picnics and dinners to bring the families of union members together. The "big dinner" they held on Labor Day was especially notable. "Can you imagine it," remarked Ed Dunivin, secretary of Local 140, "a Labor Day celebration in the zinc fields; the very first one within my knowledge, and I have been here twenty years." No longer so excited by the mining camps as Mrs. B. had been, these women saw the union as a way to fix or escape the district's shoddy housing, high rates of diseases, and general insecurity. Women bore the brunt of these conditions, as they stretched to make household finances work, struggled to raise and care for children, and faced down the looming threat that their husbands would be killed on the job or get hurt or sick with silicosis. They were also eager to enjoy prosperity. Some women organized their husbands themselves. Beggs lamented that many families could not afford a trip to the Devil's Promenade, a popular picnic spot on a bluff overlooking the Spring River a few miles east of Picher. Rather, according to Beggs, "the Devil's Promenade for the miner with a family in the Tri-State means catching the devil from his wife as she promenades an unsanitary, poorly provisioned kitchen." Beggs's line acknowledged that women were making demands on men to do something to improve their lives, often enough, it seems, for his joke to make sense. Crucially, the woman here not only expected more from her life but also blamed her husband, at least in part, for their condition.[50]

The Tri-State Council struggled, however, over safety and health policy. After a springtime of organizing speeches, the union made its first foray into action with letters of complaint to the state mine inspector. These complaints targeted alleged violations of the mine safety law. While the council seemed united around the need to reduce mining accidents, which could be easily blamed on the companies, it said nothing about dust abatement or other means of reducing silicosis. The council's only resolution at the OFL convention in Muskogee in September sounded the alarm about lung disease in the district but offered a curiously weak solution. Citing Ageton's research, it declared that "alarming conditions exist" in the Tri-State "relative to Tuberculosis," which "by its very nature causes great poverty to the families stricken." Rather than use this prime opportunity to rally OFL support for stricter regulation of mine air, the council instead asked the state to build a tuberculosis hospital in Ottawa County "for the taking care of these men." The closest one was in Talihina, over 150 miles away. The resolution did not mention any of the now well-known federal recommendations for dust abatement or indeed that dust caused lung disease. Stranger yet, the council identified tuberculosis as the problem, not silicosis or miner's consumption, both terms then in wide circulation. No one in the union, meanwhile, questioned the health effects of the shovelers' piece rate or the sliding scale. The council sought to protect the welfare of the families of sick miners and to provide palliative care for sick men but not to mitigate the conditions that caused the disease. While miners may have mistrusted the motives of government doctors, they also seemed unwilling to trust union organizers whose message echoed the recommendations of the doctors. Mine Mill had made partial progress but could not convince the men who joined it, let alone the thousands who did not, to abandon the culture of physical risk taking that guided their pursuit of individual piece rates.[51]

Despite Mine Mill's assertions of conservatism, companies opposed the union. "The mines of the Tri-State district have always been operated on the open shop plan," a company notice declared in July 1924, and would "operate on that basis in the future." Rather than start with punitive measures, the companies sought to seduce miners with assurances of white social equality and promises that the good times would return if they rejected the union. "The miners of this district are the best class of labor ever found in any mining district," the notice read. "Every man in the Tri-State district is as good as every other man so long as he conducts himself as a man." This rough bonhomie survived because miners had "grown up and worked side by side" with those who became managers, bosses, and "sometimes owners." The notice emphasized their shared language and race. "This is a white man's district so far as the mines are concerned," the notice stated, "and will always be kept as such." Rest assured, the message concluded, "there is no place under the shining sun where ability, integrity and loyalty finds quicker recognition and the reward of promotion than in the zinc and lead mines of the Tri-State district." By conjuring up the promises of the poor man's camp, the company tempted miners to keep faith in their own self-interest and the special opportunities the district offered men like them.[52]

Tri-State miners were reminded of the potential rewards as a growing market for zinc and lead spurred district production toward new heights in late 1924. In the wake of the Great War, American manufacturers incorporated zinc and lead into a wide array of products, especially in fast-growing industries like cars, chemicals, and construction. These sectors boomed as consumers pursued advertisers' promises with plentiful, cheap credit. By early 1925, the average price of zinc ore was over fifty-five dollars per ton. After holding back production with cutbacks and shutdowns, companies opened old mines and developed untapped deposits and continued to drill for new ones. By February 1925, the OPA reported weekly ore sales of $1 million. Firms put miners back on full-time and hired more. Crucially, companies also raised wages in accord with the sliding scale. By the spring of 1925, machine men were earning five dollars per day and shovelers making 12.5 cents a can, the highest rates since the war. For many miners, the sliding scale reaffirmed both their belief in the market and in their own power. What shovelers had first demanded in 1907, and strikers had won in 1915, industry officials now trumpeted as a customary means of "profit sharing."[53]

Mine Mill rapidly lost members as plentiful work and high sliding-scale wages made the union seem unnecessary. In November 1924, less than eight months into the organizing campaign, Moyer sent union stalwart Emma Langdon to help Finley revive interest. She fared no better. Finley was dismayed that miners would abandon the union at the first sign of higher pay. In December, he lambasted them in an *Oklahoma Federationist* editorial that showed how old animosities between Tri-State miners and union organizers still cut both ways. With heavy sarcasm, Finley explained that before 1924 "they had not needed an organization because conditions were so good." Free of foreign or radical influences, Tri-State miners had been "allowed to work for whatever they please" while their "families live in little shanties ... and the children can amuse themselves by watching the ambulances pass." A miner who did not get killed could "rest assured that he will live on an average of five years," since "only about 90 per cent of them contract T.B. the first twenty-four months." Finley asked who was "to blame for the miners not having an organization to sell their labor through" and answered, "It has been the fault of the stupid miner himself." Like the federal doctors, he concluded that too many men did not want to change. Yet the OFL leader added encouragement for those still in the union. Their allegiance would prove that "a real sign of manhood has been found through the zinc fields." As with previous union scolds, however, Finley's harsh challenge probably lost more members than it won.[54]

Picher miners had reason to doubt union manhood as the OFL fell into crisis in 1925. Across Oklahoma, employers harassed and weakened AFL unions. Open-shop campaigns in the coalfields decimated Oklahoma's UMW locals. The AFL offered no help. Gompers died in December 1924 after a long illness. His successor, William Green, a member of the AFL's executive council and a national UMW officer, favored cooperative relations with employers. Green did not come to the aid of the UMW in Oklahoma. As the AFL retreated further, Finley lost control of the OFL that summer to rivals who disagreed with his strategy of organizing new workers. In Finley's absence, Mine Mill's threadbare national office could do nothing to support the Tri-State Council as its locals collapsed one after the other. Moyer was still isolated and powerless. By August 1925, only Cardin 138 continued to pay dues, and it stopped in January 1926. Although some members remained, Mine Mill was again dead in the Tri-State. It was barely alive anywhere. Moyer and the executive board were forced to resign in 1926; only six people attended the 1927 convention.[55]

As prosperity flowered, most miners forgot why they had ever needed a union. In 1925, Tri-State companies produced more than 806,000 tons of zinc ore and more than 136,000 tons of lead mineral worth a combined $57.3 million. The district had never been busier, with all firms in full production, many with two shifts. Some even opened old Missouri mines. Although average prices dipped slightly in 1926, companies produced more than 823,600 tons of zinc ore, a new record, and more than 132,000 tons of lead mineral, worth a total of $52.5 million. That year Tri-State miners produced 68 percent of the nation's and 31 percent of the world's total zinc ore. Wages remained high, over $4.50 per day for machine men and twelve cents a can for shovelers. With over 7,000 miners at work full-time in 1926, Ottawa County's combined mining payroll topped $7 million, a new record. The field was again a "rich, rip-snortin' son-of-a-gun," one miner recalled.[56]

Mining families began to enjoy prosperity in Picher. They had more access to consumer goods and leisure activities. Picher's first radio station, KGGF, began broadcasting in 1925, two years before Joplin's first station. In 1926, construction crews completed work on U.S. Route 66 from Joplin to Miami, a path that took the iconic national highway through Picher and Cardin. Picher also boasted new cinemas and spectator sports, the most popular of which were football, baseball, and boxing. In the mid-1920s, Picher's King Jacks baseball team began traveling to play other area teams. The community followed the success of boys' and girls' high school teams with avid interest. The front page of the *King Jack* featured high school sports news in nearly every issue from the mid-1920s onward. Mining families also enjoyed better health as the local economy improved. Ottawa County's infant mortality rate fell by more than half.[57]

Picher and neighboring towns remained rough, raucous working-class places where men did what they pleased. Picher in the mid-1920s had "a lot of beer joints," resident Lawrence Barr recalled, "a beer joint on every corner." The town's red-light district was busier than ever. Despite Prohibition, the police still did little to regulate these places, Barr said, because they were afraid someone would kill them. Men reveled in the hazards of their work in ways that outsiders considered selfish or destructive. According to the *Engineering and Mining Journal* in 1926, Tri-State miners were notorious among safety engineers for their "cussedness," "a general, all-around lack of principle, with utter disregard of the spirit of fairness." When not at work they drank and fought in Picher's bars, such as the Bucket of Blood, the Bloody Knuckle, or the Monkey Inn, and on the streets. "The hard-rock miner drank his whiskey straight," a local historian explained, and "settled his disputes with his bare fists or a pick handle." His was "a world of a reckless breed" and of people "as hard as the rock itself."[58]

Working people took collective pride in this place, with all of its grabbing, hustling, and violence. In May 1927, more than 12,000 people took to the streets of Picher to celebrate its twelfth anniversary. They joined or watched a mile-long parade, competed in games and contests, and caroused at a street dance that lasted until 1 A.M. In a commemorative booklet published for the event, town boosters lauded Picher's residents for their strength, energy, and guts. "Now a first class city," Charles Brown wrote, Picher "sprang into existence as a necessity instead of the result of a blue-printed utopia of an idealist. The vitality was here; the resources were here and people came before preparations were fully made to receive them." Although the town was far from perfect, he admitted, "Picher unblushingly presents no excuse or apology for apparent delinquencies." Frank Hills applauded their courage and defiance in terms that invoked the poor man's camp mythology. "They left their former dwelling places because they were enterprising enough to search for better conditions," he said. "They are red-blooded, rugged individuals, the better kind of people who are willing to sacrifice and suffer a great deal for their ideals." Hills credited their race and nationality. "They are practically all Americans and all white," he explained. "They boast of the complete absence of negro and South European labor and prove by their manner of living that the zinc and lead miner is of the highest type of American laborer." That night, to mark the occasion, someone set alight four large crosses, each at least fifteen feet high, atop the highest chat piles in Picher. Once the fire department discovered that the mills were not on fire, it let the "fiery crosses burn."[59]

Tri-State miners soon learned the limits of their power, however, as mining companies continued to consolidate control over the district. The boom of 1925–26 prompted a series of mergers. Eagle-Picher purchased two competitors that added 100,000 acres of mining land and a zinc smelter at Hockerville. Commerce Mining bought the ninth-leading producer. Golden Rod Mining acquired several smaller competitors. These firms also invested in larger mills equipped with new flotation technology that processed ore more efficiently and enabled the re-milling of old chat piles. These companies sought closer cooperation in the OPA; Eagle-Picher and Commerce Mining joined in October 1928. Soon after, the OPA absorbed the AZI chapter. The association looked to cut costs, particularly wages and insurance premiums, as zinc prices began a slow, long slide, from fifty dollars per ton in late 1926 to forty dollars per ton in early 1929. Citing the sliding scale, operators cut wages for machine men to $4.25 per day and for shovelers to eleven cents a can. To control insurance costs, the OPA drew on the direct assistance of the U.S. Bureau of Mines and the PHS to fully implement Ageton's safety and health recommendations. The OPA drive not only sought to change how miners behaved on the job but also considered preventing those with certain injuries or illnesses from working altogether. With no union allies and an uncertain economy after 1926, Tri-State miners navigated these treacherous new circumstances alone.[60]

In early 1925, district operators began taking Ageton's safety recommendations more seriously when the Oklahoma State Insurance Board proposed raising workmen's compensation insurance rates from 6 to 9 percent of total payroll. With production high, companies faced ballooning insurance costs. The OPA and the AZI chapter lobbied the board for an exemption. In a day of hearings in Miami, their representatives emphasized the steps already taken to reduce accidents and compensation claims. "The operators have been fully alive to it," OPA executive secretary J. D. Conover explained. The board agreed to freeze maximum insurance premiums at 6 percent for a year but warned companies to take command of the risks. "So it is really up to the operators," the board's chair concluded, "to so conduct their mines and so supervise them, and so regulate them, as to produce fewer losses, and then they will get a lower rate."[61]

Mining companies gave ground bosses new incentives and authority to reduce accidents. The OPA started a "New Broom" club that recognized ground bosses in mines that recorded a month without a lost-time accident by nailing "a new broom" above their office doors. Eagle-Picher, meanwhile, started a cash bonus system for ground bosses that paid twenty dollars if their crews went accident free for a month. With the power to hire and fire, ground bosses could better compel reluctant miners to obey precaution or to not report injuries. Many companies also hired their own safety engineers to implement Ageton's recommendations. According to his surveys, these measures worked. From 1924 to 1926, OPA companies claimed to reduce the frequency and severity of reported accidents.[62]

In 1926, the Bureau of Mines and the OPA decided to expand the silicosis clinic. For over a year, Frederick Flinn and his successor, F. V. Meriwether, a PHS doctor who had recently helped identify black lung in Alabama, offered voluntary health exams to miners and their families, free of charge, in order to gather more data. The federal doctors could only examine a fraction of the workforce in the small facility provide by the AZI chapter. If they wanted a true picture of the problem, Meriwether argued, they needed a bigger clinic that could handle examinations for every miner in the district. The OPA and Ageton approved. So did Royd Sayers, who convinced A. J. Lanza, now head of Metropolitan Life's Industrial Health and Hygiene Service, to enlist his company's support. In May 1927, the Bureau of Mines, Metropolitan Life, and the OPA agreed to jointly support a much larger, more ambitious operation. The new plan called for a clinic of five doctors with full-time X-ray, laboratory, and dust-sampling technicians. They would provide free physical examinations, including chest X-rays, to all mine employees, with a particular focus on "controlling the silicosis-tuberculosis situation." Despite Meriwether's advice that the exams be compulsory, the OPA would only encourage them on a voluntary basis. The companies would use the results to ensure that men were given only jobs they could fully perform, thus reducing injuries, accidents, and crucially, company exposure to workmen's compensation claims. Men with serious issues, such as silicosis, tuberculosis, or heart problems, would be advised to seek treatment and given jobs "where they will be a minimum risk." As Ageton explained to a local mine owner, our "physical examination of employees is primarily intended as a means of increasing efficiency and production." The bureau agreed to provide $8,000 per year for operating expenses, supply all of the equipment and supplies, and assign PHS-affiliated medical staff. The bureau placed Meriwether in charge. Metropolitan Life agreed to match the bureau's $8,000 annual grant. The OPA contributed $16,000 per year.[63]

In the first year, Meriwether's team examined 7,722 employed men, 642 men seeking work, and 261 women and children. The exams for the men were unusually thorough in a place that had long lacked adequate medical care. Patients first completed a survey about their "personal, family, and occupational history" and then had their photograph taken, received a chest X-ray, gave spit and blood samples, and went through a full examination of their "eyes, teeth, hearing, nose, throat, chest, abdomen, rectum, hernia, genitals, and limbs." The lab tested their blood for syphilis and their spit for tuberculosis. Clinic staff used the results to form a diagnosis and issued a card to each patient that included his photograph and signature and a list of "all physical defects noted and a rating of his working efficiency." The ratings ranged from A to G, with Class A reserved for "perfect specimens of manhood"; Class B for the "average" man with only minor defects, such as slightly impaired hearing or vision; Class C for men with "a physical defect that may interfere with the man's working efficiency," such as missing fingers or first-stage silicosis; Class D for men with serious physical defects or venereal disease; Class E for men with advanced silicosis; Class F for men with early-stage tuberculosis or advanced syphilis; and Class G for men with advanced tuberculosis. Since the exams were voluntary, the OPA allowed individual companies to decide how or if to use them. Meriwether, however, considered the C grade "a warning." "The entire thought," he explained, "is to keep the workman informed of his physical condition hoping that he will take the necessary steps to have corrected such ailments as found" and to convince a miner with "tuberculosis and silicosis in its incipient state ... to go to a high, dry climate, where his chances for recovery are decidedly greater." He continued to advise the OPA that the exams should be mandatory.[64]

Meriwether's first annual report revealed the terrible physical toll that Tri-State miners had paid for their work. In the twelve months from July 1, 1927, the clinic gave A grades to 1.65 percent of patients, B grades to 63.90 percent, C grades to 16.77 percent, D grades to 7.92 percent, E grades to 0.41 percent, F grades to 6.13 percent, and G grades to 3.14 percent. Of the 7,722 miners examined, 26 percent had some stage of silicosis or tuberculosis. Of those who had worked in the district for ten years or more, 52 percent were so afflicted. For all men who labored in the ground, the overall rate was 38 percent, with machine men and shovelers reporting the highest rates of incidence, 43 and 39 percent, respectively. Shovelers, however, developed silicosis much faster. Those with first-stage silicosis had worked in the mines for an average of 8.4 years, while first-stage-silicotic machine men had labored in the ground for an average of 11.4 years. Meriwether advised men with first-stage silicosis to "leave the mines and seek occupation in the open air," under the assumption that recovery was possible. However, workers with first-stage silicosis remained productive. First-stage shovelers averaged forty-three cans per day. These men were unlikely to leave the mines voluntarily. Once miners developed second-stage silicosis, their productivity dropped considerably, to an average of thirty-three cans per day. Meriwether urged the OPA to adopt clear, tough restrictions on hiring them. "As soon as the second stage develops," he advised, "the men are not recommended for work in the mines." Meriwether was also alarmed that 20 percent of the men examined had syphilis, which some actuaries and doctors identified as a cause of serious accidents, most notably Thomas Parran, who would later oversee the Tuskegee syphilis experiment as U.S. surgeon general. Meriwether recommended that these men be excluded from work until they had completed treatment.[65]

Picher clinic grade card, 1929. Picher Mining Collection, folder 960, box 88, Pittsburg State University. Courtesy Baxter Springs Heritage Center.

Miners distrusted the new examination regime, especially the individual grades. "The men thought that the work was mainly for the protection and benefit of the operators and of little, if any, benefit to them," Meriwether reported in 1928. His weekly meetings with insurance adjustors, who took a keen interest in the exam results, belied the OPA's public claims that the clinic was for the benefit of the miners. To allay suspicion, the OPA launched a publicity campaign to emphasize that the exams were entirely voluntary and would not result in anyone losing their job. In late 1927, the OPA posted large notices at all mines stating, "You are not required to take this examination. It is available for you if you desire it. You are free to act as you wish in the matter." The exam grades could still complicate employment for men who were new to the district or those looking to change employers. The clinic's expansion coincided with reduced production in 1927 and early 1928. Mining companies reinstituted periodic shutdowns for days or weeks at a time that laid off hundreds of miners, 3,000 at the worst point in early 1928. Uncertain employment made it harder for miners to protest by switching jobs, as they had in the past, especially so for men with C or D cards. Miners learned to alter the first examination grade cards by switching out photographs with those on cards with higher grades or erasing and replacing the original information entirely. After several unsuccessful attempts to prevent fraud, the clinic began imprinting the cards with a Bureau of Mines seal that embossed the letter grade and the photograph. Meriwether was confident that the federal agency's imprimatur made the grades indelible.[66]

Meriwether gradually increased the clinic's influence. In early 1929, he requested and received funds to treat sexually transmitted diseases, which exams had shown to be rampant. In its first six months, Meriwether's "VD section" treated 363 cases of syphilis and 188 cases of gonorrhea. By his count, however, almost 300 additional syphilitics continued to work without treatment. Influenced by Parran's interest in the social and economic effects of syphilis, Meriwether argued that these infections represented a serious compensation risk that rivaled the costs of tuberculosis. "There is no excuse for the men with such diseases continuing to claim the right for employment," he reported to the OPA. Meriwether's vigilance against sexual infections gave the clinic new invasive powers, literally. He instructed clinic physicians to perform prostatic massages to test the seminal fluid of any male patient who seemed likely to have gonorrhea. The OPA enthusiastically backed this broadened mission. Its welfare nurse reported uncooperative women with syphilis or gonorrhea to the local police, who forcibly treated them in the county jail.[67]

The OPA, Bureau of Mines, and Metropolitan Life renewed the clinic agreement for another two years in 1929. Meriwether's second annual report showed that the number of men with first-stage silicosis had not changed much but that the number of those with advanced silicosis or tuberculosis had fallen. He noted that "companies are enforcing physical examinations and are carefully studying cards before employing men." "It has taken some time to bring this about," Meriwether explained, "without causing serious labor disturbances in the field." It had paid in lower insurance costs, down to 3.6 percent of payroll. Still, the report reiterated that all employers should resolve to eliminate "tuberculosis and second and third degree silicosis from the mines of the district" by "hiring [only] men free from such diseases," a standard that would exclude those with a D grade or lower. Meriwether wanted them to go even further. Anyone who presented "increased compensation risks," he advised, "should be encouraged and persuaded or coerced to change their occupations." Sayers decided not to publish this report. In private correspondence, however, Meriwether reiterated his call for the OPA and the clinic to commit to "the very closest cooperation in elimination of undesirable and unfit men" from the Tri-State.[68]

Beginning in 1915, miners had rushed into the new Picher field with renewed faith that brutal, risky work for individual incentives would yield them prosperity and respect. For more than a decade, their muscle again delivered record production and profits for mining companies. Their achievements in the ground shaped a rough working-class community around the prerogatives and appetites of forceful men. While many women seized opportunities in this male-dominated world, all were subject to its dangers. They were confident that white nationalist privileges would not only ensure good fortune but also offer protection. By the mid-1920s, despite periodic slumps, Tri-State miners and their families expected to enjoy the same standard of living that other white Americans seemed to enjoy. For a time, union organizers from Mine Mill and the OFL tried to convince them that collective action offered the only means to achieve better, more secure lives, often conceding to their racist and nativist views in doing so. Although some might see the value of unionism, most miners remained too eager to chase high wages in the hopes that the next boom would be different and would last.

By 1929, however, their claims on prosperity offered no defense against a new scientific business regime that sought efficiency, stability, and above all, control. In fact, miners' faith in the fortunes of strong men who ran hard risks in pursuit of commensurate rewards defined the terms by which mining companies would control them. The firms that developed the Picher field, led by Eagle-Picher, Commerce Mining, and others, were bigger and richer than any companies in the history of the district and benefited from the direct assistance of various federal agencies. With government and insurance industry help, companies began to see miners who were once valued for their risk-taking productivity as themselves primary risks to future profits. By 1929, the district's white working-class male ideal led into the civilizing cage of the examination clinic and its grade cards. Most miners could still get work at wages determined by the sliding scale but in the knowledge that any injury or illness could mark them for oblivion as unfit and unemployable. Men who had power and prestige in 1917 now had neither just as work became harder for everyone to find as metal prices and mine production plummeted to the lowest levels in living memory.

CHAPTER 7
BACK TO WORK

For miners whose livelihoods depended on racial and gendered claims to metal market prosperity, the Great Depression posed an existential threat. American manufacturing and construction ground to a halt as capitalism descended into crisis. Zinc ore prices fell from forty dollars per ton in August 1929 to seventeen dollars per ton in 1932, the lowest price in real terms since the early 1880s. Tri-State mine production fell from 598,000 tons of zinc ore in 1929 to 169,705 tons in 1932, the district's lowest tonnage since 1896. Many small companies folded. Large firms closed for months at a time. At the outset, more than 7,000 men had jobs in the mining district, mostly at full-time. By the summer of 1932, only 1,500 men had work, seldom more than half-time. While jobs and shifts disappeared, mining companies also slashed wages: by 1932, miners earned $2.00 to $2.50 per day, if they were lucky to have work, about half of what they had made in 1925. As the district failed, some miners left the area. Those who stayed suffered.[1]

Men once celebrated for their aggressive, indispensable power felt emasculated in idleness. They lost status and a sense of say-so, as companies used clinic grade cards and ground bosses to weed out workers now considered undesirable or unsafe. Meanwhile, women struggled more than ever to keep family economies functioning, now subject to the approval of charity workers funded and directed by district elites. As Tri-State mining families tried to make sense of what had happened to their once defiant communities, they blamed the new fetters applied by the Tri-State Zinc and Lead Ore Producers Association (OPA) in the 1920s—assertive ground bosses, the clinic, and the grade cards—not capitalism.

In 1933 President Franklin Roosevelt's New Deal offered a variety of measures designed to help working-class white men like those in the Tri-State. In deep economic and social crisis, miners rallied to the provisions of the National Recovery Administration (NRA) to boost employment, raise wages, and give employees more power in relation to their employers, in the last case by encouraging workers to join unions. Hundreds of miners responded with renewed interest in the International Union of Mine, Mill and Smelter Workers (hereafter referred to as Mine Mill) and its American Federation of Labor (AFL) allies when organizers returned to the district. Unlike many others in the nation's growing union movement, indeed many in Mine Mill, they looked backward, to a restoration of what had been lost—the jobs, pay, and swagger befitting white American men—not a radical new departure. Dependent on federal support, the union used popular anger at low wages and the clinic to mobilize the district's still nonunion majority. As the recalcitrant companies of the OPA stalled full implementation of the NRA, however, Mine Mill's leaders made a desperate scramble to deliver something that would keep the Tri-State movement alive. That gambit ended in 1935 with a bold but minority-led strike that shut down all mining operations as the NRA hung in the balance of the Supreme Court.

The union staked its legitimacy in the Tri-State on New Deal promises for labor that soon proved unreliable. In the months after the strike began, federal courts invalidated the NRA and then paralyzed the subsequent National Labor Relations Act (NLRA). As the strike failed, nonunion miners turned against Mine Mill. They began to blame the union for their suffering and see it as the enemy. Thousands of men pushed to go back to work and to break the strike they now believed had betrayed them. Angry and afraid of defeat, they listened eagerly to counterstrikers who offered strength, scope for action, and a sense of superiority. The operators, working together, harnessed these men in a new company-controlled movement, the Blue Card Union.

In exchange for loyalty, the Blue Card Union offered strikebreakers the world they had lost. It assured them of work on the old terms of the sliding scale and white nativist privilege. The Blue Card Union freed its members from the clinic's grade cards. It encouraged public violence against Mine Mill that recalled the exploits of Joplin's strikebreakers and the vigilantism of the 1910s. The war against Mine Mill had finally come home. Attacking weak unions, weak men, and the bureaucrats and lawyers who tried to protect them, the Blue Card Union won the allegiance of a majority of Tri-State miners with traditional assurances—of the primacy of strong men, racial nationalism, and market incentives—that took on new power in an age of authoritarian assertions.[2]

Far from an isolated story, Mine Mill's strike and its defeat showed how white working-class conservatism snaked through the heart of the American labor movement in the 1930s. Still an AFL affiliate when the strike began, Mine Mill sided with those unions dedicated to democratic industrial organizing strategies that formed the breakaway Committee for Industrial Organization in late 1935 and the Congress of Industrial Organizations in 1938 (CIO). Tri-State miners again found themselves in the middle of internecine war between the AFL and the more progressive leaders of Mine Mill and now its CIO allies. Once again, the AFL leadership hoped to use the miners against radical unionism by chartering the Blue Card Union in 1937 to replace and ultimately destroy Mine Mill. The approval of hundreds of thousands of AFL affiliate members nationwide for this merger revealed the limits of truly democratic working-class ideals.

The New Deal labor regime wavered but it did not fall. CIO advocates of democratic unionism held on in the Tri-State, ultimately prevailing against the Blue Card Union in cases before the National Labor Relations Board (NLRB). Clinging to federal law, Mine Mill holdouts continued to champion policies for real collective bargaining, grievance procedures, and safer workplaces, particularly when it came to the silicosis scourge. Legal victories, however, did not mean that the working-class majority would ever share their vision. The New Deal could side in Mine Mill's favor but it could not overrule the reactionary commitments in the hearts of working-class white men, no matter the cost of those commitments. That enormous task remained as the nation launched another war mobilization in 1941 that would rely on Tri-State zinc and lead.

As the worst economic crisis in American history decimated Tri-State mining communities, President Roosevelt's New Deal promised a restoration of lost prosperity. Miners and their families hoped for a return to work, higher wages, and the respect they had once commanded. In the depths of the Great Depression, they had learned that this return would be neither simple nor easy. Mining companies had gathered more power than ever, in size and scope, in the OPA and in the health clinic partnership with the federal government and the insurance industry. Nothing made the dominant authority of the mining companies clearer than their control of charitable poor relief in the worst months of the crisis. Many miners tried to reassert themselves in wildcat fashion against that authority but soon concluded they would need help.

The OPA led initial relief efforts for the unemployed in conjunction with local government and charities. In January 1930, city and company officials in Picher organized a community chest to collect goods for those in need, but the mayor had to curtail the program a month later when overwhelming demand exhausted its resources. The OPA's welfare department took up the cause by buying and distributing food, clothing, and fuel. As the crisis worsened, however, companies in the OPA struggled to fund relief. Executives for the Eagle-Picher Lead Company considered the market collapse the "most severe" in its history, worse even than in the 1890s, and instituted "rigid economies" to cut every possible cost. The OPA curtailed welfare spending. By the end of 1930, the OPA's welfare department sought and distributed donations, mostly of fuel, but no longer provided food or clothing. As with charitable relief efforts elsewhere, neither the OPA nor the local government could meet the district's escalating needs as unemployment neared 50 percent.[3]

Tri-State communities struggled to understand why work had become so scarce. In early 1930, miners trained their rising anger on ground bosses, a key instrument mining companies used to impose control. Unemployed Picher miners accused their immediate superiors of running a widespread "buddy car" racket that locked them out of jobs. Since the mid-1920s, miners and mill workers who lived in Missouri or Kansas had commuted to work in private buses. Now, the unemployed claimed, many ground bosses were running their own buses and would only employ men who paid them to get to work. Picher miners believed they lost jobs to these outsiders. Kansas miners made similar allegations against ground bosses who favored men from Missouri and Oklahoma. Some threatened violence to stop the scheme. There was truth in the complaint. Ground bosses had accrued considerable power over hiring and firing in the field, particularly in the mines of the big companies. Some no doubt used that power to favor workers who rode in their buses. While the buddy car scheme did not explain the economic crisis, or cause the surge in unemployment in early 1930, the popular anger was real.[4]

Seeking containment, local officials held a series of mass meetings so the unemployed could air their grievances. Picher police chief Joe Nolan, the son of a Webb City shoveler, organized this show of respect between white men in early April in conjunction with the mayor and the OPA. Hundreds of miners attended "to protest against alleged discrimination against Oklahoma workmen." "One after another miners who had stories to tell regarding the 'buddy car' evil" spoke out, the Miami paper reported. The state's district mine inspector, Riley Clark, validated their claims. At one meeting, Clark declared that the buddy car racket had "thrown hundreds of miners with families out of work." Nolan and Mayor J. H. Klinefelter pledged to help. They called on county and state officials to enforce a law that required permits for such services. They also called on mining company leaders to rein in ground bosses "in behalf of the miners and mill workers."[5]

The companies seemed sympathetic. Representatives of the OPA attended the meetings. They publicly accepted that some ground bosses were extorting employees but denied any knowledge of it. They pledged to resolve "the grievances of local miners" in conjunction with a committee set up to "rectify the purported tyrannical attitude of the under-bosses in hiring and firing miners because of their place of residence." Company officials certainly knew that the buddy car issue was not the cause of rising unemployment yet seemed happy to let ground bosses take the blame.[6]

At these meetings, however, miners also voiced anger at other perceived culprits, particularly the clinic doctors and their grade cards. Chief physician F. V. Meriwether informed Royd Sayers, his U.S. Bureau of Mines superior, in May 1930 that "it was proposed by some of the miners that they dynamite the clinic and run the employees out." No one objected, he added, even though the mayor and chief of police were present. By then, most miners in the district had been examined. Their suspicions about the grades were not unfounded. Meriwether noted in April 1930 that men with C cards, particularly first-stage silicotics, found getting hired "exceedingly difficult." Protests against the buddy cars and the clinic revealed resentment against the new balance of power in the district. Unemployed miners understood, if only partially, that these arms of company authority had something to do with their misfortune, not just as individuals but as a group. But Nolan's meetings worked—in the short term. The buddy car scandal subsided. The OPA stationed an armed guard at the clinic for two weeks, which seemed to quiet the threats.[7]

As ore prices continued to fall, companies relied more on the clinic's grading system to cut costs, despite the risk of renewed protest. In late 1930, labor reformers in Kansas and Oklahoma pushed to expand workmen's compensation laws to cover industrial diseases, including silicosis, which would expose district companies to significant claims. The OPA, backed by Meriwether, lobbied hard to defeat the proposal in Oklahoma, although company officials worried that proponents would try again. To mitigate that threat, Eagle-Picher began requiring all workers to show a new rustling card that proved their eligibility for work. These cards listed the worker's clinic grade. Men with C grades or better got yellow cards, which meant they could work wherever the ground boss wanted to hire them. Those with D grades or worse got red cards that limited them to certain jobs or barred their hire. Men who may have avoided examination could do so no longer. Other firms also adopted rustling cards. Meriwether and A. J. Lanza, who represented the Metropolitan Life Insurance Company, were pleased. They had achieved, Lanza boasted in early 1932, universal recognition of the principle of "Compulsory Physical Examination Before Employment."[8]

The rustling cards exacerbated unemployed workers' anger toward the clinic. Meriwether reported to Sayers that miners were circulating a petition to call on the governor to investigate the clinic and its grading system. A month later, in April 1931, Communist Party organizers from the coalfields arrived in Picher to encourage the unemployed to protest on May 1. Their discussions with local workers focused on the clinic. Already nervous about potential attacks on him and other staff, Meriwether asked the OPA for another armed guard. Popular unrest, he reported in April, had "reached an alarming state." Although local police hounded the Communists out of town before May Day, miners continued to speak against the clinic. In June, OPA executive secretary M. D. Harbaugh informed Sayers that he had "heard reverberations of the rather sullen and violent attitude of a great many of the workmen in the district toward the Clinic." They believed "that it is something set up to bar them from a job," he explained. Both Harbaugh and Meriwether made public appeals in the months that followed to reassure local workers that the examinations were meant to benefit them as well as the companies. The reassurances fell flat. "The suffering of the poorer class in the field is certainly intense," Meriwether informed Sayers in early 1932, "and is causing considerable worry at the present moment."[9]

Despite this, firms continued to use the clinic to purge payrolls of sick or injured men. According to Harbaugh, the cards revealed "an alarming number of men physically unfit for this class of work." "The tendency thereafter," he explained, "was to improve the physical standard of the employees" through "rigid preemployment and periodic physical examination." This got easier in July 1932 when the OPA took full control of the clinic after spending cuts forced both the Bureau of Mines and the Metropolitan Life Insurance Company to withdraw support. The OPA incorporated the renamed Tri-State Industrial Examining Bureau as a separate legal entity to limit liability. Now led by a local, OPA-picked doctor, the bureau eliminated all services except grading exams. "The purpose of this Bureau should be to furnish complete pre-employment physical examination of men for work in the mines, and periodic examination of employees in the mines," an internal OPA report declared. It recommended but did not offer treatment; sick or injured men had to arrange their own care. Now the OPA itself oversaw the grades that company rustling cards required and pledged to make "full use of clinical data," including records about sexually transmitted diseases. The bureau further extended its reach by requiring "members of the immediate families of" miners to be checked and treated for these infections when a company ordered it as a condition of employment. In addition, the OPA agreed to share individual medical records with companies facing workmen's compensation cases. According to Harbaugh, they sought "the elimination of low grade (physically) men" from employment. This policy, he admitted, "resulted in the growth in the mining field of a considerable number of permanently unemployable men." "It was the company's plan to get rid of employees before they became too great a liability," explained Mr. U, a miner who was graded out in 1932. While the bureau did not cause the unemployment crisis, it determined that only men it deemed healthy were employable and that those marked weak or infirm were not.[10]

Once bristling with pride, mining communities in the Picher field were by 1932 pummeled to despair as company policies conflated the twin crises of unemployment and occupational injury and disease. More than 6,000 miners had lost their jobs since 1929. As few as 500 worked more than half-time. Since 1927, clinic doctors had diagnosed 6,108 miners with some form of silicosis and 642 miners with syphilis. Many of these men had left the district or were dead by the time the depression bottomed out. A 1933 study of the causes of death in the Tri-State counties revealed that 788 men had died from tuberculosis or silicosis since the clinic opened, men like Edward Legg, a fifty-one-year-old, Virginia-born hoisterman who had worked in the district since at least 1900. According to his son, Legg received an F card in 1932. He sought treatment at a state hospital, but his family "had a terrible time having food or a place to live." Legg died in 1933. Many injured men remained in the field, at home and jobless. Most could not afford medical care. Women in mining families bore the burden of providing whatever income they could, often while also caring for sick or dying husbands and sons. "It's a pitiful way of living," Mrs. I. said.[11]

Mining families had nowhere to turn as companies further consolidated power. In addition to the examination bureau, the OPA controlled the provision of unemployment relief in Ottawa County through its own welfare department and the Red Cross chapter, which Harbaugh chaired. Of all the OPA companies, Eagle-Picher had the most sway. Its leaders took advantage of the crisis to buy rights to more land from shuttered or struggling competitors. In 1932, Eagle-Picher opened a central mill at Cardin that could process 3,600 tons per day—the largest zinc-concentrating mill in the world at the time—to create a single processing site for all of its mines and those of smaller firms.[12]

The presidential campaign of 1932 offered solutions that appealed to the district's past, as Ottawa County Democrats and Republicans alike campaigned in favor of maintaining high protective tariffs. Calling for voters to reelect Herbert Hoover, who won the county in 1928 with 64 percent of the vote, Republicans insisted "that the great protector of the mining payroll is one thing—It Is the Tariff Wall." Democrats, meanwhile, assured voters that Roosevelt's call for selective tariff protection would apply to zinc. The problem was not the tariff, they reasoned, but Hoover. Promising a return to better days, even though their presidential candidate had not won the

Miner's home with chat piles in the background, Ottawa County, Oklahoma, 1936.
Library of Congress, Prints & Photographs Division, FSA/OWI Collection,
LC-USF34-004136-E [P&P] LOT 530.

county since 1916, Democrats urged votes for Roosevelt: "Let's Get Back What We Had!" Roosevelt routed Hoover in Ottawa County with 71.8 percent of the vote; in Webb City and Carterville, he claimed 61.5 percent of ballots.[13]

In its first 100 days, Roosevelt's New Deal pumped life into the Tri-State mining economy. By the time of the March inauguration, zinc ore prices had fallen to sixteen dollars per ton, with the banking system teetering on the brink of collapse. From March to June, however, Roosevelt's decisive intervention in banking, finance, agriculture, and unemployment relief boosted public confidence and, crucially for miners, spurred industrial demand for metal. By late July, zinc ore prices had climbed to thirty-five dollars per ton. Companies increased payrolls and pay. That summer 3,100 miners had jobs. They also enjoyed rising wages, up to $3.25 per day for machine men and 7.5

to 9.0 cents per can for shovelers. The administration also created jobs for the unemployed in Ottawa County through the Federal Emergency Relief Administration (FERA) and the Civil Works Administration. With the New Deal, a local mining reporter predicted, it will soon be "like old times."[14]

Of all New Deal measures in 1933, the National Industrial Recovery Act, passed in June, sparked the most hope in the Tri-State. With an ambition and scope that invoked the Great War mobilization, the law created the NRA to stop the deflationary spiral, with measures aimed at raising prices, increasing employment, and boosting wages. Modeled on the business-friendly War Industries Board, the NRA asked companies to voluntarily write and obey codes of fair competition within their industries. These codes exempted participants from antitrust law so that they could set minimum prices to eliminate ruinous undercutting and promote reinflation. In exchange for collusive pricing power, the NRA required that the codes prohibit child labor, set minimum wages, and fix a maximum number of working hours in a week in order to give more people jobs and protect them from starvation pay. To encourage support, the NRA created a Blue Eagle symbol for participants to display as a sign of their patriotism and urged Americans to shop only where they saw it. The local press was confident that the "operators and miners of this great Tri-State district, may be depended upon, now as always, to do their part in this local and national recovery program as outlined by President Roosevelt," which it described as "a mass attack by voluntary agreement to lift wages and shorten hours."[15]

The OPA, itself a product of government assistance during the Great War, looked to shape the NRA to serve company interests. In early August, a majority of companies in the district, including Eagle-Picher, Commerce Mining and Royalty Company, and Federal Mining and Smelting Company, agreed to a temporary code covering employment, known as the President's Re-Employment Agreement (PRA), as negotiations over the terms of a final zinc code continued. Under the PRA, companies negotiated special terms that set the workweek at forty-two hours and the minimum wage at thirty cents an hour for common labor, which they preferred over the thirty-five-hour week and forty-cent minimum wage in the recommended temporary code. The Tri-State PRA retained the sliding scale for wages above the minimum.[16]

The NRA also included important provisions for union organizing that brought Mine Mill and the AFL back to the Tri-State. To counterbalance company code-making power, section 7(a) of the law stipulated that workers in all industries under codes or reemployment agreements "shall have the right to organize and bargain collectively through representatives of their own choosing," free of employer coercion or restraint, and that workers could not be forced to join company unions. The NRA established, for the first time, a federal right to organize, with logic that suggested that labor unions would have a formal role in the recovery effort. Union leaders across the country launched a wave of organizing drives in the summer of 1933 to take advantage of the law. Mine Mill had only 1,500 members in six locals, only three of which were really active. New president Thomas Brown wasted no time, however, sending organizers to rebuild the union. He had high hopes for workers in the Tri-State, who in the mid-1920s had been among the last to show interest in Mine Mill. Brown dispatched Roy Brady to Picher in July. Brady organized Picher Local 15 on August 16. With the help of new members, he soon organized Local 17 in Galena. OPA leaders knew about Brady's organizing drive but were confident in their power to contain it.[17]

In October, a Blue Eagle Day celebration in Miami demonstrated broad public support for the NRA's intervention, as did similar parades across the country that autumn. In a show of enthusiasm not seen since Picher's anniversary party, more than 10,000 spectators from all corners of Ottawa County watched 1,500 people march along streets festooned with American flags and Blue Eagle banners in "vivid tribute to NRA." The celebration was organized by the mayor of Miami and a steering committee of local and county officials. Conservative Democratic congressman Wesley Disney attended. Alongside school bands and civic clubs, twelve mining companies joined the parade, with miners, mill workers, and supervisors walking under company banners. Eagle-Picher's contingent had an "attractive float." Everyone who worked at Commerce Mining's Wilbur mine marched, including a sick miner who had to be carried, and won first prize for participation. After the parade, people enjoyed a series of contests, including a sack race, a fat man race, and a shoveling contest. Mark Trask won fifteen dollars for shoveling one ton of chat into a can in seventy-four seconds, two seconds faster than the runner-up. According to the *Miami News-Record*, "the thousands who attended the affair fully realized the spirit of the day—that it was all in observance of the NRA, improvised by President Roosevelt to aid the great masses of American people." The paper did not list Mine Mill among the day's participants or comment on its organizing drive, but its coverage left little doubt that the NRA enjoyed broad acceptance among workers, mining companies, and politicians, seemingly without reservation.[18]

Brady took advantage of this popular, patriotic enthusiasm for the NRA. According to W. P. McGinnis and M. R. Corwin, who both joined Local 15 in 1933, Brady said "that the President of the U.S. wanted all working men to join a union." Brady told others that the NRA required them to join Mine

Mill in order to get a job under the zinc code. Neither statement was true and would not be unless Mine Mill negotiated a closed-shop union contract, something that had never happened before in the Tri-State. Both of these claims seemed convincing, however, amid the hope and confusion of the NRA's improvisational implementation. Roosevelt called his reemployment agreement a plan to raise wages and "create employment." The *Miami News-Record* described it as "a covenant" "to reemploy the idle jobless." The Blue Eagle Day celebration reinforced these messages with calls for nationalistic cooperation between workers and companies. Miners hired after companies signed up for the reemployment agreement were called "NRA swing men" or "NRA extra men" in district parlance. Brady encouraged the perception of official ties between the NRA and Mine Mill. He told a meeting in Galena that the "union was authorized by the NRA headquarters to explain the effect of the codes on the miners." The NRA seemed to recognize Mine Mill as the official representative of Tri-State workers by seating Brady on "the NRA labor advisory board" in negotiations over the zinc code in Washington, D.C., in December 1933. Brady limited his statement solely "to conditions that exist in the mines of the Tri-State district." With hopes high for the NRA to boost the recovery, and companies hiring more and more miners as ore prices and production rebounded that winter, many men joined Mine Mill for fear of being left out of whatever deal the union was on its way to achieving. Few did so out of allegiance to Mine Mill.[19]

Brady built support for the union by channeling local grievances. He aimed first at the OPA's examination bureau. Brady assured miners that Mine Mill would stop the bureau's "increasingly rigid" exam policy by closing it. These attacks on the bureau elicited strong support among unemployed miners with health problems. By 1934 the OPA justified the bureau only in terms of cost reduction, no longer claiming that miners stood to benefit. It blamed liberal workmen's compensation judgments and an alleged increase in "the 'malingerer' or false claimant for compensation." The "employer's only protection from him is to be extremely and, at times, almost unreasonably careful in the selection of his employees," often including "the necessary rejection for employment of a considerable number of able-bodied workmen" with minor untreated ailments. Miners were especially angry because the bureau blocked their access to jobs even as the industry continued to recover in 1934, when 4,300 men in the field had work. Thousands of miners with C cards or worse were left unemployed, they feared permanently. The bureau's strict avoidance of expensive compensation claims "is working hardships upon hundreds of bona fide workmen," Local 15 explained. Men who lost their jobs in the 1930s already felt like inadequate men; to be unemployable because of physical damage threatened to shred whatever sense of manhood they had left. "This was hard to take by those miners who were self-respecting and not malingerers," one observer recalled. According to Harbaugh, a hostile chronicler, "these men became fertile ground for labor agitators and organizers who appeared upon the scene in 1933, when the new deal for the down-trodden man loomed in the offing."[20]

Union organizers also called for long-term health and safety reforms. Mine Mill organizers demanded stricter mine inspection and more robust ventilation requirements. Local 15 sought lobbying help from the Oklahoma State Federation of Labor (OFL). At its annual convention in September 1934, the OFL passed a resolution submitted by Local 15 calling for coverage of silicosis and lead poisoning under the Oklahoma workmen's compensation law. If in the short run Mine Mill argued that sick and injured men should be allowed to work, it argued in the long run that miners needed better protections to reduce the number who got sick or hurt on the job.[21]

Mine Mill's immediate goal was to raise the pay of Tri-State miners through the NRA. The OPA's reemployment agreement had delivered less than promised. When ore prices fell back below thirty dollars per ton in 1934, some companies reduced wages below the agreed minimum wage but held fast to the maximum hours limit. Meanwhile, those hired to round out shifts, the "NRA extra men," still worked half-time or less. Brady pressed these grievances at every opportunity, particularly at hearings on the zinc code. He called for higher minimum wages that reflected the dangers of the work. "The miner, during his employment, considering the risk involved and the labor involved, must have a fair minimum wage in order that he may provide for himself and family," Brady told NRA officials. Despite proposing safer working conditions in the future, Mine Mill mainly presented itself to Tri-State miners as the champion of strong, risk-taking men against domineering operators. By late 1934, the union had built stable locals in Picher and Galena, each with "a substantial membership" of several hundred members, perhaps more than 1,000 men combined. It now sought official recognition from the operators as the representative of all Tri-State miners.[22]

The OPA had no intention of bargaining with Mine Mill. It had molded the final zinc code to benefit Tri-State operators: winning a special lower wage scale for the district and blocking an industry-wide division to resolve wage disputes. The OPA could not, however, prevent the code from reiterating the right to organize and bargain collectively. It registered strong opposition nonetheless. "Mining operations of this district have never been unionized, employees are all white native Americans, and there has never been any suggestion of labor difficulties between employers and employees," Harbaugh informed American Zinc Institute negotiators. Wary that the New Deal was giving "undue power" to organized labor, the OPA pledged to resist anything that threatened to "disturb the existing satisfactory employer-employee relations."[23]

With Mine Mill gaining momentum and the OPA steadfast, something had to give. The union seemed to give way first. After attending the Mine Mill convention in August 1934, Brady absconded with the treasury of Local 15, a betrayal that echoed decades of local denunciations about union crooks. The theft forced the union to stop providing food relief to unemployed members. If never enough, union relief had provided vital sustenance and dignity for people with little recourse aside from OPA charity and New Deal relief that proved weaker than advertised. State officials hostile to New Deal spending had cut required matching funds for federal work programs. In October 1934, FERA programs in Ottawa County provided 3,200 men and women each a meager twelve hours of work a week. County leaders slashed that number to 2,300 at the end of the year and to zero in January 1935. Only a few hundred people worked at the county's state-level FERA projects: a mattress factory in Picher and a canning facility in Miami. The canning facility made beef broth available to families in need. To collect it, however, people in Picher had to bring an empty bucket to city hall. After Brady's treachery, Mine Mill needed bold action to save Locals 15 and 17 from dissolution.[24]

In January 1935, union president Thomas Brown came to Picher to lead the reorganization himself. Mine Mill had gained national strength in the year after the creation of the NRA, now boasting more than 15,000 dues-paying members in ninety-four locals, mostly in old western strongholds but also in new areas such as Alabama's iron mines and mills. The union was adamant about realizing the New Deal's plan to improve working conditions, especially the right to organize, and fought hard against employer opposition with strikes in Montana and Alabama. "President Roosevelt had told us to organize and had advocated shorter hours, increased wages and higher standards of living," one miner declared at the 1934 convention, so "it was up to us to stand behind him and help to keep organizing going." With the help of dedicated local members, Brown restored confidence in Picher and Galena and organized five new locals: Webb City 106, Baxter Springs 107, Joplin 108, Miami 110, and Treece 111. These locals became the basis for Mine Mill's new District Four.[25]

Local leaders emerged to replace Brady. M. E. Cartwright, who had led the district's union council in 1924, became president of the Miami local. Ed Cassell, a miner known for "espousing all local liberal causes," and Ted Schasteen, a thirty-year-old mill mechanic born in Jasper County, led the Treece local. J. A. Long, who had mined in Picher since 1915, took over as president of Local 15. Tony McTeer, who had worked in Picher since 1919, also became a talented leader in Local 15.[26]

They were anxious to make good on the union's promises. In March 1935, District Four filed a formal complaint with the NRA labor compliance board against Tri-State companies for allegedly paying less than stipulated under the PRA. The complaint was rigorous in detail, with individual statements from more than 1,000 miners. With the zinc code under final review, Mine Mill sought leverage to force the operators to bargain. If the compliance board found in favor of the union, it could refer the complaint to the newly created NLRB, which had power, albeit untested, to compel resolution. With the complaint pending, District Four asked the companies in a joint letter to negotiate a union contract. The companies ignored the letter. Union leaders next approached the OPA to negotiate, but Harbaugh referred them back to the individual companies. After the approval of the NRA zinc code on March 26, Mine Mill appealed to the U.S. Department of Labor Conciliation Service for help but to no avail—the agent who investigated saw no basis for negotiation because company officials refused to meet. Meanwhile, the OPA provided enough statistical documentation to satisfy NRA investigators, who dismissed Mine Mill's complaint in April.[27]

Union leaders recommended a strike as the only way to realize the New Deal's promise. "The operators have ignored all attempts of the union to negotiate for collective bargaining, and establishing industrial relations between employer and employee as provided under NRA," Brown explained. Having "exhausted all reasonable means," another leader asserted, the union considered a strike its last resort. Urging support from John L. Lewis and the United Mine Workers of America (UMW), national Mine Mill secretary James Robinson claimed it would be "a crime to prevent these men from making a last desperate effort to improve their conditions." Each local held a vote at the end of April, with only employed members eligible to cast ballots. Of the 700 men polled, nearly 600 voted to strike on May 8, 1935. Although a majority in the union, the strike supporters were a small minority among the district's more than 4,800 miners and mill workers. Union leaders held a series of meetings to explain the reasons for the strike and finalize the picketing strategy. On May 7, more than 2,000 people met in Picher where J. A. Long promised a "peaceful strike" and urged the participation of all non-union workers. "We realize," he said, "that no organization is large enough to buck public sentiment." Their main goal was employer recognition of Mine Mill as the bargaining agent for district mine workers, in accordance with rights granted by the NRA and the zinc code. With that, Long pledged, the union would push for "better working conditions, a shorter work week and adherence to American standards of a living wage."[28]

Mine Mill's strike paralyzed the district. On May 8, hundreds of workers did not report for work at mines, mills, and smelters from Picher to Joplin. Union pickets targeted production bottlenecks, a key vulnerability of the district's sprawling firms. They blockaded the two biggest mills, Eagle-Picher's Central and Commerce Mining's Bird Dog, which processed most of the ore mined in Ottawa County. Pickets also closed the Eagle-Picher smelter in Galena. Roving pickets enforced the strike at smaller mines and mills. Aside from a few fights, there was no violence. Local Mine Mill leaders co-ordinated the strike from union halls in Picher and Galena; Brown set up an office in Joplin. They announced aims consistent with the long-standing claims and priorities of the district's mining communities. Strikers wanted to achieve the NRA's promises for recovery, District Four explained. They "are 100% American, white, and are very open-minded and willing to co-operate, and this the operating companies have taken advantage of." While awaiting funds and supplies, the union urged all strikers to apply for fed-eral and state unemployment relief to sustain them until it was over. Nearly 5,000 men who had worked the day before were now idle.[29]

The operators responded by announcing a district-wide shutdown. With ore prices low and large stocks on hand, Harbaugh explained, firms hoped the closure would raise prices. Their main goal, of course, was to challenge Mine Mill's support among nonunion workers. Company officials hoped to break the strike by starving people into submission. Federal and state relief coordinators in Oklahoma and Missouri supported that strategy, at least in-directly, by declaring that strikers would not receive government assistance. John Campbell, Eagle-Picher's personnel manager, expected that after a week or two this approach would bring "most of them to destitution" and turn them against Mine Mill.[30]

The strategy worked. Within a week, some miners began organizing among themselves to end the strike. On May 17, T. L. Armer, a forty-seven-year-old blacksmith and welder for Commerce Mining who had recently moved back to the district from the Kansas oil fields, convened an im-promptu meeting near Cardin to talk about returning to work. Several hun-dred men attended, both union and nonunion among them. Although Mine Mill members spoke in favor of the strike and their goals of union recog-nition and contracts, the majority "was overwhelmingly in favor of going back to work immediately, or as soon as the mine operators would reopen their mines." That night, 200 men signed a petition asking companies to re-open the mines under the same conditions that prevailed before the strike. Armer scheduled another meeting for the following day. Nearly 2,000 men attended, about 200 of them Mine Mill members. Jimmie Hall, of Local 15, repeated the union's goals but was rebuked with new ferocity when he answered a question about how people were expected to live during the strike by repeating the union's advice to apply for unemployment relief. Several men shouted that they did "not want relief, but wanted to work." More than 1,000 men signed Armer's back-to-work petition.[31]

The OPA moved quickly to claim the back-to-work movement. At a third meeting, held on May 19 at the Miami fairground, several company leaders, ground bosses, and local officials, including former Picher sheriff Joe Nolan, flanked Armer as he announced plans to create a new organization dedicated to ending the strike. "It is going to be permanent and it will be composed of such men as are at this meeting," he declared, "who will protect our interests from here out," unlike those in Mine Mill, now deemed outsiders, who launched the strike in the first place. Nearly 600 more men signed the petition after that meeting, bringing the total to around 1,800. This time, Armer did not permit strike advocates to speak. The back-to-work movement now belonged to the OPA.[32]

Mine Mill rallied to bolster the strike. On May 20, 600 union men stymied a back-to-work meeting in Baxter Springs. They booed Armer and Nolan from the stage, but were themselves prevented from speaking by nonunion men. The following day, over 1,000 Mine Mill supporters, including several dozen women, marched in a series of parades bearing signs that read "Better Wages," "Better Cars," and "Goodbye Clinic" in Picher, Cardin, Commerce, and Miami, where a crowd of 3,000 rallied in the afternoon to hear strike leaders and allies. Brown, Cartwright, and McTeer spoke, alongside Ira Finley, the former OFL president. Brown reiterated the union's goal of recognition and collective bargaining under the terms of the NRA. He claimed that the strike was necessary to make companies adhere to the wage and hour provisions of the zinc code. Finley, who now led the Veterans of Industry of America, an Oklahoma-based organization of as many as 50,000 unemployed workers, delivered a more radical line. He declared that the government should confiscate the property of any company that violated the terms of the NRA and urged the strikers to hold out for contract terms that would provide "a home, all electrical appliances, sufficient money for the education of your children, and for travel"—the American standard of life that he had advocated since the 1920s. The union held another rally that night in Picher, where 2,000 men and women heard the same speakers. Strike "until you whip hell out of this bunch," Finley told them. Despite a poor strike strategy and mounting opposition, diehard Mine Mill supporters remained.[33]

They could not stop the growing back-to-work movement, however. On May 25, Armer and other leading strike opponents, including several mine managers, formally organized. In the context of the NRA, they decided that a rival union, nominally independent but under company control, was the surest means to end the strike. A new leader, Frederick W. "Mike" Evans, took charge. Born in Iowa in 1885, Evans came to Ottawa County in the 1910s, and soon became locally famous for gambling, womanizing, and picking rich mining leases in Ottawa County. Known as the "greatest risk taker," he had developed a bonanza mine and a large bootlegging operation in the 1920s and now ran a hotel, a pool room, and a dance hall in Picher. Evans also owned a mining company that leased the Big Chief mine from Eagle-Picher and belonged to the OPA. People who knew Evans described him as charming, authoritative, and persuasive, with "an uncountable number of friends from all walks"—the very embodiment of poor man's camp aspirations in an era of charismatic demagogues. Evans designed the new organization with local lawyer Kelsey Norman, Nolan, and Glenn Hickman, a twenty-five-year-old ground boss who had briefly joined Mine Mill in 1933. It would promote the "general welfare" of the district's miners, millers, and smelter workers by negotiating their return to work with the operators, with whom the new group would "establish mutual confidence and create and maintain harmonious relations." Members were required to cut all ties to Mine Mill and, in return, would be given preferential access to jobs. In two days, more than 3,100 men signed a new petition pledging allegiance to the Evans organization.[34]

The Tri-State Metal Mine and Smelter Workers Union was officially founded on May 27 at a mass meeting of more than 4,000 people in Miami. Evans was elected president and Hickman secretary-treasurer. The twelve-member executive board included eight mine managers and four workers. Mining companies, especially Eagle-Picher, directed and financed their work. Members were given distinctive blue cards that lent the organization its informal name, the Blue Card Union. Above all, its members wanted to go back to work. Mr. M. joined "because I thought my wife and kid 'ud starve if I didn't an' a lot more o' them that joined felt like me." Few thought the Blue Card was independent, but desperation dictated fealty.[35]

However corrupt in origin, the Blue Card offered members an empowering crusade against the strike that indulged district traditions of male violence. Strikers had tried to stop Evans from attending the Miami rally. When police, including the county sheriff, arrived to escort him, Mine Mill men attacked them with rocks and pipes. They did not stop Evans but badly wounded several police officers. As tempers flared at the Miami meeting, Nolan mobilized several hundred Blue Card men to return to Picher in a show of force to "take back something that belongs to us." Along the way, they armed themselves with guns, pieces of pipe, and pick handles made available by the operators. From the Picher High School football field, Nolan's pick-handle army marched in growing columns into the heart of town where the members searched for union supporters to attack. Now numbering 2,000 or more, they rallied in a lot near the Local 15 union hall. "Who'll go back to work with me in forty-eight hours?" Blue Card leader June Walker shouted. "We'll all go," a chorus responded. "We'll bust this strike wide open." To keep the peace, Governor E. W. Marland sent a contingent of Oklahoma National Guardsmen with bayoneted rifles that night. While pledging not to intervene on either side, the unit's commander positioned a machine gun in front of the Connell Hotel on Picher's central thoroughfare. This was Evans's hotel and the Blue Card Union's headquarters. Meanwhile, the state police arrested five Mine Mill men, including M. E. Cartwright, for the morning's attack.[36]

Over the next few weeks, the Blue Card Union broke the strike. Evans and Nolan organized squad cars of armed members to intimidate Mine Mill men. These enforcers broke up pickets, stopped cars suspected of carrying strikers, and harassed Mine Mill members at their homes in the night. "The record abounds in incidents of lawless violence committed by the squad-car men," federal investigators reported. Evans paid them with funds provided by the mining companies. Eagle-Picher gave at least $17,500 to the Blue Card Union to cover these expenses and to buy weapons. Within a week, they had driven strikers out of sight in Picher, Cardin, and Commerce and to the edges of the district in Baxter Springs and Galena. The Oklahoma National Guard did nothing to stop them. Evans and Hickman held a series of meetings with company officials in the union's first week and reported that operators would restart operations as soon as they "were given definite assurance they would not be interrupted." On June 5, the OPA signed a general agreement with the Blue Card Union that acknowledged it as the sole bargaining agent of the district's mine and smelter workers and pledged to employ only Blue Card members.[37]

Born in opposition to a Mine Mill strike, the Blue Card Union now claimed a closed shop. Several hundred miners were back at work later that week. The Blue Card contract set wages on the traditional sliding scale at levels established in the Tri-State PRA. By June 15, the OPA reported that district mines were running at 85 percent of prestrike capacity and employed more than 3,000 workers, all of them Blue Card members. Most of the miners were happy to go back to work. Some had never agreed with the decision to strike; if they had, it was now convenient to forget. Merle Chambers "was not in favor of a strike and went back to work when the mill started." Tom Hood also opposed the strike and was relieved to go "back to work as soon as I could." The Blue Card Union not only seemed to succeed where Mine Mill failed but consolidated support by inviting its men to attack strikers who now appeared to be their enemies.[38]

Facing defeat in early June, strikers fought back. In Baxter Springs, a Mine Mill picket attacked truck drivers to keep the Beck Mining Company closed. In Treece, Mine Mill and Blue Card miners exchanged gunfire, which wounded Ted Schasteen. Meanwhile, saboteurs, presumably strikers, dynamited the electrical lines serving Cherokee and Ottawa Counties. Fearful of escalating violence, Cherokee County officials convinced Kansas governor Alf Landon to dispatch the National Guard. In both counties, the National Guard focused on Mine Mill pickets; by all accounts, Blue Card squad cars continued to harass strike sympathizers unimpeded. Confident that order prevailed, state officials removed both military contingents in late June. Within days, however, strikers again attacked men on their way to work, this time at the Eagle-Picher smelter in Galena. In Picher, Mine Mill and Blue Card men again clashed. Landon and Marland sent the National Guard back in. In Cherokee County, the guard commander declared martial law. The commanding officer in Ottawa County, meanwhile, confiscated all firearms, including several seized during raids of Mine Mill offices. With the military again in control, operators restored normal production by the end of July.[39]

The operators wanted to eradicate Mine Mill from the district. In early June, they had rejected an offer from Brown to end the strike if companies would rehire union members without discrimination. A Department of Labor conciliator recommended the deal, as did the commanding National Guard officers, who urged peaceful resolution. Company representatives explained that "other arrangements had been made for re-employing men" and "that the agreement which had been signed with the union," meaning the Blue Card, precluded any recognition whatsoever of Mine Mill. The claim was in bad faith, since the OPA controlled the Blue Card Union, which was headed by Evans, whose company was an OPA member. By this point, the OPA no longer feared federal intervention because the U.S. Supreme Court declared the NRA unconstitutional on May 27, the same day the Blue Card Union was founded. The OPA knew that the NLRA to revive federal labor protections passed the U.S. Senate on May 16 but was assured by lawyers that it, too, would be found unconstitutional.[40]

The Blue Card Union consolidated its authority. The union set up locals in Picher, Galena, and Joplin that met every week, collected dues, and in August started publishing its own newspaper. Now the gateway to employment in the district, the union also implemented strict membership requirements to ensure loyalty to the companies. Potential members first had to surrender their Mine Mill card, if they had one. Their names were published in the newspaper for the whole membership to consider. Those with known past affiliations or sympathies with Mine Mill appeared with an asterisk. Potential members then went before their local for a vote of approval, with all members casting ballots. Candidates appeared in person to answer questions about Mine Mill, their role in the strike, and whether they would now "defend the Tri-State Union with pick-handles." The union rejected those who received three or more no votes. This gave Blue Card members the power of exclusion and a sense of control. Finally, the union's executive officers either accepted or rejected the individual for membership. Blue Card officers imposed strict ideological discipline by threatening to expel anyone who advocated a strike. "It will take vigilance to keep the undesirables out," a Blue Card editorial declared.[41]

Meanwhile, the strikers received meager support from Mine Mill. President Brown oversaw the beginning of the strike from Joplin but left the district after the companies rejected his peace offer. Mine Mill was too weak to do more. Its resurgence had stalled after the demise of the NRA, as emboldened companies counterattacked. At the union's convention in Salt Lake City in August 1935, Brown reported "little progress" in new organizing since May and that many existing members had left. Mine Mill had only $4,979 in reserves. Brown and other leaders spent much of that summer appealing to John L. Lewis and the UMW for a loan of $50,000, with no success. To make matters worse, disgruntled members ousted Brown from office in Salt Lake City, weakening union leadership. The convention voiced support for the Tri-State strikers through resolutions but nothing more, despite passionate appeals from Schasteen and McTeer, who both attended. Mine Mill's boldest statement on behalf of the strikers was a resolution asking the AFL to contribute $50,000 to support them.[42]

Mine Mill's relationship with the AFL, long the key to organizing in the Tri-State, soon faltered. Union leaders advocated on behalf of the Tri-State strikers at the AFL convention in Atlantic City in October 1935. They argued that the strike was still winnable. They submitted a resolution to the committee on industrial relations calling for AFL unions to boycott Eagle-Picher products. Their interest in the strike was secondary, however, to their concern over the AFL's stance on industrial organizing strategies. The 1935 convention roiled with debates over how the AFL should organize workers in mass production industries. Led by Lewis, leaders of existing industrial unions such as the UMW called for a concerted campaign to organize them into new mass unions that superseded existing craft union jurisdictions. They squared off against guardians of the old craft order in an increasingly acrimonious dispute. Mine Mill's Paul Peterson used the Tri-State strike example to press the case for a more ambitious industrial strategy. Mine Mill's fate depended on what happened in the Tri-State, Peterson declared, a district that had "provided the scabs which for years have broken" its strikes. He claimed that Mine Mill's persistent organizing efforts had brought "these scabs" into the union. Peterson called on the AFL, Mine Mill's ally, to help keep them organized. The industrial relations committee voiced its support but refused to authorize a boycott. This was a small defeat among many in Atlantic City, as the AFL famously rejected the demands of the industrial unionists. The following month they founded their own group, the CIO, with Lewis as president. Mine Mill was a founding member.[43]

Despite unsteady national allies, a few hundred strikers refused to concede. Locals in Picher, Galena, Treece, Baxter Springs, and Joplin remained active, despite losing members. Those in Picher, Galena, and Treece were the largest, with over 100 members each in early 1936. They looked for help from regional union affiliates, particularly the OFL. "This strike isn't over and is a long way from being lost for you cant whip any bunch of union men that wont quit fighting or accept their cause is lost," McTeer reassured OFL leaders. In September, the state federation offered support at its convention in Muskogee. President G. Ed Warren, a Tulsa lawyer who had worked with the AFL since the 1910s, said he had "never seen better loyalty than displayed by them under the most adverse conditions. They have suffered every indignity known to the exploiting employer and have been menaced and abused by professional gunmen, strike breakers, and thugs." The convention raised $68.04 for strike relief. The OFL, along with the Kansas State Federation of Labor, also distributed a circular letter calling for a boycott of products containing Tri-State zinc and lead, such as Dutch Boy and Sherwin-Williams paints. The Mine Mill stalwarts drew confidence from the new NLRA, which President Roosevelt signed into law in July 1935. The NLRA created stronger protections for workers to join unions of their own choosing, bargain collectively with their employers, and go on strike. The law also prohibited a set of unfair labor practices by employers, including interfering with independent organizing, discriminating against union members, and creating company-dominated unions—everything, it seemed, that had happened in the Tri-State. The NLRA also reformed and expanded the NLRB

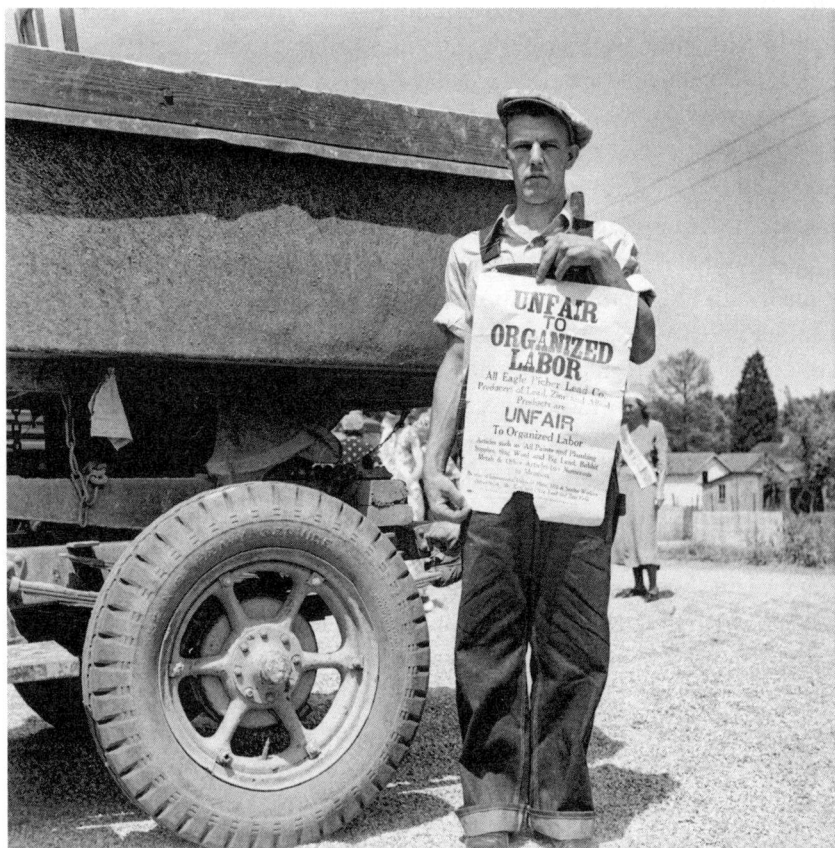

Striking zinc miner, Columbus, Kansas, 1936. Library of Congress, Prints & Photographs Division, FSA/OWI Collection, LC-USF34-004166-E [P&P] LOT 428.

with new power to enforce these provisions. "We are free men and have the right to organize and bargain collectively for hours, wages, and working conditions," Local 15 declared that November.[44]

Local strike leaders were their own best advocates. In June 1935, they had delivered several affidavits to the first NLRB detailing company support for the Blue Card Union and the violent tactics of the pick-handle brigades. The NLRB sent George Pratt, who directed its regional office in Kansas City, Missouri, to investigate, but nothing happened as the NLRB reorganized. Strikers resubmitted their complaints to the new NLRB in the autumn of 1935. They also appealed to the White House for help as anti–New Deal forces challenged the law in court. Those challenges seemed formidable. In late December, Judge Merrill Otis, a Coolidge appointee on the U.S. District Court in Kansas City, declared the NLRA unconstitutional. The day after Otis's verdict, striker John Millner wrote to Roosevelt to remind him that he had "given us the right to organize without fear of being fired." The strikers had trusted that right when they acted to end conditions that "are not American but more of the lord manor and the surf [*sic*]." The operators had ignored the law. "We placed our confidence in the Wagner labor disputes bill passed by Congress but little action has been taken yet," Millner explained. He implored the president "to make these laws stand."[45]

To build national support, strike leaders began emphasizing the district's human suffering from occupational injury and disease. "The average life of a miner is ten years," Millner informed Roosevelt, mainly due to silicosis, but "if a man so much as hints he will stand up for his rights and demands better working conditions he gets fired." "We have some of the worst working and living conditions to be found anywhere in the country," Local 15 told readers of the *Oklahoma Federationist*. "The average life of a zinc miner in this field is eight years," it declared, and "our wives and children are subject to silicosis" from "the high silicate content of the air from the chat piles." By emphasizing humanitarian motives, the strikers cast their struggle in terms that appealed to growing widespread concern for the plight of poor people, a concern that was becoming politically potent as journalists and photographers documented those fleeing the Dust Bowl or living hand to mouth in sharecropper cabins. While doing so, the union dropped its criticism of the examination bureau and any suggestion that damaged miners wanted to risk their bodies again. This change mobilized a rhetoric of responsible, breadwinner manhood that had little basis in the recent history of the Tri-State but effectively shaded the Blue Card Union as reckless, destructive, and crucially, athwart the working-class politics of the New Deal.[46]

Mine Mill acknowledged that the Blue Card Union had grassroots support. "This district is noted for its scabby people," one appeal for support declared. Victory, the strikers now admitted, would require defeating the company union and the strikebreakers who had joined it. If the Blue Card Union prevailed, they warned, it would threaten union workers outside of the district. This appeal recalled the antiscab invective that earlier Western Federation of Miners leaders had used to demonize Joplin strikebreakers. While encouraging help from allies, it also deepened the divide between Mine Mill and Blue Card members.[47]

The New Deal gave the strikers a boost in early 1936 when the NLRB renewed its investigation. Pratt returned to gather evidence in February. Although the union filed complaints against several companies, Pratt recommended charges first against Eagle-Picher. The NLRB scheduled a hearing for May. Pratt believed that the strikers had a strong case but reported that they were "almost completely beaten." NLRB action, however, brought fresh support from national allies. Mine Mill leaders returned to the Tri-State to help the locals prepare evidence and testimony. More significant was support from the new CIO. John Brophy, the CIO's first executive director, advocated on behalf of the strikers with the NLRB. Lewis also monitored the case.[48]

New Deal labor law was not yet strong enough, however. In May, just before the hearing against Eagle-Picher was to begin, Judge Franklin E. Kennamer, of the U.S. District Court in Tulsa, granted a restraining order and then an injunction against the NLRB pending higher court rulings on the constitutionality of the NLRA. The strikers again verged on total defeat.[49]

While the strikers waited on the New Deal, the Blue Card Union intensified its demonization of Mine Mill, combining long-standing animosity toward foreigners and radical unions with the rhetorical weapons of fascism. Answering union cries of "scab," Evans and Hickman called the strikers "bohunks," "yellow-bellies," "'smelly' internationalites," and "undesirables." They were "the ragged remnants of that cesspool of communistic propagandists," dupes lured by "the abscess of corruption in the breasts of hired agitators who live off unrest and unemployment." According to Blue Card logic, these people deserved expulsion or worse. Blue Card members also targeted the AFL, with which Mine Mill was still affiliated. "The A. F. of L. is infested with reds and communists," the Blue Card Union's paper declared in late 1935. AFL president William Green "and his dirty blood-sucking leeches" had "the interest of only one small minority at heart, namely, the racketeers who run the American Federation of Labor." These attacks drew from an older language of anti-immigrant thought, now adorned with fresh anti-Semitic invective. While Green was not Jewish, the Blue Card Union played on perceptions that many union leaders were, particularly those in the CIO.[50]

Blue Card leaders backed up their rhetoric with incitement to violence. In late 1935, Evans announced that "hunting season was open and that it was time for the blue card members to go hunting for" Mine Mill holdouts. For Blue Card members, the pick handle symbolized their willingness to attack enemies. "A pick-handle is something like a badge," member Burt Craig explained; it showed the carrier was a Blue Card man. "These pickhandles we have are our signs," a back-to-work poem read, "we'll meet you strikers at the mines." Joe Nolan, who had led the antistrike march on Picher, became known as the "Pick-Handle King." He maintained stocks of pick handles at strategic sites to arm members when needed.[51]

Blue Card leaders told members that strong white men could justly use violence to defend America and capitalism. They were "loyal, true and hard working American citizens" who wanted to provide for their families. Forced out of work by strikers, Blue Card men fought to "banish slavery" to the dictates of Mine Mill and the AFL. Since they "purged" the district "of the horrible stench," the union's newspaper exulted, "white folks can go about their business of personal gain." OPA leaders portrayed the Blue Card as the latest iteration of the market-driven white nationalism of poor man's camp lore. "It is doubtful if there is any part of the country where the fundamentals of independent and self-reliant Americanism still are so deeply rooted as they are among the people of this region," Harbaugh explained in 1936. They were all "white Americans—no foreign or colored labor," he emphasized, which meant they shared "a fair conception of the wage that can be paid, based on the market price of product." The strikers, by contrast, were weak headed, corrupt, physically unfit, and dependent on government help. Mine Mill was "composed of unemployed relief workers and dole takers," Harbaugh claimed, as well as men "whose physical condition relegated them to the ranks of the unemployables." James Wadleigh, mining editor of the *Joplin Globe*, concurred, writing that Mine Mill attracted "physical and mental deficient followers," "semi-broken men" who wanted to "make a living without doing much manual labor." These charges repeated old associations of racial and gendered inferiority, especially in relation to radical unionism, but with new cruelty toward men who had been denied work, and thus driven to strike, because of workplace injuries and illness that made them actuarially unemployable.[52]

The Blue Card Union fostered a movement culture that emphasized male aggression in ways that recalled the rough prosperity of Picher's heyday. Nolan sponsored weekly wrestling and boxing matches at the Picher American Legion Hall. Like his pick-handle brigades, these bouts also glorified violent masculinity. In October 1936, for example, "an enthusiastic crowd" watched as "'Wild Red' Berry, the bull of the wrestling ring, sometimes referred to as the Pittsburg Panther, manhandled Pete Baltram at Picher Friday night in a grueling battle ... staged for the Blue Card union at American Legion hall." The union also sponsored dances with live music for the wider community. The Blue Card displayed its economic and social power at a picnic to celebrate its first anniversary in May 1936, just days after Kennamer's injunction halted the NLRB challenge. An estimated crowd of 15,000 men, women, and children gathered at the Miami fairgrounds to enjoy free food, drinks, and games provided by the OPA. Only Blue Card members and their families were allowed in, mocking the desperate plight of those still on strike. The union slaughtered seventeen cows and twenty hogs for the barbeque and provided a full slate of activities, including amusement rides, races, dance competitions, and skill contests such as rolling-pin throwing for women and nail driving and shoveling for men. The rally also included three boxing matches. "All we want is the right to peacefully earn a living for ourselves and family," Evans told the crowd during his "state of the union" address. "We want to work with our picks and pick who we work with, but if we are forced to do this with our pickhandles, thank God, we've still got the courage to do it."[53]

The strikers sought but received little outside help in the summer of 1936 as the CIO and the AFL drifted toward open warfare, despite a presidential election campaign that gave new urgency to their cause. At Mine Mill's convention in August, Schasteen and McTeer passed a resolution that lambasted Republican nominee Alf Landon for his role during the strike. Meanwhile, Local 111 sent another circular letter to AFL affiliates. By boycotting Eagle-Picher and Landon, Schasteen wrote, AFL members would help Local 111 "assert our rights as free working men" against "strikebreakers" backed by thuggish "attempts to terrorize our people." "As you know," he exaggerated, "90 per cent of the strikes in the United States have been broken by scabs from this district." A strike victory, Local 111 promised, would give the AFL "control over these men." At the OFL convention in mid-September, McTeer offered resolutions on behalf of Local 15 opposing Landon, calling for a boycott of Eagle-Picher, and proposing a congressional investigation into silicosis and other occupational diseases. While the convention supported Local 15's resolutions, the relationship between the two was doomed. Ten days earlier, the AFL executive council had suspended the CIO unions, including Mine Mill. Despite support for the strikers from the convention floor, OFL leaders meeting in executive session cut all ties with Mine Mill.[54]

The remaining strikers faced the bleak prospect of never working in the mines again. The OPA used its control of the Blue Card Union and its self-described "control of employees through physical examinations" at the bureau to cleanse payrolls of suspect men. Several hundred strike supporters were blackballed. Some men were excluded for being friends or neighbors with Mine Mill members. The Blue Card continued to monitor those who renounced their membership. If a worker "said anything against the Blue Card Union, or said anything that sounded like you favored the 'International boys,' you got sent to the clinic for an examination an' came out with an 'F' card and no job," Mr. M. explained in 1939. Here the rhetorical assertions that the strikers were damaged men facilitated employer use of the bureau to enforce obedience and conformity to the Blue Card. If Mine Mill appealed to men who were physically unable to work, and it had with criticism of the examination bureau in 1934, then sympathy with Mine Mill was said to indicate physical and mental weakness and thus mark them as an ongoing risk in terms of ideology and insurance liability. One miner who could never fully submit to the Blue Card, even though he was not a diehard Mine Mill man, faced years of discrimination. "Somehow I could never pass the physical examination" after 1936, he recalled in the 1950s, "and without my card I could not work." Meanwhile, Blue Card members who otherwise would have failed their examinations received cards with passing grades. According to the *Blue Card Record*, Nolan, whose father had died of silicosis in 1919, interceded on behalf of "1300 of these men who received hospital cards in spite of the fact that they had injuries and deformities ranging from rupture to crooked back bones." They were safe employees who knew not to file a compensation claim.[55]

Blue Card members, meanwhile, reaped material rewards from their allegiance to Evans and the OPA as the economic recovery, spurred by New Deal measures, boosted activity in Tri-State mines. In 1936, with prices above thirty dollars per ton for the first time in months, mining companies produced 428,524 tons of zinc ore and 52,256 tons of lead mineral worth more than $16 million, the highest totals since 1929. Production continued to boom as ore prices moved over forty dollars per ton in early 1937. According to the OPA, 5,200 men had jobs in the district's mines, mills, and smelters, more than at any time since the 1920s. These workers also received higher pay as companies raised wages on the sliding scale seven times between November 1936 and March 1937. Machine men now earned 69.125 cents per hour and shovelers worked for 14.5 cents per large can, the highest rates since the Great War. Many were longtime residents who understood the significance of these raises. Some of them had gone on strike in 1935. Others were new to the district, such as Elven "Mutt" Mantle, an Oklahoma farmer who took work as a shoveler at Commerce Mining after moving to Ottawa County with his wife and son, Mickey, in 1935. Now, they all carried blue cards and fat pay packets.[56]

Many Tri-State miners looked at conditions in the spring of 1937 and saw a validation of white working-class ideals they had long held. Their economic prospects were the best in years. As Blue Card members, they had status in an organization that rewarded loyalty to employers and the market, their race and nativity, and their sense of themselves as strong men. They were proud of these commitments. From their perspective, Roosevelt's resounding victory in November 1936 on a platform that championed the common man offered further evidence, at least in their reading. More than 61 percent of Ottawa County voters chose Roosevelt for reelection; they swept Republicans from every county office. Democrats had lived up to their 1932 promise to restore prosperity. Most Tri-State miners considered these triumphs hard won, the result of their willingness to fight, and not of the generosity of elites. Blue Card men told themselves that they broke the 1935 strike on behalf of the district's majority, a view that revived the long local tradition of fighting against radical minorities and strangers. They considered Mine Mill a threat from outside and the strike the illegitimate scheme of sick, lazy subversives. Now, with OPA support, Tri-State miners thought they again had power. But Mine Mill did not give up. The few remaining strikers saw their own redemption in Roosevelt's triumph and the CIO's surge in major industries. Blue Card miners welcomed the fight. To many, Blue Card and Mine Mill alike, the outcome had never been more urgent.

Following Roosevelt's crushing victory, Mine Mill restarted its organizing campaign in the Tri-State in December 1936. Cut off from the AFL, CIO unions launched a series of such drives to harness working-class enthusiasm for a more confrontational, democratic union movement. In steel, automobile, and rubber factories, the CIO targeted the unorganized and those who were unhappy with conservative, exclusive, company-friendly AFL unions. Mine Mill's new president, Reid Robinson, believed that the men of the Blue Card Union could be convinced to join the CIO and that the strikers could still win. Robinson explained that "it was decided the only way to get justice for those still striking was to reorganize the district, taking into the organization those who had been forced to return to work." McTeer and Schasteen led the campaign but kept meetings small and secret. They looked to the Roosevelt administration for help. Schasteen wrote to the president at the start of the push to ask for a federal "investigation of conditions in this District" and urged Roosevelt "to give this matter your personal attention."[57]

Blue Card leaders were on high alert, however, as the CIO made stunning advances elsewhere. In February 1937, sit-down strikers in Flint, Michigan, forced General Motors to grant a pay raise and recognize the United Auto Workers, a major victory that gave the CIO new credibility. Three weeks later, in early March, CIO steelworkers signed a collective bargaining agreement with U.S. Steel, a corporate behemoth long opposed to unionization. The *Blue Card Record* denounced the sit-down strike tactic as foreign-inspired radicalism. Tri-State workers were "Americans and Americans don't do business that way," an editorial commented. In early March, the paper reported that "trash of the international" were trying to infiltrate the Blue Card Union. Targeting this "danger from within," the paper's editor declared that "the spirit of May 27, 1935, is still alive," "the spirit which inspired some 2,000 men to shoulder pickhandles and march down the streets of Picher with the determination to knock hell out of anyone who opposed the reopening of their jobs." The paper warned "that there are many more now than there were then willing to shoulder pickhandles" to keep the district "free from the cancerous corruption of rednecked radicals." The Blue Card Union thrived on these kinds of threats, which echoed, almost word for word, the language Joplin strikebreakers had used in Idaho and elsewhere almost forty years earlier. After four months, McTeer and Schasteen reported little progress. Robinson blamed "the fear of the workers not only of losing their jobs but of physical violence from the hands of the Blue Card Company Union."[58]

Mine Mill decided "to bring the program out in the open" with a rally that would enlist the growing power of the CIO. Robinson, McTeer, and Schasteen planned to hold it in Picher on April 11. Robinson invited Lewis to speak. Not only had the CIO just defeated two powerful foes, but now the courts seemed ready to support federal labor rights. On March 29, following Roosevelt's proposal to reform the federal judiciary, the Supreme Court for the first time issued rulings in favor of the New Deal. Crucially for the CIO and Mine Mill, the court's ruling in *Virginia Railway Company v. System Federation No. 40 et al.* indicated that it would uphold the NLRA. Mine Mill announced the Picher rally days later. Although Lewis declined the invitation, he expressed strong support for the campaign and his hope "that it will meet with deserved success among the metal miners of this district."[59]

Blue Card leaders responded as promised, with ferocity. On April 10, Evans convened a meeting of around 400 ground bosses and foremen. He told them to make sure that all Blue Card members took to the streets the following day for a "show of strength" against Mine Mill. Evans explained that everyone was expected to participate and that those on Sunday shifts would have the day off to do so. He also promised to provide plenty of food and booze as well as pick handles, "our emblem." That night, Nolan told another large gathering of Blue Card men that Robinson and Mine Mill had to be stopped if they wanted to keep working. He promised to arm them with pick handles and, if necessary, machine guns. He assured the crowd that there would be no legal consequences for their actions. In case anyone doubted, Picher police chief Al Maness was standing behind him. Mine Mill leaders, including Robinson, who was in Joplin preparing to address the rally, warned county and state officials that Evans and Nolan were planning violence. Their concerns were dismissed.[60]

Several thousand Blue Card men gathered in Picher on the morning of April 11, most of them carrying pick handles, many already "liquored up." They roamed the streets looking for Mine Mill members. "It wasn't long before C. I. O. sympathizers' heads were bouncing off the cudgels," the *Blue*

Card Record later exulted. A Blue Card mob attacked Clifford Doak and Lester Wakefield with brass knuckles and pick handles for wearing CIO pins. Another mob raided Local 15's meeting hall, where they found no people but destroyed union files and property. Similar fights and scuffles erupted throughout Picher and surrounding towns that morning. The authorities did nothing to stop the Blue Card riot. By noon, Robinson had canceled the rally and everyone allied with Mine Mill had fled Picher. To trumpet their dominance, "Blue Card workers kept up a constant pounding on the pavements and sidewalks with their pick handles."[61]

Blue Card leaders now looked to destroy the strike once and for all. With mobs in control of Picher, Nolan falsely announced from a sound car that CIO members were regrouping in Treece. Hundreds of drunk, pick-handle-wielding Blue Card members made the two-mile journey across the state line to Local 111's meeting hall, where they thrashed the few remaining Mine Mill members, ransacked the property, and stole all of Local 111's files. With Treece under Blue Card control, Evans and Nolan announced, again falsely, that Mine Mill was regrouping at Local 17's headquarters in Galena. As the pick-handle force made its way, Blue Card leaders broadcast an appeal for more supporters on a Joplin radio station. More than 500 armed Blue Card men arrived in Galena that afternoon brandishing pick handles and guns.[62]

This time, however, Mine Mill members stayed to defend themselves. As the mob gathered in front of Local 17's Main Street meeting hall, Lavoice Miller, a Blue Card member, shattered one of the windows. The Mine Mill men fought back. "It was man against man and at stake was work to put food on our tables," one later explained. They shot into the crowd. Several Blue Card men were hit, including Miller. The shooting continued sporadically until some in the pick-handle mob detonated smoke bombs. The mob took away nine wounded while the Mine Mill men slipped out the back. As the smoke cleared, the district's mines and mills resumed operations the following day with no disruption. Robinson praised the Local 17 men for defending themselves even as he regretted the resort to violence. He blamed local authorities for not heeding his warnings. Meanwhile, Evans pledged that the Blue Card Union "will continue our attempts to prevent C. I. O. unionization of this territory."[63]

Blue Card leaders next took the extraordinary step of enlisting the AFL against their CIO foes. They were concerned about the resilience of Mine Mill's reorganization campaign, not because they believed it would prevail but because they worried that the Supreme Court would uphold the NLRA and reopen Mine Mill's charge of unfair practices with the NLRB. Mine Mill's former OFL ally G. Ed Warren had arranged talks on March 29 be-

Blue Card Union pick-handle rally, Galena, Kansas, 1937. Photos 223, folder 23023, box 544, Tamiment Library, New York University, by permission of the Communist Party USA.

tween Blue Card leaders and William Green. They thought AFL affiliation would make the Blue Card Union legitimate according to the law and thus nullify Mine Mill's claims. On April 12, the day after the Blue Card riot, the Supreme Court realized their fears by ruling the NLRA constitutional. Green gave the negotiations with the Blue Card his full support on April 13. Three days later, Evans and Warren signed an agreement in Tulsa. The AFL incorporated the Blue Card locals as directly-affiliated federal labor unions while it waited on the 1937 convention to ratify a new charter. The Blue Card Union internal structure remained unchanged, with its current leaders in place and its sliding-scale contracts with district companies in force, now with AFL approval. Both sides referred to the recent violence in Galena as the impetus for the deal, which promised, according to an official announcement, "to bring industrial peace to the Tri-State District." Evans declared that the Blue Card and the AFL were in "absolute accord" and would work together to "drive out of all industry the lawless labor element that threatens the peace and security of the nation." Robinson, meanwhile, was stunned that the AFL "would stoop so low."[64]

The Blue Card Union ratified the affiliation at a mass meeting of more than 5,000 people, including many women and children, in Miami on April 18 that indulged the district's long-standing working-class assertions of white, nativist self-interest. The *Blue Card Record* billed it as "the greatest protest demonstration against the C. I. O. ever to be staged in the United States." Speaking first, Evans explained that the Supreme Court's decision on the NLRA forced them "to do something" or otherwise risk Mine Mill winning through the NLRB. Kelsey Norman assured members that the AFL was aligned with Blue Card principles because it "is opposed to strikes," especially the un-American sit-down tactic, and classified CIO members as "scabs" whom it would not allow to work in Tri-State mines under the Blue Card contract. In addition, he stressed, the AFL also wanted to make the Blue Card Union the "sole bargaining agency" for all metal mine and smelter workers in North America at its next convention. "The downfall of John L. Lewis has already begun." Norman promised that the AFL would see "that you get everything on earth that American labor is entitled to in this district." In a question-and-answer session that followed, Norman gave two key responses. He repeated the union's commitments to current contract terms, including the sliding scale. Another miner asked if the new affiliation meant that "we will have to work with niggers and wops." Norman reassured the thousands gathered that "there is not any place in this organization for them."[65]

Blue Card members approved the affiliation with authoritarian bravado. "If you believe in this new affiliation, every man who carries a Blue Card please stand up and raise his right hand," Nolan instructed. "Five thousand right arms, uplifted as one," responded, according to the local press, and "gestured defiance" to the CIO. Then, after the crowd sat, Nolan ordered all those "contrary to this agreement stand up and get knocked down." No one stood. Amid the ensuing backslapping and congratulations, organizers played "Happy Days Are Here Again," President Roosevelt's 1932 campaign theme song, over loudspeakers in a building constructed by FERA. Many no doubt thought they were acting in accord with the true New Deal. Five days later, Green welcomed Evans, Nolan, and Hickman at his Washington office to charter the three Blue Card federal labor unions.[66]

In late May, the Blue Card leaders attended the AFL's special executive council meeting in Cincinnati, where Green planned to launch a national counteroffensive against the CIO. The Blue Card Union was its first victory. Green reaffirmed his goal to give the union jurisdiction over all "workers in the zinc lead mining industry." Evans asked what would happen if Mine Mill returned to the AFL fold. Green pledged "assurance from this Executive Council that nothing will ever be done that will disturb the relationship of your group with us." The following month, the Blue Card also affiliated with the OFL.[67]

In their drive to destroy the CIO, AFL leaders capitulated to the federation's most reactionary tendencies by admitting the Blue Card Union. The AFL gained 8,000 dues-paying members by taking in the company union, an important counter to CIO gains. In time, Green hoped to build it into a national union that would replace and ultimately destroy Mine Mill, an old idea that recalled Samuel Gompers's efforts against the Western Federation of Miners in 1899. Green considered Mine Mill the most vulnerable CIO union, and both Green and Warren seemed to believe that the Blue Card Union was bona fide.[68]

Although a few AFL locals denounced the deal, Green dismissed their accusations of cynicism as CIO-inspired propaganda. "We responded to the request of the metal miners in the Tri district to become organized into the American Federation of Labor," he explained. In private correspondence, Green also reiterated plans "to make the organization established among the metal miners in the Tri district a functional organization, serving the economic and industrial needs of the workers and established upon a sound and secure" AFL basis. Meanwhile, Warren greeted the Blue Card's delegates to the 1937 OFL convention "as brothers in the great army of organized labor." The AFL ratified the Blue Card charter at its annual convention in Denver that October, with Evans, Nolan, and Hickman in attendance. "We are confident that through said affiliation and the establishment of a cooperative relationship, that the economic and industrial interests of these metal miners can be promoted in a most substantial and satisfactory way," the official proceedings declared.[69]

Now with AFL backing, the Blue Card Union reveled in its founding commitments. "Technically we are now operating under the general policy laid down by the" AFL, the *Blue Card Record* stated in May 1937, but "really we are still the Blue Card Union in every respect and as such we intend to remain." The union reaffirmed its apparent triumph with a second-anniversary picnic in Miami on May 27. More than 20,000 people, the largest gathering in the history of the county, enjoyed another day of contests, music, free barbeque, and speeches. Green sent a welcome message and his "personal felicitations." By joining together, he explained, the AFL and the Blue Card Union "stand for the preservation of our democratic form of government and for the protection of American institutions against every onslaught which may be made upon them by subversive forces." Warren addressed the throng in person. He congratulated the Blue Card for becoming "the first independent organization to come in the A. F. of L." in its fight against the CIO. By all accounts, everyone enjoyed the "continuous entertainment" with no fights and no arrests for drunkenness.[70]

The Blue Card counterattack decimated Mine Mill in the Tri-State. The riots of April 11 halted the union's reorganization drive. Worse still, local authorities charged ten Mine Mill members with murder after Lavoice Miller died from his wounds. McTeer and Cassell, with Robinson's help, formed the Galena Defense Committee to raise defense funds. Meanwhile, the Blue Card's AFL charter, its leaders' visits to Washington, and public support from Green and Warren made Mine Mill seem pathetic and illegitimate. "We regard the C. I. O. men as non-union workers," Warren told the press. The strikers were on the edge of total defeat. "We have suffered a crushing set-back here and our situation is desperate," Cassell informed Lewis.[71]

The CIO offered little direct help. After a winter of amazing gains in 1936–37, the international suffered a series of startling setbacks as companies pressed an array of harsh tactics. The Ford Motor Company and the smaller steel producers collectively known as Little Steel mobilized anti-union vigilantes to stop CIO progress. Starting that summer, in the wake of these bloody defeats, CIO unions also faced a national economic crash that hit its strongholds in mass production industries hardest. During the year-long recession, new members struggled to pay dues, which cut organizing funds. Already weak, Mine Mill barely survived. By June 1938, only 27,000 of its 46,000 members paid dues. It stayed solvent because of a $10,000 loan from the UMW. In the Tri-State, the UMW was the only union to come to the aid of Mine Mill. David Fowler, the president of UMW District 21, which covered Arkansas and Oklahoma, fought back against the AFL and Green for recognizing a company union created "to defeat the interests of organized labor." He formed a council of CIO unions in Arkansas and Oklahoma to challenge the AFL "and its affiliated scabs." If Green and Warren wanted "war," Fowler told the press, "we are willing to declare it." While the fight energized some CIO forces, Mine Mill's remaining members in the Tri-State rightly felt exposed. They "have become more or less disgruntled," Robinson admitted at Mine Mill's August 1937 convention. Although police dropped charges against the Galena ten, remaining members were angry at the CIO for not doing more to help them and consumed with "bitter feeling" toward the thousands of men in the Blue Card Union. Robinson warned that any reorganization in the Tri-State would be painstaking.[72]

The strikers' only hope was the NLRB. The Tenth Circuit Court of Appeals lifted Kennamer's injunction in July 1937, which allowed the board to re-join its scrutiny of the relationship between Eagle-Picher and the Blue Card Union. In early November, the NLRB charged Eagle-Picher with a series of unfair labor practices stemming from the strike. The board began hearings in Joplin in early December. William Ringer, who left his Indiana law prac-tice to join the NLRB, served as trial examiner. William Avrutis prosecuted the NLRB's case in close cooperation with Mine Mill's local lawyers, Louis Wolf and Sylvan Bruner. In hearings that lasted until April 1938, the NLRB collected thousands of pages of evidence and heard testimony from dozens of witnesses on all sides of the dispute. It did not see the internal records of the Blue Card Union, because those were destroyed when Evans's car mys-teriously burned. Nor did the board see the records of Mine Mill Locals 15, 17, and 111 because Blue Card rioters destroyed those on April 11. The Blue Card Union had hoped for legal cover from the AFL, which prepared to send chief counsel Charlton Ogburn until Ringer ruled that the federation was not a party to the case when the original complaint was filed. Even without AFL backing, lawyers for Eagle-Picher and the Blue Card claimed that it was a real union with grassroots support. Meanwhile, the *Blue Card Record* accused the NLRB, led by Avrutis, with his "nice sweet bolshevick name," of propping up the illegitimate CIO. Avrutis's sympathies were strong. He boasted to NLRB regional director George Pratt of his desire to attack Eagle-Picher and the Blue Card with "gleeful malice." Avrutis later called the case "one of the highlights of my entire life," because "we were in a righteous cause" to "do something about bringing, in a real constructive sense, law and order to what was a wild area." Both sides delivered closing arguments on April 28. They would wait for months for Ringer's ruling.[73]

AFL leaders continued to support the Blue Card Union after the NLRB hearing, despite damaging revelations about its links to Eagle-Picher and other companies. Its Blue Card charter had spearheaded an aggressive strategy of launching conservative unions to rival CIO counterparts; the AFL chartered similar unions among glassworkers, coal miners, and auto-workers. The main Kansas City local of the United Auto Workers–AFL, chartered in early 1939, also became known as the Blue Card Union. Green accepted Evans's resignation in November 1937 in the hopes that new leadership might deflect Mine Mill's charges; Nolan replaced him, with AFL approval. Green told the executive council in May 1938, a month after the NLRB hearings ended, that he still expected the Blue Card to become the AFL union for all metal miners. The AFL had revoked Mine Mill's charter in February and was supporting Blue Card–like countermovements among metal miners in Alabama, California, and Idaho. Like Eagle-Picher and Blue

Card leaders, the AFL executive council accused the NLRB of unfair bias in favor of the CIO. By 1938, the executive council supported or led conservative attacks on the board, often with full-throated anti-Communist invective. Whatever Ringer ruled, Blue Card leaders expected the AFL to protect "the interests of the workers who have consistently fought the C. I. O."[74]

The NLRB's intermediate report, issued in August 1938, blasted Eagle-Picher and the Blue Card Union. Ringer found Eagle-Picher guilty of violating the NLRA by promoting a company union in order to deny its workers their right, under the law, to choose their own representatives. He ordered the company to sever all connections to the Blue Card Union, to stop preventing workers from joining Mine Mill, and to reinstate more than 200 strikers with back pay. While the CIO hailed the initial outcome, Ringer's report still needed final confirmation from the NLRB in Washington. No one knew how long that might take. Pressing its advantage, Mine Mill filed similar charges with the NLRB against the smaller mining companies. In the meantime, the AFL and OFL both backed away from the Blue Card Union without explanation.[75]

CIO and Mine Mill leaders hammered the AFL with these conclusions. The editor of the *CIO News* called the federation's executive council a bunch of "misleaders, who claimed to be labor men and yet spent most of their time fighting the rank and file in the interests of the bosses" by creating fake unions. "A prize example of the phoney 'union'" was the Blue Card, he said, a group that "specialized in vicious attacks on organized labor, which it carried to the extent of arming mobs to smash union halls and beat up union workers." The line of attack bolstered the CIO in the region. Fowler's Arkansas-Oklahoma Industrial Union Council met in December 1938 with over 200 delegates and 1,000 visitors. The UMW still led the way, but now alongside growing CIO unions in the glass, furniture, retail and wholesale, and oil industries, as well as Mine Mill, which was making headway among smelter workers in Bartlesville.[76]

Back in the Tri-State, the few remaining strikers, led by McTeer and Cassell, called for federal action to address the district's health crisis. With little recourse between a wounded Mine Mill and Blue Card "fascism," Cassell appealed to the two most powerful women in the federal government, Secretary of Labor Frances Perkins and First Lady Eleanor Roosevelt, in August 1938. "Only the federal government will help us," he told a Perkins deputy. Cassell asked for the Department of Labor to launch a new health and safety investigation that privileged workers, not the companies, as the Bureau of Mines had with its clinic. "We will immediately send plenty of factual and documentary evidence that men are dieing like flies and that 8 out of every 10 women in this district are widows, 75 percent of the children orphans," he informed Roosevelt. Both women voiced support, although Perkins's Division of Labor Standards proceeded timidly as conservative opposition slowed the New Deal after the 1938 election. Meanwhile, the International Labor Defense and the National Committee for People's Rights, both Communist-affiliated advocacy groups allied with Mine Mill, began campaigning on their behalf. The National Committee for People's Rights established the Tri-State Survey Committee to investigate conditions in the district. In early 1939, the committee sent social worker Mildred Oliver to conduct fieldwork in Picher. It also enlisted Sheldon Dick, a documentary photographer who had recently been to Picher on assignment for the Farm Security Administration. In October 1939, the committee released a preliminary report that revealed—with firsthand accounts, statistics, and Dick's photographs—grim housing, sanitation, and public health conditions. Despite recent OPA attention to dust abatement, the Tri-State was a "death trap," the report stated, where a whole community suffered the "denial of the basic rights of decent living." The same month, L. S. Davidson, a former Picher teacher, corroborated the report with a semifictionalized account of the strike and its defeat, *South of Joplin*, published by W. W. Norton. Coming at once, these interventions brought new national attention to the conflict in the Tri-State.[77]

Then, in October 1939, the NLRB gave the strikers a victory by sustaining Ringer's conclusions. The board confirmed his orders against Eagle-Picher and upheld the reinstatement of as many as 200 workers, most of them Mine Mill members, with back pay that totaled over $500,000. Keen to avoid such a payout, Eagle-Picher appealed the ruling to the Eighth Circuit Court of Appeals in November. The operators were not totally defiant, however. The Blue Card Union quietly disbanded its locals. Newly confident, Mine Mill launched a reorganization drive in the Tri-State. The CIO sent Gobel Cravens and Jim Ferns, the son of Rube Ferns, who organized in the district in the 1910s, to help McTeer and the remaining local leaders.[78]

Mine Mill made workplace health and safety the basis of its new campaign. The union was determined "to show the workers how to get the things they need to free themselves from misery and death—by organization," its newspaper declared in late 1939. "With a Labor Board decision finally handed down in their favor, and with a renewed organizing drive under way, the miners and their wives and their children in the Tri-State are fighting to live like Americans and human beings." McTeer believed a focus on the common problem of silicosis could heal the divide between Mine Mill and the Blue Card men. "We know these conditions," McTeer told them. "Let us fight for health. There is death in the mines for our friends and our enemies.

There is no distinction between our lungs." "Let's all get together," he urged, "to make mining fit work for human beings." For many activists, the focus on silicosis meant more than strategy. Many of those trying to rebuild Mine Mill also suffered from the disease. According to the union, of the more than 500 men involved in the initial 1936 NLRB complaint, seventy-eight had already died, and about half of those still living were sick with silicosis. Tony McTeer was one of them.[79]

For the first time since NRA days, Mine Mill supporters in the Tri-State had both momentum and strong outside support. According to McTeer, the union added 300 new members in the month after the NLRB ruling, which gave it close to 500 total members in the district. The Blue Card Union was now gone. In early 1940, Frances Perkins announced that the Department of Labor would hold a conference in Joplin to focus attention on the silicosis crisis. During the April conference, Sheldon Dick and Lee Dick screened their experimental documentary film, *Men and Dust*, which was based on the Tri-State Survey Committee's preliminary report and narrated by actor and activist Will Geer. Eleanor Roosevelt also watched it in the White House. In her nationally syndicated My Day column, she expressed hope that the film would "awaken the interest of the people of the United States to make it easier for the unions to obtain proper working and living conditions."[80]

Yet the people of the Tri-State would themselves remain suspicious of, and in many cases hostile to, both union and government interventions. After its promising gains in late 1939, Mine Mill's campaign stalled. In April 1940, when Perkins spoke at the Local 15 union hall, "700 to 800 persons" turned up, including members, visitors, and their families. Three years before, the Blue Card Union had turned out 20,000. In June, Mine Mill received another boost when St. Louis Smelting and Refining Company, a National Lead Company affiliate that operated a big mine outside Baxter Springs, pledged to reinstate twenty strikers with back pay and abide by "the spirit as well as the substance" of the NLRA. Anticipating lucrative defense contracts, the company squared its account with the federal government rather than risk renewed NLRB scrutiny. It was "a big victory for the union, moral victory as much as otherwise," McTeer declared, "which will pave the way for organization into a real, legitimate union for the purposes of collective bargaining." To generate support, he convinced Mine Mill to hold its 1941 convention in Joplin. The union added sixty more members in the months that followed the St. Louis Smelting announcement, but it still needed more than moral victories to build a "real, legitimate union."[81]

Mine Mill advocates argued that unreliable federal backing between 1934 and 1939 made many miners too afraid to insist on their rights. "They know they have troubles aplenty, and what they are," McTeer told Perkins during the conference, "but they also know false hopes when they see them." "They have been fooled too often," he explained, "so that now they don't let themselves in for another deception." Another union member explained that it was hard "to get these boys to admit they'd like to have a union, 'cause they're afraid lettin' it be known they think that way might mean their jobs. I guess they remember, or have heard about how it used to be in the 'Blue Card Days.'" McTeer and other organizers believed that most miners wanted to join Mine Mill and that sustained federal attention would give them the courage to do so.[82]

It was dangerous to claim, however, that the long-standing conservatism of white American men could be overcome so easily. Many Tri-State miners remained loyal to the spirit of the Blue Card Union and full of enmity toward Mine Mill and the CIO. Evan Just, the new OPA executive secretary, reminded Perkins that Mine Mill, the Department of Labor's choice to speak for workers at the conference, "only represents an extremely small minority." "The real workmen in this district are not represented here today," he added. They stood aloof by choice, Just asserted, unwilling to follow work-shy radicals in the CIO. Rarely, however, did these people speak in public for themselves after 1939. At the Perkins conference, Cliff Titus, a preacher from Joplin, tried to give them voice. "The men who work in the Tri-State area are Americans, white Americans," he said. "They have quite a tradition behind them; they are pretty independent; they don't like to be bossed very much." Titus was sure of their commitment to the district's conservative traditions. "We are Americans," he declared, "and we believe in that individual opportunity, initiative and responsibility that goes to make America." These were partisan views in 1940, and Perkins did not accept them as gospel. And yet, years later, locals would hint at the resonant power of the Blue Card. "For years the story would be told of how a labor strike was broken with pick handles," historian Velma Nieberding wrote in the 1980s. The story of that strike and its end, she explained, yielded "old grudges" that "would grow, cancer-like in the hearts of many." In 1975, another miner, a Mine Mill member, insisted on anonymity when talking to a reporter about the strike. "Even after 40 years feeling runs pretty high when you get on that subject," he said. The reporter held tangible evidence of that high feeling in his hands, "a souvenir of those days," proudly preserved: "A pickhandle, inscribed in red: 'Sunday morning, April 11, 1937, Labor Demonstration, Picher, Okla. Blue Card Union, miner's strike riot.'"[83]

Despite federal intervention and Mine Mill's claimed resurgence, most miners could not deny the formidable power of the mining companies, whether they wanted to join the CIO or not. None was mightier than Eagle-Picher, which had bought its next-biggest competitor, Commerce Mining, for $10 million at the end of 1938. By 1940, the firm's production was more than double the combined output of the next two biggest companies, Federal Mining and St. Louis Smelting. Workers had reason to respect Eagle-Picher. Despite the downfall of the Blue Card Union and the contrition of St. Louis Smelting, Eagle-Picher continued to discriminate against actual or suspected Mine Mill members as it fought the NLRB ruling. "If you join the union, you lose your job," some men told occupational health scientist Alice Hamilton during the Perkins conference. To get hired, miners still needed a ground boss's good word, which required a reputation for loyalty. Eagle-Picher communicated its idea of loyalty by hiring former Blue Card officers Glenn Hickman and Joe Nolan; it also kept close ties with Mike Evans. Eagle-Picher offered some favors in exchange. Its in-house personnel decisions weakened the OPA's role in certifying prospective workers; the OPA closed the examination bureau in 1939 for lack of use. This made it easier for the company to relax health requirements for men it wanted and to exclude those it did not. Some men prospered by loyalty to the company; Mutt Mantle had risen from a shoveler to a shoveler boss to a ground boss.[84]

By 1940, most miners in the Tri-State were happy to hold jobs, with little energy or leverage to rebel, as they had in the past, because the district's industrial structure was changing, its scale shrinking. The recession of 1937–38 rocked the district. Total employment in its mines and mills fell from 6,000 to 4,500 in twelve months. "The lack of job security," Mildred Oliver reported, dominated her discussions with families in early 1939. That year's economic recovery, spurred by defense industry growth, revealed a new way of working. With no new ore discoveries since 1914, operators restructured production methods by reprocessing old chat piles, work that existing mill employees could accomplish. Between 1934 and 1939, companies produced 29 percent of the district's ore this way. Underground production ran less than half the all-time-high levels of the mid-1920s. It is telling that, of the entire 1930s, miners registered the best year of new ore production in 1930, with over 355,000 tons; that year had been the worst year by far since 1921. In 1940, miners produced only 330,000 tons of new ore. Companies also deployed more machinery. Fifteen years of mining had cut cavernous expanses that finally made mechanical loaders and diesel trucks easier to use. The job of shoveling was no longer so important and certainly not so lucrative; the best job now was machine man or in a mill, above ground. Overall, the operators now needed fewer workers, especially in the mines. In late

1941, despite rocketing wartime ore prices, they employed around 5,000 men total, a small increase from the low point of the 1937–38 recession. People were leaving. Mine Mill leader Ted Schasteen moved to California, along with many others. Picher's population fell by 25 percent during the decade. Tri-State workers lost hope in a new boom, even though another world war loomed.[85]

Whether nonunion out of fear or conviction, working-class men remained committed to an ideal of aggressive, physical strength that now informed a defensive, cruel pride, a self-satisfied meanness. Mine Mill could not shed the stain of illness, weakness, and inadequate men; the operators made sure of it. They were not "real workmen," the OPA declared. In statements to the press, Just blamed the silicosis crisis on "the remnant and offspring of a host of derelicts" who chose to live in "degradation and squalor" and were "mentally or physically unsuitable for employment in mining." The plight of Tony McTeer, who was incapacitated in a Joplin hospital by 1941, seemed to prove the point. These charges resonated among men who continued to indulge in casual violence off the job. When Lawrence Barr started working in the mines in 1938, he recalled, Picher was "a pretty tough little town," something he and other men took pride in. According to his wife, Theo Sisco Barr, whose father was a first-generation Picher miner, the work made people "mean and rough and tough, and I don't think there was nothing they wouldn't do." Long-time Picher resident Orville "Hoppy" Ray, who was born in 1925, concurred. "Most miners died before they was forty-five years old, but they were hard-rock miners—hard living." On Saturday nights, he recalled, "you'd see four, five, or six fights. Old knockdown drag-out fights. Everybody stood around and watched." John Mott, who was born in Picher in 1927, also remembered the hardness and violence of the place when he was growing up. "The miners would go down to the bars and they were tough, hard-working people," he said, "and they drank and fought. Some would rather fight than eat." Theo Sisco Barr emphasized that many men took the violence home. Miners "got drunk and cut each other up with knives," she recalled, or "beat their wives and kids." According to her, brute male force dominated community and private life in the Tri-State after the 1930s. "You could beat your wife and kids and cuss the government or do anything you was big enough to do," she said. "That's the truth."[86]

Women talked bluntly about the costs of living with Tri-State miners and the legacies of those who came before them. By the early 1940s, 206 families—more than 1,500 people total, one-fourth of Picher's population—were headed by widows or wives of miners who were too sick to work. Mrs. R.

came to Picher in 1915 from Arkansas with her husband and eight children. "When we got here," she told Mildred Oliver in 1939, "we wuz as stout a family then as you've ever seen." Since then, her husband and three sons "all took the miners' con and died." "Now there's only me an' the girls left." Her oldest son died at age thirty-one, leaving behind a widow and four children. Her second son died at age twenty-six. Her youngest son died at age eighteen. Some women knew the district would kill their husbands and urged them to leave. "Picher never struck me as a healthy place to live," Mrs. D. told Oliver. "I wuz always tryin' to git Tom to leave 'en go somewhere else but he wuz born around the mines an' liked workin' in 'em, an' he never did see my point, right up until he died of 'the con.'" Mrs. P. admitted, "Ever since I come to Picher I wished we could leave the minin' country an' go back to farmin'." But her "husband died of miners's con" after a series of gruesome hemorrhages. Now her son was sick. She had two healthy sons left and wanted "to get away from here more than I ever did 'cause I want to keep these last two out o' the mines if I can." Mrs. I., who had lived in the district since 1928, said her husband "stayed working in the mines long after he should have stopped." He was strong but would "never admit that he was sick, and he held up so well that it wasn't until almost the end that I realized he had TB." She never believed the district's promises and wanted to leave. "I always begged him to git out," she told Oliver, "but we never could put by enough money to take the risk of making a change. I never liked this mining business because, for one thing, we couldn't ever save anything while he was working." During her 1940 visit, Perkins met with a group of mining women like these, all silicosis widows. "Some of them spoke with bitterness, and some with question, and some with resignation," she said. But they spoke.[87]

Many women hoped that the federal government would bring help. Evelyn Hannon was twenty-seven years old in 1940, a wife and mother of four who had lived her whole life in Cherokee County, Kansas. "I hardly know how to begin," she wrote Perkins following the April conference. Her thirty-three-year-old husband, William, who moved to Treece from a nearby Oklahoma farm in the 1920s, was a striker still loyal to Mine Mill. But he had not worked since receiving a failing examination grade in 1935. Her father, who died in 1934, had been a shoveler. "We are afflicted with silicosis," Hannon told Perkins. "We have lived in this dust so long." They wanted to leave the district and hoped that the government could provide information about where to go. "We are willing to work if we have a chance," Hannon pleaded. Picher resident Edna Carey wrote to Perkins with information about her sister, Helen Bennett, whose husband had recently died from silicosis. Carey urged Perkins to do something because Bennett was now alone with three young daughters. "Your labor fight has been a brave one," Carey concluded, "but please, don't forget us down here."[88]

In October 1940, a full six months after the Joplin conference, the Department of Labor announced the outcome of the gathering: the formation of four committees to investigate the district's problems. The committees covered occupational hazards, social and health issues, workmen's compensation, and housing. If women like Evelyn Hannon and Edna Carey expected forceful action, this was not it. None of those appointed to serve on the committees could claim to represent the miners or their families. Those who came closest—George Maiden of the Kansas State Federation of Labor, Thelma Levering of the OFL, and David Fowler of the UMW and Oklahoma CIO—were all union officials with no firsthand knowledge of the district. No one from Mine Mill, locally or nationally, was included. The chair of the housing committee, however, was John Robinson, vice president of Eagle-Picher. The committees met for the first time four months later, in February 1941. The housing committee reported that the Farm Security Administration, the Federal Housing Administration, and the U.S. Housing Authority all rejected requests for help. Robinson concluded that any improvements would have to come from "the efforts and large financial support of local people." Mine Mill attended the first meetings of the committees on social and health problems and mine safety. Although a lot was discussed, the union reported, "no action was taken."[89]

EPILOGUE

War's return in 1941 gave new, late life to Tri-State mining. American and Allied military spending sparked heavy demand for zinc and lead. At first, federal Office of Price Administration caps on commodity prices hindered expanded mine production, particularly in places where only thin ore deposits remained, such as the Tri-State. In early 1942, the Office of Production Management authorized the Metals Reserve Company, a government entity that stockpiled strategic metal and minerals, to pay a premium price subsidy for copper, lead, and zinc from marginal mines that was at least 33 percent higher than the cap imposed by the price administration. In 1943, Tri-State operators received an average of $100 per ton for zinc concentrate, an all-time record, by far—twenty-two dollars more per ton than the previous high annual average in 1916. Government raises and extensions of the price subsidy pushed zinc to $117 per ton in 1946.[1]

These incentives made the Eagle-Picher Lead Company and the other firms "hoggish," according to Lawrence Barr. The operators took the money while it was there: more than $43 million for zinc and lead combined in 1943, when the Tri-State produced 30 percent of the nation's zinc and 7 percent of its lead. Only 1925 and 1926 had been richer. But ramped-up operations decimated already thinning ore deposits; average yields dropped from 5.00 percent in 1941 to 2.81 percent in 1946. As the money rolled in, district production declined from 478,403 tons in 1941 to 376,236 tons in 1943 and 258,373 tons in 1946—much of that from reprocessing chat piles. Eagle-Picher even mined the immense pillars it had left behind to hold up the ground, including one worth $1 million from underneath the Picher bank. This was the beginning of the end.[2]

Government mobilization gave Tri-State miners leverage again. The demands of the war created sharp labor shortages. The military draft began taking away healthy young men in late 1940. Many others went to work at new munitions plants in Baxter Springs and Parsons, Kansas, and Chouteau, Oklahoma, all within seventy-five miles of Picher. Still others found good defense jobs through the U.S. Employment Service office in Joplin. To get miners to stay, Eagle-Picher and other operators raised wages in line with ore prices according to the traditional sliding scale, up to $5.55 per day for machine men in 1941. By the summer of 1943, more than 5,400 men worked in the district. But sliding-scale raises usually could not match the high pay available elsewhere. Turnover was high.[3]

The International Union of Mine, Mill and Smelter Workers (hereafter referred to as Mine Mill) used the federal wartime labor regime to make its first major gains in the Tri-State. It reorganized by founding new locals to represent workers by company, not by town. The union focused exclusively on collective bargaining rights and pay. In May 1941, workers in Local 514 at the St. Louis Smelting and Refining Company, outside Baxter Springs, voted 254 to 131 to make Mine Mill their sole bargaining agent in a National Labor Relations Board (NLRB) certification election. The union pressed the resistant company for higher wages with a complaint to the new National War Labor Board (NWLB) in January 1942. The NWLB ruled in favor of Mine Mill's proposed wage scale in February, resulting in raises of one dollar per day; the contract made St. Louis Smelting a union shop. Mine Mill's delivery of tangible results boosted the campaign. "Our union boys point out to these unorganized workers that if they were organized they could increase their wages by at least a dollar a day," organizer Elwood Hain reported. Within the month, Local 489 won a certification election at Federal Mining and Smelting Company, the district's second-largest producer, and Local 596 obtained certification by card check at two small companies in Oronogo. All three locals won higher wages with a second NWLB complaint in early 1943. Meanwhile, smaller companies settled with the NLRB to avoid their own hearings stemming from the 1935 strike. By March 1943, sixteen firms had paid out more than $134,000 in back wages. By then, Mine Mill had more than 1,200 Tri-State members working under contract; it had over 80,000 dues-paying members nationwide. The union did not crack Eagle-Picher during the war, although the district's biggest firm could not escape its influence. To keep and attract workers, Eagle-Picher matched the union's NWLB-imposed wage scales. Mine Mill stalwart Tony McTeer did not live to see these successes; he died from silicosis two days before Christmas 1941.[4]

After the war ended, Tri-State workers finally rebelled against Eagle-Picher as part of the larger national uprising to protect wartime gains in the conversion to peace. The company gave ground only after its final appeal of the 1939 ruling ended in the U.S. Supreme Court in May 1945. In the months that followed, Eagle-Picher paid out more than $250,000 in back wages to the Mine Mill strikers. That breakthrough helped Local 108 win NLRB certification at Eagle-Picher's insulation facility in Joplin. The company was also under pressure from workers at its smelters, particularly the largest at

Henryetta, Oklahoma, where Mine Mill was pushing for a contract. Eagle-Picher signed an agreement with Local 108 in February 1946, its first union contract, but refused to bargain with the smelter workers. They went on strike in April, among more than 2 million American workers who went on strike for higher wages and stronger labor rights in 1945 and 1946. Mine Mill was at the center of that strike wave, with battles in Idaho, Montana, Alabama, and Arizona. The drive against Eagle-Picher inspired workers at its Central Mill in Cardin. They organized a new local, 861. In March, at a public rally in Picher attended by Reid Robinson, Local 861 announced a petition for an NLRB certification election. The union would first poll the mill workers, among whom its support was strongest, and then call for a vote among the underground workers. Meanwhile, Mine Mill won a series of elections at smaller mining companies in the district. The union was confident that organized labor would finally triumph in the Tri-State.[5]

The fight over price controls in the summer of 1946 galvanized Mine Mill's campaign. Conservatives in Congress fought hard to kill the Office of Price Administration, which both guarded consumers from inflation and provided lifeblood subsidies for marginal metal producers. Mine Mill was among many unions and consumer associations calling for its renewal as part of a broader postwar plan to make the American standard of living available to all. The union organized rallies in Picher to demand an extension. When Congress approved weakened price controls in June, however, President Truman vetoed the bill in the hopes of securing passage of something stronger. That veto hit hard in the Tri-State because it also stopped the premium price subsidies. Operators closed the mines to await further developments in Washington, a decision that gave the union's campaign urgency. In late July, the third week of the shutdown, Mine Mill rallied more than 700 mine and mill workers in Picher to demand the resurrection of both price controls and ore subsidies. Truman signed a renewal bill a few days later that extended consumer protections in diminished form until the end of the year and guaranteed the premium price subsidies for twelve months. The mines reopened. In August, Mine Mill won the certification election at Eagle-Picher's Central Mill, 232 to 113. It had momentum. In early September, the Henryetta smelter workers won their five-month strike with a contract and a wage increase. In November, miners at Eagle-Picher chose Mine Mill as their representative, 475 to 251. Support for the union among the miners "never was very strong," an organizer conceded, but they believed the mill workers would get a contract and did not want to be left behind. It was the union's seventeenth NLRB victory in the Tri-State since March 1946. Unwilling to risk a strike while government subsidies were still in effect,

Eagle-Picher agreed to a contract with Local 861, including both mine and mill workers, in the first week of December. It provided a union shop, grievance procedures, seniority rights, lines of progression, vacation benefits, and a wage scale that paid machine men $9.03 per day, a raise of $2.50. Eagle-Picher retained the right to send employees for physical examinations and reclassify them based on the result. By early 1947, Mine Mill represented nearly all Tri-State mine and mill workers—victory at last.[6]

A majority of Tri-State workers supported Mine Mill in 1946 because it delivered material gains, but the union could not sustain those gains without continued federal help. Despite union and company pleas, Congress allowed metal price subsidies to expire on June 30, 1947. District operators curtailed production while awaiting another potential renewal. With most mines closed, companies laid off 4,000 of the district's 6,000 mine and mill workers. Truman crushed their hopes by vetoing Congress's renewal bill in August. He advised those who lost their jobs to seek work elsewhere while the national economy was strong. Many small mining companies closed for good. District zinc production collapsed to 159,549 tons in 1948, the lowest since 1896. Mine Mill's first contract with Eagle-Picher had also expired on June 30, 1947. Local 861 prepared for "a showdown on wages" but postponed negotiations amid the closures and layoffs. That fall, union leaders tried to organize a local political action committee around demands for government operation of the mines but to little avail. In January 1948, Local 861 threatened to strike Eagle-Picher's remaining operations unless it got a wage increase, then backed down. Eagle-Picher continued to recognize Mine Mill and abide by the terms of the lapsed 1946 contract, mainly because the pact provided a rare point of stability. During contract negotiations in June 1948, Eagle-Picher offered Local 861 a raise of 10.2 cents per hour, or 9 percent. With no raise for eighteen months and inflation high, the union demanded thirty cents per hour, or 27 percent. When Eagle-Picher refused to budge, Local 861 voted to strike, 302 to 19.[7]

Mine Mill's 1948 strike reprised, on a smaller scale, the central elements of the 1935 strike. With 500 members, Local 861 blockaded Eagle-Picher's Central Mill. Unable to process ore, Eagle-Picher and twenty smaller mining companies, which together employed about half of the district's workers, ceased operations. The strike held for two months as negotiations stalled. The union insisted that Eagle-Picher could afford to raise wages after taking so much in federal subsidies since 1942. Eagle-Picher explained that the subsidies had stopped and accused Mine Mill's leadership of Communist subversion. In August, the company ran full-page spreads in district newspapers that showed evidence linking national officers Reid Robinson and

Maurice Travis to the Communist Party. As the nascent Cold War fueled the political power of anti-Communism, this was Mine Mill's principal weakness. The union had been riven with factional splits over the role and influence of Communists since the early 1940s. Mine Mill's left wing seemed to win in 1947 when Travis, a known Communist, replaced centrist Robinson as president. The issue grew graver in June 1947 when Congress responded to the recent strike wave with the Taft-Hartley amendment to the National Labor Relations Act. In addition to new restrictions on collective action, Taft-Hartley required all union officers to sign affidavits declaring that they did not belong to the Communist Party or believe in or support any organization that advocated overthrowing the U.S. government. Unions that did not submit affidavits would no longer have standing with the NLRB. Indignant at this violation of civil liberties and hopeful of repeal, Mine Mill's leaders, along with many others in the Congress of Industrial Organizations (CIO), did not sign affidavits. This would doom Mine Mill in many places but especially in the Tri-State.[8]

In early September 1948, Walter Cherry, a thirty-five-year-old welder at Eagle-Picher, launched a back-to-work movement that forced Mine Mill to end the strike. Cherry had deep roots in the district: born near Granby, Missouri, in 1913, he moved to Ottawa County a few years later when his father took work as a miner. Likely backed by the company, Cherry pledged that this group, at first called the Union of America, would be "a peaceful organization based on Americanism, not communism." Cherry assured potential members that all of its officers would sign anti-Communist affidavits, remain independent of all other unions and all companies, and crucially, petition the NLRB for a certification election to replace Mine Mill. More than 150 workers joined. Mine Mill agreed contract terms with Eagle-Picher the following day. Although the union achieved a raise of 14.2 cents an hour that other companies matched, it was losing support. Only 225 members voted on the contract, which would run until June 30, 1949. Cherry kept organizing. Claiming people were "tired of the word 'union,'" he changed his group's name to the Tri-State Mine and Mill Workers' Association. In November, Cherry's association collected enough signatures to petition the NLRB for a new election.[9]

After Cherry's gains, the American Federation of Labor (AFL) joined the challenge to Mine Mill in the Tri-State. In October, the United Cement, Lime, and Gypsum Workers International Union (hereafter referred to as AFL Cement), an AFL union chartered in 1939 to counter the CIO, launched an organizing drive to raid Mine Mill's locals. In November, Mine Mill organizers informed the national office that AFL Cement had won the allegiance of a majority of workers in Eagle-Picher's Central Mill and other surface operations. Later that month, workers at Eagle-Picher's insulation works in Joplin chose it to represent them in an NLRB election, 196 to 15; Local 108 vanished. Mine Mill was not on the ballot because it had not submitted anti-Communist affidavits to the NLRB. Meanwhile, Cherry's association gained more followers among the miners. Mine Mill organizers believed that as many as half of the underground workers in Local 861 now followed Cherry. In all cases, Elwood Hain reported, "the Communist question is quite pertinent." In April 1949, AFL Cement petitioned the NLRB for a representation election at Eagle-Picher. The board decided to hold a single election in early June with both AFL Cement and Cherry's association on the ballot. Both groups deployed blunt anti-Communist attacks on Mine Mill, which was not on the ballot. If workers wanted to keep Mine Mill, they had to vote "neither."[10]

This battle for representation took place amid a dismal backdrop of falling ore prices and production cuts. In late April, many smaller companies again closed in response to falling zinc prices, throwing 950 of the district's remaining 2,800 workers into unemployment. Workers in nonunion mines suffered pay cuts of $2.25 per day. St. Louis Smelting laid off its 250 workers. Federal Mining shut its last remaining mines in Oklahoma and Missouri, including the Klondike, the last active mine in Granby. Eagle-Picher continued to operate at the contract wage scale. Only 1,400 mine and mill workers had jobs in the district as the NLRB election approached.[11]

Mine Mill fought back against AFL Cement and Cherry with appeals to the struggle of the 1930s. "If this union-busting AFL organization is successful, it means bad working conditions, low wages and little kids going to school ragged and hungry as they did before in the old AFL Mike Evans-Joe Nolan reign of terror," one circular declared, "and it may lead to more pick-handle parades and beatings and suffering." In a radio broadcast, Mine Mill reminded listeners how hard it had been for the Western Federation of Miners (WFM), "of which your present union is descendant, to organize the Tri-State years ago." It warned of a return to the "'Blue Card' days." No doubt, it predicted, "some of you will weaken" and others "will just crawl before the boss in hopes of saving their own selfish individual skins." "This is the proven time-tested way of disaster in the Tri-State," Mine Mill declared. "There is only one way out; to stand up like the men who founded this union, like the men who built it in the Tri-State."[12]

Workers at Eagle-Picher voted for anti-Communist Americanism. AFL Cement won the election with 448 votes against 232 votes for "neither," which included both those who remained loyal to Mine Mill and those who wanted no union at all. Fifty workers voted for Cherry's association. By June 3, 1949, Local 861 was no more. In the weeks that followed, the district's agony only intensified. With the Mine Mill contract set to expire on June 30, Eagle-Picher demanded harsh new terms: a 30 percent pay cut of $3.08 per day and the reinstatement of the market-based sliding scale. Having jumped into the fire, AFL Cement rejected the offer, 139 to 98. Eagle-Picher closed its operations on July 1. More than 1,400 employees were again thrown out of work: 800 at Eagle-Picher and 600 at small companies that sold ore to its Central Mill. After three weeks, another back-to-work movement organized, this time led by Clotis Cates, an Eagle-Picher hoisterman. Cates soon had 500 Eagle-Picher employees ready to return to work on the company's terms. AFL Cement capitulated in early August: 100 members voted for the pay cut and sliding scale, 82 against. Eagle-Picher resumed operations; many other companies never did. Picher's unemployment rate for men spiked to more than 20 percent.[13]

Despite rising Cold War demand for metal, Tri-State operators could not compete under the nation's new, freer trade policies. Federal defense spending ballooned in the early 1950s after American entry into the Korean War and a massive military expansion. As the Department of Defense stockpiled strategic metals, the price of zinc ore and lead mineral soared in 1951 to $135 and $246 per ton, respectively. Rather than rely on marginal districts like the Tri-State to supply this ore, the Truman administration now drew on cheaper foreign sources. This was possible because the 1951 General Agreement on Trade and Tariffs, the latest step toward a global free trade regime to contain Communism, lowered duties on imported zinc and lead to historic levels. In 1952, imports of zinc surpassed production from American mines for the first time since 1873, the year Joplin was founded. As foreign ore flooded the market, prices fell below eighty dollars per ton. Local congressional representatives offered bills to raise the tariff on zinc and lead and to reinstitute the subsidy for domestic producers, but both bills failed. Locals accused the federal government of "serving America last." President Dwight Eisenhower, the first Republican to win Ottawa County since Hoover, maintained Truman's policy. Tri-State production fell to 71,000 tons in 1953, the lowest amount in seventy years. Picher was nearing collapse, literally. In 1951, Eagle-Picher closed a four-block area in the heart of town because engineers feared the ground would cave in after the company had mined the support pillars. The national press covering the subsequent evictions and evacuations called Picher a "dying town." Only 3,900 people still lived there.[14]

The workers who remained looked belatedly to the CIO for help in a rearguard fight to stall oblivion. The sliding-scale provision of the 1949 contract was devastating their weekly pay as imports drove down ore prices. Workers at Eagle-Picher were fed up with AFL Cement. In April 1953, they voted for the United Steelworkers of America (USW) to represent them and organized as USW Local 4915. The USW already represented the Henryetta smelter workers after raiding the Mine Mill local there in 1950. They had won a wage increase in 1952. Local 4915 hoped for the same when it demanded a raise of forty-eight cents per hour in May 1953; Eagle-Picher offered four cents. With talks in stalemate, Local 4915 went on strike in June. By again closing the Central Mill, the strikers idled the 1,100 workers who remained in the district, over 800 of them at Eagle-Picher. The strike lasted for six months. Local 4915 accepted a raise of fifteen cents per hour in December. In stark contrast to district production—which delivered around 110,000 tons of ore in the mid-1950s—they remained bullish. In July 1955, Local 4915 went on strike for three days and won a raise of ten cents per hour. That December members threatened another strike and won five cents more. The flurry of latter-day defiance continued in September 1956, when another threatened strike won an eight-cent raise. They had finally learned the lesson.[15]

The Tri-State mining district went under in 1957. Eagle-Picher ceased operations twice due to low zinc prices: in April, for five weeks, and again in July. The few remaining small companies that relied on its Central Mill closed for good. Local 4915 continued to meet with company officials through the end of the year about a new agreement that might allow it to restart operations. The recession of late 1957 ruined those chances. Local 4915 held on for another year, stubborn to keep hope alive amid crushing unemployment. The union threw a Christmas party in 1958 complete with "an appearance by Santa Claus with treats for the children." The mines never reopened.[16]

There was nothing ironic about the Tri-State miners turning to the CIO in the 1950s. After Taft-Hartley, conservative and centrist forces within the congress attacked the left-led unions that had refused to accommodate the Red Scare. By February 1950, the CIO had expelled eleven of them, including Mine Mill. The expelled unions were not only the most robust critics of capitalism but also the most committed to organizing the unorganized and to promoting racial and gender equality. The CIO, and particularly the USW, now pursued the narrow goal of delivering more pay and benefits for those who were already members. With the blessing of president Philip Murray, CIO unions joined the AFL in membership raids on their former partners, targeting more than 1,200 locals combined between 1951 and 1953. The CIO raids against labor's left wing emphasized anti-Communism and Americanism, just like William Green and the AFL; all "flag-waving, breast-beating, and Red-baiting," a contemporary reported. Some raids, such as the USW drive against Mine Mill in Alabama, also relied on overt appeals to white supremacy. In 1955, the CIO and AFL merged around these common commitments: the economic and social privileges of white working-class men, patriotic nationalism, and the defense of American capitalism.[17]

If there was any irony in the Tri-State story, it was that its miners and mill workers had found a union that fully represented them and their long history but too late to save themselves. Since at least 1896, they had pursued their own material self-interest, primarily in the form of higher wages, above all else. Tri-State miners showed little loyalty to the district's mining companies. For a half century or more, they seldom shied away from pugnacious opposition to the biggest firms, whether the American Zinc, Lead, and Smelting Company or Eagle-Picher. They switched employers without notice, often to go work as strikebreakers for higher pay in other districts. They used the courts and government agencies like the National Recovery Administration to advance their interests. Many went on strike themselves, at first in short, small wildcat actions but later, particularly in 1915, 1935, and 1948, in organized efforts with district-wide impacts. They often seemed on the verge of organizing in unions—in 1900, 1911, 1915–16, 1924, and 1933–35—but only when the unions promised material gains within the capitalist system and defended the privileges of white Americans. When independent action failed, Tri-State miners were drawn most to the AFL because its policies reflected their traditions and its devolved structure offered the most freedom from outside interference. Sometimes, significant numbers of them joined the WFM or Mine Mill but only really ever as a means to get into the AFL. Its leaders, whether Samuel Gompers, Green, or state-level officials, often reciprocated the attraction in the hopes of turning the Tri-State into a bastion of conservative metal-mining unionism. The AFL's enthusiastic acceptance of the Blue Card Union testified to the elemental power of that affinity.

Most Tri-State miners remained mistrustful of, if not openly hostile to, the broad-based solidarity of the WFM and Mine Mill for fifty years or more. After the 1896 Leadville strike, Tri-State miners regarded the militant anticapitalism and ethnic diversity of the WFM as a threat to their own prerogatives as native-born white men with entrepreneurial ambitions. Their clashes in western strike zones deepened the chasm: encouraging Tri-State miners to double down on their sense of racial and national advantage while leading the union into more confrontations to destroy the prevailing system. Men on both sides of those clashes struggled to overcome the divide, even after the WFM moderated its aims and tactics. While some Tri-State miners would join the WFM and Mine Mill, they continued to see the union as more radical and more alien than the AFL and thus always suspect.

It was easy for Tri-State miners to see the WFM as a threat in the 1890s because they had seldom doubted their chances to succeed, as workers, under capitalism. From the Granby stampede in the 1850s to the zinc boom of the 1880s, Tri-State miners had as prospectors and owner-operators created a poor man's camp with their own muscle, appetite for chance, and astute navigation of the metal market, at great risk and often greater cost. Despite a legion of failures, many men in the district continued to enjoy real, if small-scale, opportunities to work for themselves and pocket the profits when increasing numbers of American workers no longer did. They had faith in their own self-interest in the market, which set them in opposition to both radical unions and large corporations. From the 1870s on, Tri-State miners opposed monopolies that threatened to crush small-scale producers like them. Many went so far as to support the antimonopoly proposals of the Knights of Labor and the Greenback-Labor and Union Labor Parties. Yet most Tri-State miners refused to join with other miners of metal and coal who organized together, particularly in the Knights, as permanent wage-workers to confront those same monopolies. Tri-State miners would not abandon their owner-operator ambitions and traditions; they believed that they were special, that the problems others faced were not their problems. That conclusion set them on a collision course with the WFM in Colorado and Idaho, where strikebreakers unironically identified the union as another kind of monopoly trying to limit their freedom of action.

It also framed the way they and their sons understood wage labor once the poor man's camp was no more. Led first by strikebreakers and then by shovelers, Tri-State miners accommodated permanent wage labor with increasingly aggressive physical assertions tied to claims on market-based performance incentives, especially the piece rate, despite the growing silicosis crisis. At the height of their influence, in the 1915 strike, shovelers used their essential masculine power not to end the piece rate but to induce their employers to yoke the piece rate to a sliding scale that followed market prices; their triumph, then, was to further commit themselves to capitalism, just like their predecessors had with leasehold provisions on zinc ore. This approach drew upon a profound confidence in the privileges and respect due white, native-born, working-class men like them—a confidence that the Picher boom seemed to confirm.

For Tri-State miners, American nationalism was an essential white working-class vehicle that advanced both the promises of capitalism and the privileges of race and nativity. The nation-state delivered real, material benefits to them. The federal government cleared Native Americans from the land before 1840, facilitated the railroad links that connected the area to national markets, provided tariff barriers beginning in the 1890s that protected domestic ore producers from foreign competition, guarded strikebreakers across the West, cut off most foreign immigration in the 1910s and 1920s, reconstructed capitalism in the 1930s, paid premium price subsidies that kept the mines open in the 1940s, and helped silence anticapitalist radicals in two Red Scares. Most important, the United States fought wars, lots of them, which each demanded more lead and zinc than the last. These wars not only boosted market prices and wages, especially under sliding-scale contracts, but also reaffirmed who and what the nation was for. Beginning with the Spanish-American War, if not earlier, nationalists used the state to create and promote exclusionary Americanism. Tri-State miners made strong claims to this ethnonationalist project, particularly during wartime. They also asserted the prerogatives of Americanism themselves in creative, often violent ways for their own ends in the absence of war, as prospectors, strikebreakers, shovelers, and back-to-workers against supposed foreign union foes and sometimes their own employers. Between 1896 and 1920, they turned Americanism into a newly belligerent, working-class male drive for racial and nativist privilege in a political economy increasingly dominated by middle-class voters and corporations.

The Great Depression almost broke the power of white working-class Americanism in the Tri-State. The district hit peak production in 1926. The subsequent decline undercut the influence of its miners and allowed newly powerful companies like Eagle-Picher to assert control over them and their behavior with the help of federal agencies. The companies adopted a managerial logic of efficiency that threatened to remove the last means working-class white men had to command a livelihood that reflected their sense of racial and national status: their willingness to risk their bodies for economic incentives. In the Tri-State, that strategy turned doctors, insurance companies, and government-backed health plans into the new enemies of poor man's camp miners.

Tri-State miners looked with hope to the New Deal for a restoration of past glories, not a revolution. For some, the New Deal made unionism make sense without challenging the tenets of white working-class Americanism. Some rallied to CIO social democracy, but only a minority. The majority remained much like they were in the 1920s, now more open to an organized response to the economic crisis that pledged to empower working-class men—but in a changing, uncertain political climate. Exploiting confusion exacerbated by shaky federal laws, companies co-opted the district's violent, antiradical, nationalist traditions with the Blue Card Union to foment a popular backlash against Mine Mill and the new CIO's model of social democratic unionism. Despite immigration restriction, foreign threats still mattered in the 1930s as anti-Communism and anti-Semitism gave new expression to older forms of nativism and xenophobia. The New Deal labor regime rallied to kill the Blue Card Union but not before the AFL reinforced the white working-class conservative ideas at its core. Tri-State miners did not turn to those ideas in simple reaction to the CIO; they had cultivated them, in word and deed, since the nineteenth century.

With supporters like those in the Tri-State, the New Deal, and especially its wartime extension, was made in part by the conservatism of white working-class men. While voters across the district supported Roosevelt in 1932 and 1936, they did not remake themselves to do so. Hoover had won a landslide in Jasper and Ottawa Counties in 1928, with 71 percent and 64 percent of the vote, respectively. That support reflected the district's prosperity in the mid-1920s as well as the power of the Republican consensus on economic protection and nativist exclusion. Roosevelt won over a swath of these voters with the promise of an economic recovery that would again include them. The Tri-State story of the 1930s was mainly a battle to define what that would mean, with a minority vision of a new future challenging the majority's long-held certainties. Votes for the New Deal only told so much: members of both Mine Mill and the Blue Card Union selected Roosevelt in 1936—the latter in far greater numbers. The march of social democracy was not held back by southern racists in Congress alone. White working-class conservatives like those in Tri-State helped define the limits of the New Deal from within the coalition by resisting the CIO, supporting its AFL rivals, and defending white supremacy and narrow Americanism, often with violence.

The federal wartime state proved its New Deal by reaffirming the privileges of white, working-class men. Once again, a war boom provided jobs to replace those lost as district production waned. The remaining workers benefited from government help through price subsidies that supplied profits beyond what the market would bear. Mine Mill took advantage of its standing with the NWLB to push for higher wages that beat the sliding scale, which brought miners back to its locals. To maintain that standing, Mine Mill turned itself into an agent of Americanism by backing the war with a pledge not to disrupt production with strikes. Despite its democratic and progressive heritage, and important new locals with diverse memberships in Alabama and New Mexico, the union did not challenge the racial and ethnic exclusions that Tri-State miners had demanded since 1900; nor did it restrain its white members elsewhere, such as Butte, Montana, where a 1942 hate strike targeted African Americans. Tri-State miners looked and sounded like most other white union workers in the late 1940s as the war mobilization entrenched the New Deal order in its most conservative essential form: strong state support for capitalism and a collective bargaining regime that delivered material gains for white working-class men.[18]

While trade liberalization dealt the death blow to Picher in the 1950s, the Tri-State mining district was already long into a decline that had begun in the late 1920s. The population of Picher fell by half between 1930 and 1950 and then by another 35 percent in the next decade. As the workforce shrank, and mechanical loaders replaced shovelers, the intergenerational transfers of skill and meaning slowed. When World War II ended, most young men looked elsewhere for work. Lawrence Barr "decided that a man wasn't supposed to live his life underground." He left the mines after the war to start an air-conditioning business in Tulsa. Other would-be miners went to Joplin, now home to a growing pack of trucking companies. Mutt Mantle refused to have his son follow him into the mines. Every day after work, he threw batting practice to Mickey in their yard in Commerce. People in the Tri-State experienced the dislocations of deindustrialization earlier than most other Americans, but they would not be alone.[19]

Local historian Genevieve Stovall Craig worried that people would forget about the mines as the nation enjoyed postwar prosperity. In a 1955 essay for *Ford Times*, the car company's monthly travel magazine, she described the town's past and present for tourists who might travel that stretch of Route 66 and wonder what had happened there. Craig imagined the view from atop one of the chat piles: "The clamorous mills glitter as their lights begin to flicker; derricks and hoppers punctuate the prairies." "You recognize the wealth the earth yields so that we can have so gentle a thing as baby powder, as well as paint, dyes, linoleum, ink, adhesive tape, wash tubs, barn roofs, pipe organs and smoke screens." Her readers would also see "the high-fenced, desolate heart of Picher where, because of threatened cave-ins, blocks of its proudest buildings are leveled into a rubbish graveyard dense in weeds and cotton wood trees." Once the metal had been stripped out of the mines, all that was left were the chat piles, mountains of waste. Men, armies of them, had moved those mountains, metal and all, out of the ground with little more than shovels. Now, Craig wrote, Picher clung to its "perishing wealth," with no future. She could not forget the men who made it happen or the costs of the making. Her father had been a shoveler in Carterville. He died from silicosis in 1916, when Genevieve Craig was six years old. Now, she informed her readers, Picher's children were "uninterested in its history." Craig understood that "inevitably the end must come to Mother Earth's generosity, the day when the last vein is pillaged, when there is no further discovery."[20]

Some miners still believed. Prospecting scavengers, known as gougers, worked abandoned leaseholds. These men carried on in the tradition of the poor man's camp a century after the Granby stampede. The gougers were "a unique breed of men," Craig explained, "the 'old-time Saturday night hell raisers' who gave Picher a name in the lusty boom days." "Descending as far as three hundred feet underground with pick and carbide lamp, the 'old-timers' haul the sparse lead gleanings to the top in old pots and pails, using an ancient motor as a hoist." The gouger dug "against the certain belief that around the next pillar, in the next drift, there's the rich, the solid vein, just for him."[21]

Even at its end, the Tri-State district lived on in its favorite son, Mickey Mantle, who exemplified the power and pitfalls of white working-class masculinity in 1950s America. After a season with the Joplin Miners in 1950, Mantle debuted with the New York Yankees the following year at age nineteen. By 1956, the last full year of production in Picher, Mantle was the nation's most famous baseball player on its best team. He was strong, white, and native born, an Okie from a poor town called Commerce. "The story of Mickey Mantle reads almost like a deadpan parody of the red-blooded, All-American boy reaching fame and fortune," a midseason profile declared. He went on to win the Triple Crown and a World Series title that year and was named the American League's most valuable player. What impressed people most was his prodigious strength. "Take it from me," declared his coach Bill Dickey, who had played with Babe Ruth, "Mantle hits a ball harder and farther than any man I've ever seen." He swung with reckless might, what one coach later described as "pure, blue-collar farm-boy aggressiveness." In an earlier time, Mickey Mantle would have been a hell of a shoveler.[22]

Mantle never forgot where he came from and could not escape its darkest legacies. For all of his physical power, Mickey feared dying young like many of the men he had known growing up, including his grandfather, two uncles, and father. Mutt Mantle died of cancer in 1952; he is buried in Miami, Oklahoma, a few hundred feet from Joe Nolan. Mickey was convinced that Mutt died of silicosis. Mining "was killing" him, Mickey later explained. "Every time he took a breath, the dust and dampness went into his lungs. Coughed up gobs of phlegm and never saw a doctor" because Mutt reckoned "he'd only be told it was 'miner's disease.'" "So what the hell," Mutt would say, "live while you can." "He was always so strong," Mickey recalled, and then he

Picher shoveler, Cardin, Oklahoma, 1943. Library of Congress, Prints &
Photographs Division, FSA/OWI Collection, LC-DIG-fsa-8b08304.

weakened fast and was dead. "I always wished my dad could be somebody
other than a miner."[23]

Most people who stayed in Picher had no patience for regretful thinking. Velma Nieberding, a local historian who had moved to Miami in 1915,
anticipated that many readers of her 1983 history of Ottawa County might
wonder whether it all had been worth it. "Would you have the land as it
was before the mines?" she asked readers. Would you erase the whole history of the Tri-State district if you could? Nieberding must have harbored
her own doubts about living in such a place. Her first husband, a Miami
police officer, was gunned down by thieves in 1934. Since the mines closed,
Nieberding and her neighbors lived amid the environmental fallout from
the mines—subsidence, residual lead contamination, and acidified ground-

water. The new Environmental Protection Agency listed Picher as one of its worst Superfund sites in 1983. Nieberding made it clear that they wanted no pity. "Do not use the words waste, pain, tragedy and death. These are the words of failure," she instructed. "The era of King Jack was lusty, exciting, fearful, prodigious and cruelly strong. But it was not failure, not exploitation, not unbridled waste. It was exultant progress."[24]

ACKNOWLEDGMENTS

I benefited from the help of many people while working on this book. Virginia Laas not only inspired the project with her story about a woman's unlikely attachment to a beautiful chat pile, but she also shared research material and knowledge about Joplin and the Tri-State district and read the whole manuscript. Her influence runs even deeper than that. Virginia took me under her wing during my undergraduate days at Missouri Southern State College and helped set me on the path to becoming a professional historian. Without her help and advice I would probably be managing some department in a Big Box store right now. Kimberly Harper generously opened her files about Joplin to me and read parts of the work, for which I am grateful. Hanna Jameson and Jessica Wilkerson read most of this book at a late stage and helped me to improve it considerably. They are great readers and even greater colleagues. I have also benefited from discussions with David Anderson, Gergely Baics, Larry Cebula, Greg Downs, Rosemary Feurer, Erik Gellman, David Roediger, and David Rosner about the book's arguments. Sanjay Paul was an excellent fact-checker and interlocutor; he is an impressive historian in the making. Nat Case made the wonderful maps. Monica Campbell saved her PhD supervisor during a research trip of her own by photographing some material that I had overlooked. Carol Roll helped with a trip to our hometown library on my behalf (thanks, Mom!). I am indebted to them all.

The editorial and production teams at the University of North Carolina Press have been terrific to work with. I am especially grateful to Brandon Proia, whose enthusiasm and advice gave me the energy to get this book finished. Many thanks also to Leon Fink and Ken Fones-Wolf whose evaluations of the manuscript at various stages made it better. Their scholarship has been inspirational to me as a labor historian, and I feel lucky to have had them as readers.

I had the good fortune of presenting parts of this project to a number of critical and encouraging audiences over the past six years or so: the Annenberg Seminar in History at the University of Pennsylvania, the Shelby Cullom Davis Center Seminar at Princeton University, the Center for the Study of Southern Culture and the Arch Dalrymple III Department of History at the University of Mississippi, the Labor and Working-Class History Association annual meeting, and the Newberry Seminar in Labor History. Your responses made this book stronger. The errors and faults that remain are all my own.

Some of this research appeared previously in "Sympathy for the Devil: The Notorious Career of Missouri's Strikebreaking Metal Miners, 1896–1910," *Labor* 11, no. 4 (Winter 2014): 11–37; and "Faith Powers and Gambling Spirits in Late Gilded Age Metal Mining," in *The Pew and the Picket Line: Christianity and the American Working Class*, ed. Christopher D. Cantwell, Heath W. Carter, and Janine Giordano Drake (Urbana: University of Illinois Press, 2016), 74–95. Thanks to everyone involved in the review and publication of those pieces—your assistance shaped the book project for the better.

I have worked in academic departments with some great people who helped me in ways big and small. Robert Cook and Richard Follett provided considerable support as department chairs and colleagues at the University of Sussex during the project's early stages. Joseph Ward and Noell Howell Wilson continued to aid and abet the project after I moved to the University of Mississippi. More broadly, these people especially have made work interesting and fun over the past ten years: Mikaëla Adams, Paul Betts, Jesse Cromwell, Shennette Garrett-Scott, Jonathan Gienapp, Darren Grem, Garrett Felber, David Fragoso Gonzalez, Douglas Haynes, Zachary Kagan Guthrie, Theresa Levitt, Rebecca Marchiel, John Ondrovcik, Ted Ownby, Eva Payne, Paul Polgar, Chuck Ross, Brian Sherry, Peter Thilly, Anne Twitty, Jeff Watt, and Clive Webb.

I could never have attempted the archival scope of this project without contributions of money and time from the following institutions: the British Academy; the Shelby Cullom Davis Center for Historical Studies at Princeton University; the School of History, Art History, and Philosophy at the University of Sussex; the Arch Dalrymple III Department of History and the College of Liberal Arts at the University of Mississippi; and the National Endowment for the Humanities.

Once I got to those archives, I benefited from the abundant knowledge and assistance of excellent archivists and librarians: John Bradbury and Kathleen Seale at the State Historical Society of Missouri Research Center in Rolla; David Hays at the University of Colorado Boulder; Randy Roberts and Steve Cox at Pittsburg State University; Charles Nodler at Missouri Southern State University; Mary Billington at the Baxter Springs Heritage Center and Museum; and Christopher Wiseman at the Joplin History and Mineral Museum. Mary, Christopher, Kathleen, and Steve were especially helpful and generous with image reproductions and permissions. I am especially indebted to Randy Roberts because twenty years ago he let me write a finding aid for the Julius A. Wayland/*Appeal to Reason* archival collection. That summer in Pittsburg convinced me that there were many important stories close to home that had not yet been told.

This book tells one of those stories, painful but necessary to hear. I dedicate it to R. S. and M. S. R., who are always close to my mind and heart no matter how far away we often are in person. One of these days we will hit our pay dirt.

NOTES

ABBREVIATIONS

AZR   R10, American Zinc, Lead, and Smelting Company Records, 1901–1965, State Historical Society of Missouri Research Center, Rolla, Missouri

b.   box

Baxter   Small Manuscripts Collection, Baxter Springs Heritage Center and Museum, Baxter Springs, Kansas

CUA   Catholic University of America, American Catholic History Research Center and University Archives, Washington, D.C.

Dun   R. G. Dun & Co. Credit Report Volumes, Harvard University Business School, Baker Library, Cambridge, Massachusetts

Easton   Stanley Easton (1873–1961) Papers, 1900–1916, Manuscript Group 5, University of Idaho, Special Collections and Archives, Moscow, Idaho

EMJ   *Engineering and Mining Journal*

f.   folder

FPP   Frances Perkins Papers, 1895–1965, Columbia University Rare Book and Manuscript Library, New York, New York

JG   *Joplin Globe/Joplin Daily Globe*

JH   *Joplin Morning Herald/Joplin Daily Herald*

JN   *Joplin Daily News*

JNH   *Joplin News Herald*

JUL   *Journal of United Labor*

LSR   Leadville Strike Reports Collection, 1896–1899, MS334, History Colorado, Denver, Colorado

MBLS   Missouri Bureau of Labor Statistics

MBM   Missouri Bureau of Mines

Meany   George Meany Memorial AFL-CIO Archives, University of Maryland Libraries, Special Collections and University Archives, College Park, Maryland

MHM   Missouri History Museum Library and Research Center, St. Louis, Missouri

MM   International Union of Mine, Mill and Smelter Workers

MNR   *Miami News-Record/Miami Daily News-Record*

MSA   Missouri State Archives, Jefferson City, Missouri

MSS   Missouri Secretary of State

NARA-CP   National Archives and Records Administration, College Park, Maryland

NARA-FW   National Archives and Records Administration, Fort Worth, Texas

OFLMC   Oklahoma State Federation of Labor Manuscript Collection, 1907–1958, University of Oklahoma, Western History Collections, Norman, Oklahoma

OHC   Oklahoma History Center, Oklahoma City, Oklahoma

OR	*War of the Rebellion: A Compilation of the Official Records of the Union and Confederate Armies*. 128 vols. Washington, D.C.: Government Printing Office, 1880–1901. Unless otherwise indicated, all references are to Series 1. *OR* citations take the following form: volume(part number where applicable):page number.
PMC	Picher Mining Collection, Pittsburg State University, Axe Library Special Collections and Archives, Pittsburg, Kansas
Rolla	State Historical Society of Missouri Research Center, Rolla, Missouri
SHSK	Kansas State Historical Society, Topeka, Kansas
SP	George R. Suggs Papers, Southeast Missouri State University, Special Collections and Archives, Cape Girardeau, Missouri
Tamiment	New York University, Tamiment Library/Robert F. Wagner Archives, New York, New York
UCL	University of Colorado Boulder Libraries, Special Collections and Archives, Boulder, Colorado
UMW-P	President's Office Correspondence with Districts, 1894–1983, United Mine Workers of America, Pennsylvania State University, Eberly Family Special Collections Library, State College, Pennsylvania
WCF	William J. Cassidy Files on the Tri-State Mining District of Missouri, Kansas, and Oklahoma, TAM 063, New York University, Tamiment Library/Robert F. Wagner Archives, New York, New York
WCR	*Webb City Register/Webb City Daily Register*
WFM/IUMM&SW	
	Western Federation of Miners/International Union of Mine, Mill and Smelter Workers Collection, University of Colorado Boulder Libraries, Special Collections and Archives, Boulder, Colorado

INTRODUCTION

1. In order of quotation: *Miners Magazine*, May 1901, 6–7; executive board report, July 18, 1910, 3:5, WFM/IUMM&SW; "Zinc Miners Gaining," *Oklahoma Federationist*, December 5, 1924; Mills, "Joplin Zinc," 657–58; Harrington et al., "Dust-Ventilation Investigation," 32, RG70, NARA-FW.

2. This framing governs the major syntheses of the field: Lichtenstein, *State of the Union*; Dubofsky and McCartin, *Labor in America*; Zieger, Minchin, and Gall, *American Workers, American Unions*; and Storch, *Working Hard*, 303.

3. Norwood, *Strike-Breaking and Intimidation*; Gaventa, *Power and Powerlessness*; Kelly, *Race, Class, and Power*; Harris, *Bloodless Victories*; Tomlins, *State and the Unions*; Simon, *Fabric of Defeat*; Simon, "Rethinking"; Pearson, *Reform or Repression*; Feurer and Pearson, *Against Labor*; Du Bois, *Black Reconstruction*; W. E. B. Du Bois, "The Black Man and the Unions," *Crisis*, March 1918, 216–17; Foner, *From the Founding*; Taillon, *Good, Reliable, White Men*; Mink, *Old Labor*; Saxton, *Indispensable Enemy*; Simon, "Appeal of Cole Blease"; Brattain, *Politics of Whiteness*; Nelson, *Divided We Stand*; Stein, *Running Steel, Running America*; Ervin, *Gateway to Equality*; Lombardo, *Blue-Collar Conservatism*. For this interpretation of the Tri-State district in the 1930s, see Suggs, *Union Busting*.

4. Honey, *Southern Labor*; Cohen, *Making a New Deal*; Goldfield, "Race and the CIO," 1–32; Andrews, *Killing for Coal*; Arnesen, *Waterfront Workers*; Gellman and Roll, *Gospel*; Cowie, *Great Exception*.

5. Derickson, *Workers' Health, Workers' Democracy*; Rosner and Markowitz, *Deadly Dust*; Jameson, *All That Glitters*; Wilkerson, *To Live Here*.

6. Greene, *Pure and Simple Politics*; Montgomery, *Fall of the House*; Taillon, *Good, Reliable, White Men*; Brattain, *Politics of Whiteness*; Kimeldorf, *Battling for American Labor*; McKillen, *Making the World Safe*. For the conservative interpretation, see Perlman, *History of Trade Unionism*, 303–5 (quote p. 303).

7. Andrews, *Killing for Coal*; Letwin, *Challenge of Interracial Unionism*; Jameson, *All That Glitters*; Emmons, *Butte Irish*; Corbin, *Life, Work, and Rebellion*; Wyman, *Hard Rock Epic*; Jensen, *Heritage of Conflict*; Green, *Devil is Here*.

8. Montgomery, *Fall of the House*, 2 (second and fourth quotes), 171 (first, third, and fifth quotes); Montgomery, "Mythical Man," 56–62 (sixth quote p. 59, seventh quote p. 60). For an emphasis on the AFL roots of working-class conservatism, see Moody, *Injury to All*, xiv–xvi.

9. See Richards, *Union-Free America*, 1–6; Minchin, *What Do We Need*; Case, "Losing the Middle Ground"; Fones-Wolf and Fones-Wolf, *Struggle for the Soul*; and Caldemeyer, *Acting Against the Order*. For similar works that take a longer view, see Shane Hamilton, *Trucking Country*; Lou Martin, *Smokestacks in the Hills*; and Perkins, *Hillbilly Hellraisers*.

10. This book is the first labor history of the Tri-State miners from the district's beginnings in the 1850s to its end in the 1950s. It builds on the excellent work of earlier scholars who examined particular aspects of that history, including Gibson, *Wilderness Bonanza*; Wyman, *Hard Rock Epic*, 52–57; Suggs, *Union Busting*; Rosner and Markowitz, *Deadly Dust*, 135–77; Robertson, *Hard as the Rock Itself*, 121–83; Laas, "Nineteenth-Century Rugged Individualism"; Knerr, *Eagle-Picher Industries*; and Norris, *AZn*.

11. This argument is indebted to scholarship on the historical relationships among categories of class, race, and gender. See esp. Hunter, *To 'Joy My Freedom*; Glenn, *Unequal Freedom*, 6–17, 56–92; Nelson, *Divided We Stand*, xviii–xliv; Bederman, *Manliness and Civilization*; Merritt, *Masterless Men*; and Baron, "Masculinity."

12. Gerstle, *American Crucible*, 159–60; Basso, *Meet Joe Copper*, 6–14, 28, 277; Taillon, *Good, Reliable, White Men*, 3–7, 69.

13. Zieger, *CIO*, 1–3 (third quote p. 1); Cowie, *Great Exception*, 9–33, 91–151 (first quote p. 91, second and fourth quotes p. 9).

14. Cowie, *Great Exception*, 123–24 (first quote p. 123); Gerstle, *American Crucible*, 130–31, 180–83 (second quote p. 130).

15. Blevins, *History of the Ozarks*, 156–96; Perkins, *Hillbilly Hellraisers*, 13–19; Moreton, *To Serve God*, 13–16; Postel, *Populist Vision*, 4, 9, 32; Hoganson, *Heartland*, 74–166.

16. Corbin, *Life, Work, and Rebellion*, 25–35; Lou Martin, *Smokestacks in the Hills*; Caldemeyer, "Run of the Mine," 31–36, 268–70; Caldemeyer, *Acting Against the Order*; Shane Hamilton, *Trucking Country*, 2, 43–51, 106–8, 188–90; Derickson, *Dangerously Sleepy*, 108–10; Currarino, *Labor Question in America*, 139–51; Davis, *Prisoners*, viii, 21–40.

17. Salvatore, *Eugene V. Debs*, 37–40; Bryan and Bryan, *Memoirs*, 106 (quote), 111; Sanders, *Roots of Reform*, 143; Scranton, *Figured Tapestry*, 1–2, 21–22; Rodgers, *Work Ethic*, xiv, 14–15, 168.

18. Hahn, *A Nation Without Borders*, 406–8; Laurie, *Artisans into Workers*, 218; Fink, *Workingmen's Democracy*, 12–13; Grob, *Workers and Utopia*, 189.

19. Meyer, *Manhood on the Line*, 3–35; Kimmel, *Manhood in America*, 69–76; Simon, "Appeal of Cole Blease"; Mary Murphy, *Mining Cultures*, 18–19; Baron, "Masculinity," 144–52 (quote p. 146); Fink, "Culture's Last Stand," 183–87; Kessler-Harris, "Treating the Male," 193–201.

20. Painter, *History of White People*, 291–310; Davis, *Prisoners*, 38–40; Basso, *Meet Joe Copper*, 159–93; Gerstle, *American Crucible*, 182–86; Churchwell, *Behold, America*, 43–65; Bederman, *Manliness and Civilization*, 1–44. On claims to white racial superiority within the WFM, see Jameson, *All That Glitters*, 140–60.

21. On the alternative vision of responsible manhood pursued by unions, see Basso, *Meet Joe Copper*, 35–41; Mary Murphy, *Mining Cultures*, 107–31; Jameson, *All That Glitters*, 93–113; and Derickson, *Workers' Health, Workers' Democracy*, 57–85, 155–88. These effects of white masculinity persist. See Metzl, *Dying of Whiteness*.

22. Dawley, *Struggles for Justice*, 297–333; Derickson, *Workers' Health, Workers' Democracy*, 189–219.

23. Cowie, *Great Exception*, 91–121 (quote p. 92); Cowie and Salvatore, "Long Exception," 3–32; MacLean, *Freedom is Not Enough*, 13–34; Self, *All in the Family*, 10, 43.

24. Harper, *White Man's Heaven*; Lynn, *Preserving*, 178–80.

25. Harper, *White Man's Heaven*, 32 (quote); U.S. Bureau of the Census, *Statistics for Missouri*, 604, 613, 652–87, 722–23.

26. See Roll, "Faith Powers," 74–95.

27. My use of the word "belief" here and throughout draws on the work of scholars of religion who seek to understand human faith as the intersection of intellectual and material practice. See Morgan, "Materiality," 73; Callahan, *Work and Faith*, 8–10, 129–51; and Rutherford, "Enchantments of Secular Belief."

28. See, for example, Cowie, *Great Exception*; Durr, *Behind the Backlash*; and Lombardo, *Blue-Collar Conservatism*.

29. Nelson, *Divided We Stand*, xxxiii–xxxiv; Lichtenstein, *State of the Union*, 66–67; Nelson, "Working-Class Agency," 414–20; Windham, *Knocking on Labor's Door*, 40–43.

30. Bruner address in MM, *38th Convention*, 783–84.

CHAPTER 1

1. Susan B. Carter et al., *Economic Sectors*, 308–15; Appleby, *Relentless Revolution*, 189; Laurie, *Artisans into Workers*, 34–46.

2. Gibson, *Wilderness Bonanza*, 14–18; Matthews, *Promise Land*, 84–86.

3. Bezis-Selfa, *Forging America*, 13–40, 167–218; Dew, "Disciplining Slave Ironworkers," 393–418; Paul, *Mining Frontiers*, 28–36, 270–77; Lingenfelter, *Hardrock Miners*, 3–4; Gates, *Michigan Copper*, 8–38, 98–101; Krause, *Making*, 205–34; Laurie, *Artisans into Workers*, 15–46; Larson, *Market Revolution*, 166.

4. Foley, *History of Missouri*, 6–11, 59–60, 127–28; Gracy, *Moses Austin*, 53–138; Lucy Eldersveld Murphy, *Gathering of Rivers*, 79–100; Aron, *American Confluence*, 98–100; Rowe, *Hard-Rock Men*, 37–61; Wyman, *Wisconsin Frontier*, 127–56; Ingalls, *Lead and Zinc*, 94–107, 119–35, 200–205; Twitty, *Before Dred Scott*, 50–52; Mutti Burke, *On Slavery's Border*, 21.

5. Laurie, *Artisans into Laborers*, 47–112; Wilentz, *Chants Democratic*, 363–89.

6. Schoolcraft, *View*, 254 (quote); Gibson, *Wilderness Bonanza*, 14–19.

7. Renner, *Joplin*, 15–18; U.S. Census Office, *Seventh Census*, 655; "John Cox," District 41, Jasper County, and "William Tingle," District 63, Newton County, both in Missouri, slave schedules, 1850 federal manuscript census.

8. Renner, *Joplin*, 18–19; Matthews, *Promise Land*, 94–95; North, *History of Jasper County*, 834–35; Draper and Draper, *Old Grubstake Days*, 3–6; Jobe, *Reprint of Centennial History*, 15; Gibson, *Wilderness Bonanza*, 19–21; *EMJ*, December 1, 1894, 508; James, *"Earth's Hidden Treasures,"* 1:1–2; "Amos Spurgeon," Shoal Creek, Newton County, Missouri; "William Tingle," Shoal Creek, Newton County, Missouri; and "John Cox," District 41, Jasper County, Missouri, all in 1850 federal manuscript census.

9. Winslow, *Missouri Geological Survey*, 6:602–7; Gibson, *Wilderness Bonanza*, 11.

10. "Lead in Southwest Missouri," 119–22; W. Alexander and Street, *Metals*, 172–74.

11. "Journal of Mining and Mines," 411–12; "Lead in Southwest Missouri," 119–22 (quote

p. 120); "John Cox," District 41, Jasper County, Missouri, slave schedules, 1850 federal manuscript census; North, *History of Jasper County*, 487–88, 817, 835–36.

12. Certificates 7530, 7577, 7602, 7876, 8426, 9170, 12350, 17341, 17352, 17353, and 21934, Springfield Office, in U.S. General Land Office Records; "John Cox," Mineral Township, Jasper County; "Amos Spurgeon," Shoal Creek Township, Newton County; and "John Spurgeon," Shoal Creek Township, Newton County, all in Missouri, 1860 federal manuscript census; "John Cox," District 41, Jasper County, Missouri, 1850 federal manuscript census; North, *History of Jasper County*, 835.

13. Ingalls, *Lead and Zinc*, 200–205.

14. "Journal of Mining and Mines," 411–12 (quotes); Margaret S. Carter, *New Diggings*, 33; Winslow, *Missouri Geological Survey*, 6:201–2; "William S. Moseley, Neosho," Missouri, 28: 365, Dun; "George W. Moseley," Neosho, Newton County, Missouri, and "William S. Moseley," District 62, New Madrid County, Missouri, both in 1850 federal manuscript census; Certificates 10454 and 10459, Springfield Office, in U.S. General Land Office Records.

15. "Journal of Mining and Mines," 411–12; "Lead in Southwest Missouri," 119–22; Gibson, *Wilderness Bonanza*, 116; *EMJ*, December 1, 1894, 508.

16. "Lead in Southwest Missouri," 119–22; Swallow, *First and Second Annual Reports*, 160–64; "William Moseley," 352 (quote); "McKee and Tingle, Grand Falls," 365; and "William S. Moseley, Neosho," 365, all in Missouri, vol. 28, Dun; Gibson, *Wilderness Bonanza*, 115–17.

17. "Journal of Mining and Mines," 411–12; "Lead in Southwest Missouri," 119–22; Swallow, *First and Second Annual Reports*, 160–64 (quotes p. 164); Certificate 10424, Springfield Office, in U.S. General Land Office Records; "Francis Reando" and "David Sunday," both in District 41, Jasper County, Missouri, 1850 federal census; *Centennial History*, 15–17; Broadhead, *Report*, 89; Ingalls, *Lead and Zinc*, 313–15.

18. Certificates 11624, 11625, 11626, 11627, 12180, and 12181, Springfield Office, in U.S. General Land Office Records; "Ferd Kennett" and "John Casey," both in Union Township, Washington County, Missouri, 1850 federal manuscript census; "Ferdinand Kennett" and "John Casey," both in Washington County, slave schedules, 1850 federal manuscript census; Leonard, *Industries of St. Louis*, 101; *Year Book*, 125–27.

19. Swallow, *Geological Report*, iii–viii; Act Granting the Right of Way to the State of Missouri, Pub. L. No. 32–19, 10 Stat. 8 (1852); Miner, *St. Louis–San Francisco*, 8–21.

20. *Centennial History*, 15–17; Broadhead, *Report*, 89; Swallow, *First and Second Annual Reports*, 160–63 (first quote p. 161, second quote p. 160); "Lead Mining in Southern Missouri," 389–91.

21. Ingalls, *Lead and Zinc*, 102–4; Gracy, *Moses Austin*, 83–84; Rowe, *Hard-Rock Men*, 40–42.

22. Rowe, *Hard-Rock Men*, 40 (first quote); Rohrbough, *Trans-Appalachian Frontier*, 474–76 (second quote p. 474, third quote p. 476); Everett Dick, *Lure of the Land*, 86–91.

23. Swallow, *First and Second Annual Reports*, 161–63.

24. Swallow, 162–63 (first and second quotes p. 164, third quote p. 162). See Schoolcraft, *View*, 69.

25. Bishop, *Loyalty on the Frontier*, 38 (first quote); "Granby," *Daily Missouri Republican*, March 15, 1858 (final quote); "Missouri Lead Mines," 2; Conrad, *Encyclopedia*, 3:84.

26. Demographic statistics compiled from Granby Township, Newton County, Missouri, 1860 federal manuscript census.

27. "George and Sarah Benge"; "William and Sarah Linton"; "A. B. and Catherine Fowler"; "Jonathan and Agnes Tisdall"; and "David and Mary Holland," all in Granby Township, Newton County, Missouri, 1860 federal manuscript census; Kimmel, *Manhood in America*, 11–53.

28. "Granby in 1857," in Swallow, *Geological Report*, between pp. 36 and 37 (quote p. 37); Winslow, *Missouri Geological Survey*, 7:601–6; James, *"Earth's Hidden Treasures,"* 1:4–5.

29. Parker, *Missouri as It Is*, 97 (quotes); Swallow, *Geological Report*, 36–37; "Granby," *Daily Missouri Republican*, March 15, 1858; Winslow, *Missouri Geological Survey*, 7:501.

30. Buckley and Buehler, *Geology*, 1–2; "William Moseley," Missouri, 28:352, Dun (first quote); "Missouri Lead Mines," 2 (second quote); "Granby," *Daily Missouri Republican*, March 15, 1858 (third quote); James, *"Earth's Hidden Treasures,"* 2:2.

31. Ingalls, *Lead and Zinc*, viii, 28–30, 86, 105–7, 200; Holley, *Lead and Zinc Pigments*, 17–20; Dacus and Buel, *Tour of St. Louis*, 242–43.

32. Parker, *Missouri as It Is*, 98; Scharf, *History of St. Louis*, 608; Primm, *Lion of the Valley*, 174, 192; "Henry Taylor Blow," in Christensen et al., *Dictionary of Missouri Biography*, 85–86; "Charles, Blow and Co., drugs," Missouri, 38:184, Dun; "Ferdinand Kennett," Jefferson County, and "Peter Blow," Newton County, both in Missouri, slave schedules, 1860 federal manuscript census; "Granby," *Daily Missouri Republican*, March 15, 1858; "Kennett, Ferdinand, et al, vs. Plummer, John, et al," 1857, 5–7, f. 9, b. 232, Supreme Court of Missouri Historical Records, MSA.

33. Winslow, *Missouri Geological Survey*, 6:206–7; "Information Sheet," R369, Rolla; Swallow, *Geological Report*, 92; Parker, *Missouri as It Is*, 98 (quote).

34. Parker, *Missouri as It Is*, 97–98.

35. "Kennett, Ferdinand, et al, vs. Plummer, John, et al," 1857, f. 9, b. 232, Supreme Court of Missouri Historical Records, MSA; Parker, *Missouri as It Is*, 98 (second quote); "Royalty Enrages Miners," in Haase, "Granby," 12, Jasper County Public Library, Joplin, Missouri (first quote); "John Plummer" and "Eli Powers," both in Neosho, Newton County, Missouri, 1860 federal manuscript census; "John Plummer," Neosho, Newton County, Missouri, slave schedules, 1860 federal manuscript census.

36. Richardson, *Beyond the Mississippi*, 210 (first quote); Swallow, *Geological Report*, 36–37 (remaining quotes); "Lead at Granby," *Daily Missouri Republican*, September 20, 1858.

37. Parker, *Missouri as It Is*, 100; Swallow, *Geological Report*, 37; "Joseph Hopkins" and "B. K. Hersey," both in Granby, Newton County, Missouri, 1860 federal manuscript census.

38. Richardson, *Beyond the Mississippi*, 212–13.

39. "W. W. Saunders and Co. (L. P. Roberts and W. W. Frazer), Granby Mines," Missouri, 28:378, Dun (quotes); "George and Sarah Benge"; "William and Sarah Linton"; "A. B. and Catherine Fowler;" "David and Mary Holland"; "Jacob Blackwell"; and a random sample of ninety-four other "miners" listed, all in Granby Township, Newton County, Missouri, 1860 federal manuscript census; Long, *Wages and Earnings*, 39–49.

40. Swallow, *Geological Report*, 37–39 (quotes p. 39); "William Orchard," Granby Township, Newton County, Missouri, 1860 federal manuscript census; Winslow, *Missouri Geological Survey*, 7:518; Tunnard, *Southern Record*, 82; Ingalls, *Lead and Zinc*, 108–10.

41. Steele and Cottrell, *Civil War*, 11–38.

42. Winslow, *Missouri Geological Survey*, 6:131; Tunnard, *Southern Record*, 81–82 (first quote); Thomas Richeson to H. W. Halleck, March 14, 1862, file 2116, reel F1584, Missouri Union Provost Marshal's Papers, State Historical Society of Missouri Research Center, Columbia, Missouri (second quote); "Peter E. Blow, drugs," Missouri, 38:262, Dun; "Missouri Lead Mines in the Hands," 626–27 (last two quotes).

43. Benjamin to B. McCulloch, October 11, 1861, and G. W. Clark to Benjamin, October 14, 1861, both in *OR*, 3:717–18; "Lead in Arkansas," *Weekly Raleigh Register*, October 30, 1861.

44. "Committee from Missouri" to H. W. Halleck, November 1861, and R. H. Chilton to L. Hebert, December 31, 1861, both in *OR*, 8:370–71, 726–27; Steele and Cottrell, *Civil War*, 39–50.

45. Clark Wright to Headquarters, April 9, 1862, *OR*, 8:363; report of Brig. Gen. Egbert B. Brown, July 17, 1862, *OR*, 13:163; T. W. Ridgely to Blow & Kennett, July 10, 1862, b. 2, Blow Family Papers, MHM (quote); Blow & Kennett to Lincoln, December 1, 1862, Lincoln Papers, Library of Congress, Washington, D.C.; Conrad, *Encyclopedia*, 2:84; Steele and Cottrell, *Civil War*, 53–54.

46. Thomas T. Crittenden to William F. Cloud, May 18, 1863, *OR*, 22(1):328; Gibson, *Wilderness Bonanza*, 26.

47. "Letter from Granby," *Missouri Weekly Patriot*, July 13, 1865; "The Granby Mines," *Daily Missouri Republican*, September 19, 1865 (quote).

48. "Industrial Enterprise in Missouri," *Daily National Intelligencer*, September 15, 1865; "Granby Mining and Smelting Company," April 5, 1865, Charter 5810 1/2, Missouri Online Business Filings, MSA; Reavis, *St. Louis*, 412–14, 435–39; Robert W. Jackson, *Rails across the Mississippi*, 12–28; Missouri General Assembly, *Journal of the Senate*, 268–69, 305–7; Miner, *St. Louis–San Francisco*, 37–51.

49. "Granby Mining and Smelting Co.," *Daily Missouri Republican*, September 20, 1865.

50. "From Newton County" (first quote) and "Southwest Missouri" (second quote), both in *Missouri Weekly Patriot*, October 19, 1865.

CHAPTER 2

1. Sang, "Corrosion of Iron and Steel," 493 (first quote); Ingalls, *Lead and Zinc*, 204; U.S. Centennial Commission, *International Exhibition*, 144; "Missouri at the Centennial," *Missouri Cash-Book*, September 1, 1875; *Engineering and Contracting*, February 22, 1922, 169 (second quote).

2. Ingalls, *Lead and Zinc*, 204; Susan B. Carter et al., *Economic Sectors*, 312.

3. On inside contracting, see Clawson, *Bureaucracy*, 71–125; and Buttrick, "Inside Contract System," 205–21.

4. Ingalls, *Lead Smelting and Refining*, 25 (quotes); *St. Louis Globe-Democrat*, December 10, 1875, 8; "Statistical Statement," October 14, 1864, f. 7, b. 51, Fox Collection, New-York Historical Society Museum and Library, New York, New York.

5. "From Newton County," *Missouri Patriot*, August 10, 1865; "Granby Mining and Smelting Co.," *Daily Missouri Republican*, September 20, 1865; "Miners Wanted!," *Carthage Banner*, June 11, 1868 (quotes); "200 Miners and Six Good Smelters Wanted," *Daily Missouri Republican*, April 18, 1868; Gibson, *Wilderness Bonanza*, 130.

6. "From Newton County," *Missouri Patriot*, November 23, 1865 (first quote); "Martin Jarrett," Granby Township, Newton County, Missouri, 1860 federal manuscript census; "Jasper Moon," District 42, Jefferson County, Missouri, 1850 federal manuscript census; "Southwest Items," *Missouri Patriot*, September 20, 1866; "Southwest Items," *Missouri Patriot*, October 18, 1866 (second quote); Blow to Eads, July 16 (third quote), December 22 (fourth quote), 1868, f. 6, b. 1, Eads Collection, MHM; "In Mineral Wealth Missouri is Incomparable," *Carthage Banner*, April 23, 1868; "Southwest Missouri," *Carthage Banner*, November 14, 1867 (last two quotes).

7. General survey of miners and information regarding "Bazil Meek," "Robert Meek," "John Meek," "John Trevaskis," and "Thomas Walker," all from Granby Township, Newton County, Missouri, 1870 federal manuscript census; U.S Census Office, *Statistics of the Population*, 399.

8. General survey of miners and information regarding "E. Moffett," "Michael Brady," and "James Cummins," all in Mineral Township, Jasper County, Missouri, 1870 federal manuscript census; McGregor, *Biographical History*, 32–33; "Miners Wanted!," *Carthage Banner*, June 11, 1868 (first quote); "Minersville, Galesburg, and Georgia City," *Carthage Banner*, March 10, 1870 (second quote); Gibson, *Wilderness Bonanza*, 31.

9. McGregor, *Biographical History*, 32–33; Shaner, *Story of Joplin*, 5 (quote); Renner, *Joplin*, 25; "Moffett and Sergeant," Missouri, 18:710, Dun; Gibson, *Wilderness Bonanza*, 31.

10. Shaner, *Story of Joplin*, 6 (first quote); *Illustrated Historical Atlas Map*, 73; "A Trip to the Mineral Region," *Carthage Banner*, July 20, 1871 (remaining quotes).

11. Lloyd and Baumann, *Mineral Wealth*, 17, 21; Shaner, *Story of Joplin*, 9–11.

12. "The Joplyn Mines," *Carthage Banner*, August 10, 1871 (first, second, and third quotes); "Lead Mining," *Carthage Banner*, June 22, 1871 (last two quotes).

13. McGregor, *Biographical Record*, 66; "The Joplyn Mines," *Carthage Banner*, August 3, 1871 (quote); "W. P. Davis and P. Murphy, Jasper Mining and Smelting," Missouri, 18:674, Dun; "Missouri's Mineral," *St. Louis Globe-Democrat*, October 18, 1887.

14. Lloyd and Baumann, *Mineral Wealth*, 13–41; McGregor, *Biographical Record*, 427–28; Shaner, *Story of Joplin*, 12–33; Broadhead, *Report*, 445–69.

15. "Joplin Mass Meeting," *Carthage Banner*, March 7, 1872 (quotes); Paul, *Mining Frontiers*, 23–24.

16. "Joplin Mass Meeting," *Carthage Banner*, March 7, 1872 (quotes); Arnold, *Fueling the Gilded Age*, x, 28–29; Wyman, *Hard Rock Epic*, 29–30; Corbin, *Life, Work, and Rebellion*, 30–39.

17. "Joplin Mass Meeting," *Carthage Banner*, March 7, 1872 (quotes); Livingston, *History of Jasper County*, 157; Andrew Jackson, State of the Union address, December 8, 1829.

18. Lloyd and Baumann, *Mineral Wealth*, 8, 35, 38, 41 (last quote); Broadhead, *Report*, 490 (first and second quotes), 496, 500; "Notice to Miners," *Carthage Banner*, August 15, 1872; "$10,000 in Cash!," *Carthage Banner*, May 8, 1873.

19. "Joplin Items," *Carthage Banner*, May 8, 1873 (first two quotes); "The Greatest Strike in the Mines!," *Carthage Banner*, May 15, 1873 (last three quotes).

20. *St. Louis Globe-Democrat*, February 14, 1873, quoted in Livingston, *History of Jasper County*, 166–67 (quotes p. 167); Lloyd and Baumann, *Mineral Wealth*, 47, 49; "Jasper County Lead Mines," *Carthage Banner*, September 26, 1872; "Joplin Mining and Smelting Company, Joplin," 743, 753, and "S. B. Corn, Joplin," 760, both in Missouri, vol. 28, Dun.

21. Shaner, *Story of Joplin*, 12–33; Livingston, *History of Jasper County*, 79 (first quote); "In the Mines," *St. Louis Globe-Democrat*, November 26, 1875; "William A. Daugherty," in Conrad, *Encyclopedia*, 2:227; "Neighboring Mines," *Carthage Banner*, August 6, 1874 (final quotes). Those who named Oronogo might have been influenced by Oronoco, Minnesota, an 1850s gold mining settlement whose founders probably referenced the Oronoco River in South America.

22. "Mining Items," *Carthage Banner*, July 24, 1873.

23. "Lead Mining," *Carthage Banner*, June 22, 1871 (first quote); "In the Mines," *St. Louis Globe-Democrat*, November 26, 1875 (second quote); Lloyd and Baumann, *Mineral Wealth*, 9 (last quote).

24. Jolly, *U.S. Zinc Industry*, 22–23, 57–63; Livingston, *History of Jasper County*, 163–64; Siebenthal, "Early Days," 342–43; Ingalls, *Lead and Zinc*, 286–89.

25. "Mining Items," *Carthage Banner*, July 24, 1873 (first quote); "Observations of the State Geologist in Jasper County," *People's Tribune*, February 26, 1873 (second quote); Gibson, *Wilderness Bonanza*, 124–25.

26. Susan B. Carter et al., *Economic Sectors*, 312; Douglas, "History of Manufactures," 113–14; McGregor, *Biographical Record*, 27; Winslow, *Missouri Geological Survey*, 7:503; "Zinc Furnaces," *People's Tribune*, April 29, 1873 (quotes).

27. Winslow, *Missouri Geological Survey*, 7:503; "Mining Items," *Carthage Banner*, October 16, 1873 (first two quotes); "Jasper County," *Carthage Banner*, January 1, 1874 (last quote).

28. "Communism," *Carthage Banner*, July 23, 1874 (first and second quotes); MBLS, *Sec-*

ond *Annual Report*, 124 (last quote); Broadhead, *Report*, 495–96; Livingston, *History of Jasper County*, 178.

29. Holibaugh, *Lead and Zinc*, 7–8.

30. "State News," *People's Tribune*, June 17, 1874 (first quote); "Mining Items," *Carthage Banner*, October 16, 1873 (second quote); Broadhead, *Report*, 496 (third through sixth quotes).

31. Foner, *From Colonial Times*, 439–74; Jensen, *Heritage of Conflict*, 10–24; Laurie, *Artisans into Workers*, 141–49.

32. "Railroads," *Carthage Banner*, April 30, 1874; "Joplin Items," *Carthage Banner*, June 11, 1874; Haase, "Granby," 16, Jasper County Public Library, Joplin, Missouri; Violette, *History of Missouri*, 426–28.

33. "Blowing Up Philadelphia Engine," *Carthage Banner*, July 29, 1875.

34. "Joplin Items, "*Carthage Banner*, June 11, 1874 (second quote); *EMJ*, November 9, 1907, 864–65; Livingston, *History of Jasper County*, 178; "Communism," *Carthage Banner*, July 23, 1874 (first and third quotes).

35. "Burning of Pitcher's Furnace," *Neosho Times*, July 30, 1874; "Communism," *Carthage Banner*, July 23, 1874 (first quote); "A Pleasant Call," *Neosho Times*, August 27, 1874; "We Are Theirs!," *Carthage Banner*, November 5, 1874 (third quote); "Boys, They're Ours!" and "Newton County is in the Hands of the Righteous" (second quote), both in *Neosho Times*, November 5, 1874.

36. Ingalls, *Lead and Zinc*, 203; "Notice," *Carthage Banner*, September 10, 1874; "Joplin Items," *Carthage Banner*, October 15, 1874; Knerr, *Eagle-Picher Industries*, 52–53.

37. All in *Carthage Banner*: "Joplin Items," July 1, 1875; "Joplin Items," July 8, 1875 (first quote); "The Mines," September 2, 1875 (second quote); "The Mines," October 7, 1875 (last quote); "The Mines," October 28, 1875 (Coyle quotes).

38. "From Oronogo," *Carthage Banner*, October 15, 1874; "Joplin Items," *Carthage Banner*, March 25, 1875 (second quote); "Rich Lead Discovery," *Carthage Banner*, September 2, 1875 (third and sixth quotes); "Our New Bonanza," *Carthage Banner*, September 23, 1875 (first, fourth, and fifth quotes); "Something about the Webbville Mines," *Cherokee Index*, July 6, 1877.

39. "How Jasper County Celebrated," *St. Louis Globe-Democrat*, July 6, 1876; MSS, *State Almanac*, 88–89; Astor, *Rebels on the Border*, 180–247.

40. "Lots of Lead," *Bates County Record*, June 2, 1877; "The Lead Mines of Kansas," *Emporia News*, July 13, 1877; "Among the Miners," *Galena Miner*, June 9, 1877 (quote); Holibaugh, *Lead and Zinc*, 21–22; Haworth, *University Geological Survey*, 20–26; Ingalls, *Lead and Zinc*, 203.

41. "Lots of Lead," *Bates County Record*, June 2, 1877 (quotes); "From Joplin," *Carthage Banner*, June 13, 1878; Winslow, *Missouri Geological Survey*, 7:596–600.

42. U.S. Census Office, *Statistics of the Population*, 175, 242, 244, 421; Miner, *St. Louis-San Francisco*, 169–73; Winslow, *Missouri Geological Survey*, 7:518, 528; Ingalls, *Lead and Zinc*, x, 205.

43. MBLS, *Second Annual Report*, 122–25 (second quote p. 124); "Lead Mining," *Carthage Banner*, June 22, 1871 (first quote); "Joplin Mines," *St. Louis Globe-Democrat*, November 21, 1880.

44. MBLS, *Second Annual Report*, 122–25 (first quote p. 125, second and third quotes p. 123); "Notes in Southwest Missouri," *Boston Evening Journal*, July 19, 1879 (last quote).

45. MBLS, *First Annual Report*, 92 (first quote); U.S. Census Office, *Report on the Mining Industries*, 805; "Notes in Southwest Missouri," *Boston Evening Journal*, July 19, 1879 (final quotes).

46. U.S. Geological Survey, *Mineral Resources* (pub. 1883), 370 (first quote), 371 (remaining quotes); MBLS, *First Annual Report*, 92 (second quote); U.S. Census Office, *Report on the Mining Industries*, 804.

CHAPTER 3

1. "The Miner," *State Line Herald*, January 6, 1881.

2. Ingalls, *Lead and Zinc*, 201–3; Winslow, *Missouri Geological Survey*, 7:519; Oscar E. Schmidt, quoted in Knerr, *Eagle-Picher Industries*, 54–61 (quote p. 59); *EMJ*, February 2, 1878, 73; "Notes in Southwest Missouri," *Boston Evening Journal*, July 19, 1879.

3. "Joplin Mines," *Carthage Banner*, August 9, 1877 (first quote); Ingalls, *Lead and Zinc*, 203; "Reducing Salaries at the Granby Mines," *St. Louis Globe-Democrat*, November 23, 1877; MBLS, *Second Annual Report*, 123; MBLS, *First Annual Report*, 92–94; George Case to D. Baumann, August 28, 1878, reprinted in full in *Carthage Banner*, January 9, 1879 (second quote).

4. "The Granby Mining War," *Carthage Banner*, October 3, 1878; "The Granby Lead Mines," *St. Louis Globe-Democrat*, September 12, 1878; "Greenbacks," *St. Joseph Gazette-Chronicle*, June 21, 1878 (first quote); "Greenback Convention," *Carthage Banner*, August 29, 1878 (second quote); MSS, *State Almanac*, 94; MSS, *Official Directory, 1881*, 48, 55; Hild, *Greenbackers*, 9–44; Hahn, *Nation Without Borders*, 406–7.

5. "Baumann's Vindication," *Carthage Banner*, January 9, 1879 (first two quotes); "The Business Outlook," *Colorado Weekly Chieftain*, February 6, 1879; "Throughout the State," *Boulder County Courier*, October 18, 1878 (third quote); "J. H. Wilbraham, Joplin Lodging House," in *Clark, Root, and Co.*, 161.

6. Jensen, *Heritage of Conflict*, 19–21; Wyman, *Hard Rock Epic*, 42; Fink, *Workingmen's Democracy*, 3–15.

7. "From Leadville," *JH*, June 6, 1880 (first quote); "Leadville," *JH*, August 23, 1880 (second quote); "Leadville," *JH*, September 26, 1880; Enyeart, *Quest*, 43–44; Wyman, *Hard Rock Epic*, 162–65; Jensen, *Heritage of Conflict*, 19–24; Paul, *Mining Frontiers*, 127–28.

8. "Another Side to the Story," *Carthage Banner*, February 20, 1879.

9. "Another Side to the Story," *Carthage Banner*, February 20, 1879 (last two quotes); "From Leadville," *JH*, June 6, 1880 (first and second quotes).

10. Gibson, *Wilderness Bonanza*, 124–25; Ingalls, *Lead and Zinc*, x, 203, 340; Winslow, *Missouri Geological Survey*, 7:503, 518–21, 546; U.S. Geological Survey, *Mineral Resources, 1883 and 1884*, 426–27; Holibaugh, *Lead and Zinc*, 10; U.S. Geological Survey, *Mineral Resources, 1886*, 156; Knerr, *Eagle-Picher Industries*, 52–61.

11. Holibaugh, *Lead and Zinc*, 19–22 (first two quotes p. 19); U.S. Geological Survey, *Mineral Resources* (pub. 1883), 369 (third and fifth quotes); MBLS, *Tenth Annual Report*, 101 (fourth quote); *Frank Leslie's Illustrated Newspaper*, June 7, 1890, 391.

12. Holibaugh, *Lead and Zinc*, 10; Haworth, *University Geological Survey*, 106; "First Annual Report," 255.

13. "First Annual Report," 229–31, 242–45; "Report of the State Mine Inspector," in MBLS, *Eleventh Annual Report*, 489 (first quote), 507 (second quote).

14. MBLS, *Tenth Annual Report*, 104 (first quote); MBM, *Report of the State Mine Inspector, 1890*, 44–56; Haworth, *University Geological Survey*, 106; Holibaugh, *Lead and Zinc*, 6–10.

15. "First Annual Report," 229–54; MBM, *Report of the State Mine Inspector, 1890*, 44–61 (quote p. 45).

16. MBM, *Report of the State Mine Inspector, 1890*, 44–61; MBLS, *Tenth Annual Report*, 117 (quote).

17. U.S. Geological Survey, *Mineral Resources* (pub. 1883), 369–70; *Frank Leslie's Illustrated Newspaper*, June 7, 1890, 391; Gibson, *Wilderness Bonanza*, 106–8 (first quote p. 106); Ingalls, *Lead and Zinc*, 295 (final quotes).

18. MBM, *Report of the State Mine Inspector, 1890*, 36–40, 44–61.

19. "First Annual Report," 230, 255; MBM, *Report of the State Mine Inspector, 1890*, 36–40, 44–61; MBM, *Fifth Annual Report*, 162–76.

20. MBM, *Fifth Annual Report*, 159–76; Long, *Wages and Earnings*, 95–98; Wyman, *Hard Rock Epic*, 34–36; MBLS, *Eleventh Annual Report*, 337 (first quote); "Joplin on the Jump," *Atchison Daily Globe*, December 1899 (second quote); Shaner, *Story of Joplin*, 78–79 (third quote); U.S. Census Office, *Report on Population*, 143, 216, 219; "A Booming City," *St. Louis Globe-Democrat*, April 18, 1887.

21. Winslow, *Missouri Geological Survey*, 7:520, 524, 529; "Charles DeGraff," Joplin Ward 4, Jasper County, Missouri, 1900 federal manuscript census; MBM, *Report of the State Mine Inspector, 1890*, 55; Haworth, *University Geological Survey*, 106; MBLS, *Tenth Annual Report*, 104.

22. U.S. Geological Survey, *Mineral Resources* (pub. 1883), 371 (first quote); MBLS, *Second Annual Report*, 123 (second quote); MBLS, *Eleventh Annual Report*, 336 (second quote); MBM, *Fifth Annual Report*, 155–61; Aurora Centennial Committee, *Aurora Centennial*, "Underground Riches" section; "Aurora, Mo.," *Kansas City Times*, May 8, 1888 (third quote).

23. MBM, *Report of the State Mine Inspector, 1890*, 41; "Aurora, Mo.," *Kansas City Times*, May 8, 1888 (quotes).

24. MBLS, *Eleventh Annual Report*, 336–38 (remaining quotes), 379 (first two quotes).

25. MBLS, 338–39, 342–44.

26. MBLS, 340 (first quote), 345–46 (final quotes).

27. MBLS, *Tenth Annual Report*, 104 (Clerc quote), 116 (Davey quote); MBLS, *Eleventh Annual Report*, 336–37 (Woodson quotes); "Missouri's Mineral," *St. Louis Globe-Democrat*, October 18, 1887.

28. U.S. Geological Survey, *Mineral Resources*, 370–71 (Clerc quotes); MBLS, *Eleventh Annual Report*, 114–17, 456 (Woodson quote).

29. Knights of Labor, "Preamble and Declaration"; MBLS, *First Annual Report*, 5–6; Kaufman, Albert, and Fones-Wolf, *Samuel Gompers Papers*, 454–54; "The Missouri Convention of Miners," *JUL*, September 25, 1885; "First Annual Report," 16–22.

30. Garlock, *Guide*, 86–87, 132–33, 246–47.

31. John Samuel, letter to the editor, *JUL*, August 15, 1881 (quotes); "R. J. Davis" and "Edward Armstrong," both in Webb Township, Jasper County, Missouri, 1880 federal manuscript census; Second General Assembly proceedings, 1879, 65, and Eighth General Assembly proceedings, 1884, 822–25, both in reel 1, film R-7552, Tamiment; "Locals Attached to the General Assembly," *JUL*, November 1882; "Education and Discipline," *JUL*, June 2, 1888; "Locals Still Voting," *JUL*, June 9, 1888; John H. Mills to Charles H. Litchman, August 8, 1887, f. 29, b. 13, Hayes Papers, CUA.

32. Livingston, *History of Jasper County*, 225–29, 279–80; North, *History of Jasper County*, 632; MSS, *Official Directory, 1883*, 43, 50; MSS, *Official Directory, 1885*, 26.

33. Knights of Labor, "Preamble and Declaration" (first quote); Loftus to Terence Powderly, December 24, 1879 (quotes), reel 2, *Terence V. Powderly Papers*; "Correspondence," *JUL*, July 15, 1881; MBLS, *Eleventh Annual Report*, 340.

34. Loftus to Powderly, December 24, 1879 (Loftus quotes), reel 2, *Terence V. Powderly Papers*; MBLS, *Eleventh Annual Report*, 340; Fink, *Workingmen's Democracy*, 9–13; Weir, *Beyond Labor's Veil*, 1–66, 231–76; Caldemeyer, "Run of the Mine," 104–15.

35. Seventh General Assembly proceedings, 1883, 405 (first quote), and Eleventh Gen-

eral Assembly proceedings, 1887, 1354 (second quote), both in reel 1, film R-7552, Tamiment; Peter McEntee, Charles Storey, and E. N. Morton to John W. Hayes, October 10, 1890 (third quote), and McEntee to Hayes, August 11, 1891, both in f. 29, b. 13, Hayes Papers, CUA.

36. "First Annual Report," 16–22, 180–81 (quotes p. 180).

37. "First Annual Report," 16–22, 180–81 (Wolfe quotes); MBLS, *Eleventh Annual Report*, 458 (Woodson quotes); MBM, *Report of the State Mine Inspector, 1890*, 35.

38. "First Annual Report," 255 (Wolfe quotes).

39. Gibson, *Wilderness Bonanza*, 148–150; "Mines and Mining," *Southwestern Miner*, June 9, 1892, R873, Rolla (quote); Knerr, *Eagle-Picher Industries*, 63–65.

40. MBM, *16th Annual Report*, 35–38, 48–49, 51; "A Huge Mining Deal," *St. Louis Post-Dispatch*, October 5, 1890; MBM, *Eighth Annual Report*, 256–57; Holibaugh, *Lead and Zinc*, 13–17, 20–22; Rex Mining and Smelting Company, Charter 5468; Empire Zinc Company, Charter 5675; and Western Zinc Company, Charter 7050, all in Missouri Online Business Filings, MSA; Gibson, *Wilderness Bonanza*, 148–49.

41. Holibaugh, *Lead and Zinc*, 7–8 (first, second, and fourth quotes), 13 (third quote); "Jasper County," *Southwestern Miner*, June 9, 1892, R873, Rolla.

42. Holibaugh, *Lead and Zinc*, 13 (first quote); MBM, *Eleventh Annual Report*, 180; "The Galena-Joplin Lead and Zinc District," in Rothwell, *Mineral Industry*, 666–68; *EMJ*, March 18, 1899, 321 (second quote); *EMJ*, October 6, 1917, 595; Gibson, *Wilderness Bonanza*, 42–43.

43. MBM, *Fifth Annual Report*, 158–61; MBM, *Seventh Annual Report*, 174–77; MBM, *Eighth Annual Report*, 256–57; payroll ledger, 1898, J. J. Luck Mining Company Records, SHSK; *EMJ*, October 6, 1917, 595; "Granby Mining & Smelting Company," *Southwestern Miner*, October 28, 1892, R873, Rolla (quote).

44. Holibaugh, *Lead and Zinc*, 10–11 (first quote p. 10); MBM, *Seventh Annual Report*, 158 (second quote); MBM, *Tenth Annual Report*, 192.

45. "Kansas Miners Out," *JH*, May 20, 1893; "Great Coal Strike," *JH*, May 21, 1893; "No End in Sight," *JH*, May 26, 1893 (first quote); "To Spread the Strike," *JH*, May 27, 1893; "The Mines," *JH*, June 18, 1893; "The Strike," *JH*, May 31, 1893 (second and third quotes); DeMoss, "Missed Opportunity"; "Striking Miners," *Aspen Evening Chronicle*, May 27, 1893; "30,000 Miners Will Strike," *Journal of the Knights of Labor*, June 8, 1893; John W. Hayes to LA 535, Scammon, June 30, 1893, f. 35, b. 8, and Hayes to R. M. Goodman, July 26, 1893, f. 13, b. 9, both in Hayes Papers, CUA.

46. Editorial, *JH*, May 31, 1893 (first quote); editorial, *JH*, May 28, 1893 (second quote); "The Mines," *JH*, June 18, 1893 (third quote); Sapp to Joseph W. Ridge, August 3, 1893, f. "Battlefield Mining Company," b. 1, Ridge Papers, SHSK (fourth quote); Holibaugh, *Lead and Zinc*, 10.

47. All in *JH*: "South Joplin Jottings," June 6, 1893 (first quote); editorial, May 31, 1893 (second quote); and editorial, May 23, 1893 (third quote).

48. "Smelteries Forced to Close," *JH*, June 27, 1893 (first quote); "The Mines," *JH*, July 2, 1893 (second quote); "The Mines," *JH*, July 23, 1893 (third quote); "Carterville," *JH*, June 25, 1893; "The Mines," *JH*, August 13, 1893; "The Shut Down," *JH*, August 10, 1893; "Not Yet Settled," *JH*, August 17, 1893; MBM, *Ninth Annual Report*, 218.

49. Phillips to John W. Hayes, July 5, 1893, f. 29, b. 13, Hayes Papers, CUA.

50. "Prosperity" (first quote) and "They Talk It Over" (second quote), *JH*, August 18, 1893; "The Miners' Meeting," *JH*, August 19, 1893.

51. "Prosperity" (first and second quotes) and "They Talk It Over" (third, fourth, and fifth quotes), *JH*, August 18, 1893; "The Miners' Meeting," *JH*, August 19, 1893; "Miners Meeting at Carterville," *JH*, August 23, 1893 (sixth quote); Baron, "Masculinity," 146–47; Kimmel, *History of Men*, 42–43.

52. Holibaugh, *Lead and Zinc*, 10 (first quote), 32; Sapp to Joseph W. Ridge, August 11, 1893, f. "Battlefield Mining Company," b. 1, Ridge Papers, SHSK (second quote); MBM, *Seventh Annual Report*, 165–66 (third quote p. 166).

53. Holibaugh, *Lead and Zinc*, 13; MBM, *Seventh Annual Report*, 158 (first and second quotes); "The Zinc Industry," *JH*, September 7, 1893 (third quote); "Webb City," *JH*, August 19, 1893; MSS, *Official Manual, 1895–1896*, 84.

54. "Mining Conditions," *JH*, August 20, 1894; MBM, *Eighth Annual Report*, 229 (first quote); MBM, *Ninth Annual Report*, 207–8, 216, 218, 220 (final quote p. 208).

55. *EMJ*, August 22, 1896, 182 (first two quotes); "Why They Went," *JG*, November 3, 1896 (final quotes).

56. "Why They Went," *JG*, November 3, 1896.

CHAPTER 4

1. *American Federationist*, November 1901, 468; Wyman, *Hard Rock Epic*, 172–74; Jensen, *Heritage of Conflict*, 54–58.

2. Gompers to Frank Weber, March 21, 1896, vol. 14, and Gompers to Robert Askew, April 6, 1896, vol. 15, both in reel 9, *Samuel Gompers Letterbooks*; executive council minutes, March 23, 1896, reel 2, *American Federation of Labor Records*; WFM charter application, June 29, 1896, f. 19, and Northern Mineral charter application, November 30, 1895, f. 23, both in b. 4, RG28-03, Meany; Jensen, *Heritage of Conflict*, 58–60; "E. J. Smith of Evansville," *JN*, June 20, 1896; Kaufman, Albert, and Paladino, *Samuel Gompers Papers*, 253–57; Taft, *A. F. of L.*, 150–52.

3. Gazzam, "Leadville Strike," 89; Joseph Gazzam, "Leadville Version of 'Show Me,'" *St. Louis Post-Dispatch*, July 14, 1941, Gazzam Papers, MHM (quote); Philpott, *Lessons of Leadville*, 2, 30–60, 107n25; "Seeley W. Mudd," in Rickard, *Interviews with Mining Engineers*, 391–92, 396; "List of Graduates," in Washington University, *Catalogue of the Officers*, 175; Emmons, *Butte Irish*, 263–286; Wyman, *Hard Rock Epic*, 32–60; Mellinger, *Race and Labor*, 22–36.

4. Paul, *Mining Frontiers*, 111, 124–25, 128; "Leadville," *JH*, August 23, 1880.

5. "Joplin, MO., Aug. 30th, 1896" (first quote) and "Galena, Kansas, Aug. 31st, 1896" (third, fourth, and fifth quotes), both in f. 4, b. 1; and "Saturday, September 5th, 1896" (second quote), f. 47, b. 2, all in LSR.

6. "Joplin, MO., Sept. 2nd, 1896"; "Saturday, September 5th, 1896" (first quote); and "Galena, Kansas, Aug. 31st, 1896" (second quote), all in f. 4, b. 1, LSR. See also Roll, "Sympathy for the Devil," 11–37.

7. "Wednesday, September 9th, 1896," f. 47, b. 2, LSR (quotes).

8. "Scabs are Scarce," *JG*, September 19, 1896 (first quote); "Miners for Leadville," *JG*, September 22, 1896 (second quote); "A Great Bluff," *JH*, September 24, 1896 (third quote); editorial, *JH*, September 25, 1896 (fourth quote); "Consumptives and Bums," *Aspen Tribune*, October 7, 1896 (fifth quote); "The Situation," *JG*, September 24, 1896 (final two quotes).

9. "Why They Went," *JG*, November 3, 1896 (quotes); Kimmel, *Manhood in America*, 57–69; Kessler-Harris, *Women Have Always Worked*, 62–70; Baron, "Masculinity," 146–47.

10. "Will Import Miners," *Denver Republican*, September 25, 1896; "Miners Out of Jail," *Denver Republican*, September 26, 1896 (quotes); Griswold and Griswold, *History of Leadville*, 2163–64; Anthony DeStefanis, "Violence and the Colorado National Guard," in Gier and Mercier, *Mining Women*, 195–97.

11. "Soldier Boys Heard From," *Carthage Press*, December 24, 1896 (first and second quotes); "Ingersoll Has a God," *Denver Daily News*, September 28, 1896 (third quote); *EMJ*, November 21, 1896, 493 (fourth quote).

12. "Another Riot of Strikers," *New York Times*, November 13, 1896 (first quote); "Strikers as Missionaries," *Denver Republican*, October 5, 1896 (second quote); "Miners Will Fight It Out," *Denver Republican*, October 1, 1896 (final two quotes).

13. *American Federationist*, November 1896, 200, 204; *American Federationist*, December 1896, 220; *American Federationist*, March 1897, 11–12; Kaufman, Albert, and Paladino, *Samuel Gompers Papers*, 253–57; *EMJ*, October 31, 1896, 421 (first quote); *EMJ*, November 21, 1896, 493 (second quote); "Importations of Miners," *Los Angeles Times*, December 31, 1896; Philpott, *Lessons of Leadville*, 5, 58–67; *EMJ*, February 13, 1897, 169 (third quote); *EMJ*, April 3, 1897, 136 (fourth quote); "Must Be Some Mistake Here," *Leadville Herald Democrat*, February 23, 1897; "Scott McCollum," Joplin Ward 3, Jasper County, Missouri, 1900 U.S. federal manuscript census.

14. Jensen, *Heritage of Conflict*, 57–60; Foner, *Policies and Practices*, 408–9; Enyeart, *Quest*, 122–24; Boyce to Gompers, March 16, April 7, 1897, 304–5 (first and second quotes), 313 (third quote), and "The Western Federation of Miners and the AFL," 257 (fourth quote), both in Kaufman, Albert, and Paladino, *Samuel Gompers Papers*; *Miners Magazine*, April 1902, 8–9 (last quote).

15. "Miners Uneasy," *Emporia Daily Gazette*, January 5, 1897 (first quote); *EMJ*, January 30, 1897, 121 (second quote); "Shipping in Miners," *Rocky Mountain News*, February 6, 1897 (third and fourth quotes).

16. *EMJ*, February 13, 1897, 149; MSS, *Official Manual, 1897–1898*, 23; MBM, *Eleventh Annual Report*, 179 (quote).

17. Ingalls, *Lead and Zinc*, 340, 345; A. J. Martin, *Summary Statistics*, 16.

18. MBM, *16th Annual Report*, 192; U.S. Census Office, *Census Reports*, 163, 241–42, 245, 583, 590–91; Norris, *AZn*, 4–26.

19. MBM, *13th Annual Report*, 175, 179; *EMJ*, January 7, 1899, 20–21; Gibson, *Wilderness Bonanza*, 149–50, 154, 166–68; Norris, *AZn*, 4–26; "When Hell's Neck Became Thriving Neck City," undated clipping, f. 5, R57, Rolla; "The Boom in Zinc," *Denver Evening Post*, October 31, 1899.

20. *EMJ*, January 7, 1899, 21 (first three quotes); MBM, *13th Annual Report*, 179 (fourth quote); "Rich in Zinc and Lead," *Butte Weekly Miner*, June 8, 1899; "Joplin on the Jump," *Atchison Daily Globe*, December 1, 1899; MSS, *Official Manual, 1899–1900*, 11–12, 88, 93.

21. "The Zinc and Lead Industry," *Appeal to Reason*, March 11, 1899 (second quote); "Brother Norton of Greenback Fame," *Appeal to Reason*, May 6, 1899 (first quote).

22. "Laborers and Spectators," *Los Angeles Times*, July 11, 1899 (first quote); "Two Weeks' Vacation," *JN*, June 24, 1899 (second quote); "The Shut Down," *JG*, June 30, 1899 (third quote); "Missouri, Jasper County," *EMJ*, July 8, 1899, 47–48; "Gas and Coal Fields for Smelters," *Nevada Daily Mail*, July 25, 1898; Norris, "Missouri and Kansas," 322–25.

23. Jensen, *Heritage of Conflict*, 74–84; Phipps, *Bull Pen to Bargaining Table*, 24–29, 282 (quote).

24. Campbell to J. W. Kendrick, June 10, 1899, 379, f. 84, b. 1, Campbell Papers, Northwest Museum of Arts and Culture, Spokane, Washington; "Wanted, 1000 Miners," *JN*, June 16, 1899 (quote).

25. "Miners for Idaho," *JG*, June 17, 1899 (first quote); "Train Load of Miners in Camp," *Spokesman-Review*, June 23, 1899 (second quote); "Missouri Men Like Their Jobs," *Spokesman-Review*, June 24, 1899 (third and fourth quotes); "There is No Strike," *JG*, June 16, 1899.

26. "Missouri Men Like Their Jobs," *Spokesman-Review*, June 24, 1899 (first quote); "Train Load of Miners in Camp," *Spokesman-Review*, June 23, 1899 (second quote); "A Letter

from Idaho," *JN*, July 12, 1899; "The Idaho Miners," *JG*, June 27, 1899 (sixth quote); "Letter from John Cotner," *JG*, June 30, 1899; "Doe Isbell Writes," *JG*, July 1, 1899; "Letter from John Maddy," *JN*, July 13, 1899 (third, fourth, and fifth quotes); "The Idaho Mines," *JG*, July 2, 1899 (seventh quote).

27. "Order in the Idaho Mines," *San Francisco Chronicle*, June 25, 1899 (first and second quotes); "Missouri Miners Deceived," *JN*, June 30, 1899 (third, fourth, and fifth quotes); "Warning to Miners," *JN*, July 11, 1899 (sixth quote); "A Foe of Independence," *Washington Times*, November 3, 1900.

28. "Missouri Men Like Their Jobs," *Spokesman-Review*, June 24, 1899 (first and second quotes); "Letter from John Maddy," *JN*, July 13, 1899 (third, fourth, and fifth quotes); "Choosing Native Born," *Los Angeles Times*, June 24, 1899; "A Letter from Idaho," *JN*, July 12, 1899 (sixth quote).

29. "A Letter from Idaho," *JN*, July 12, 1899.

30. *EMJ*, July 22, 1899, 92; "Doe Isbell Writes," *JG*, July 1, 1899; "Joplin" and "Missourian" in American Dialect Society, *Dialect Notes*, 26; *Miners Magazine*, February 1902, 15.

31. *EMJ*, September 9, 1899, 316; "New Men Work in Every Mine," *Spokesman-Review*, July 17, 1899 (quote); Phipps, *Bull Pen to Bargaining Table*, 42–44.

32. *American Federationist*, April 1899, 44 (first quote); *American Federationist*, June 1899, 93; Charter 7500, zinc and lead miners, Oronogo, August 30, 1899, 2:32–33, RG2-10, Meany; "Sam Dodson," District 45, Joplin, Jasper County, Missouri, 1900 federal manuscript census; Jensen, *Heritage of Conflict*, 60–71; Greene, *Pure and Simple Politics*, 36–47; Foner, *Policies and Practices*, 136–37, 261–64.

33. "Miners' Union Organized," *JN*, August 26, 1899 (quotes); "Another Shut Down Probably at Joplin," *Kinsley Graphic*, September 22, 1899.

34. Charter 7567, Neck, October 14, 1899, 38–39; Charter, Central Labor Union, Joplin, October 26, 1899, 38–39; and Charter 7590, Webb City, October 30, 1899, 40–41, all in vol. 2, RG2-10, Meany; *American Federationist*, December 1899, 258 (first quote); Samuel Gompers to Edward J. Mahoney, November 3, 1899, reel 21, *Samuel Gompers Letterbooks* (second quote); MBLS, *Twenty-Third Annual Report*, 273.

35. Gompers to Ed Boyce, March 26, 1897, in Kaufman, Albert, and Paladino, *Samuel Gompers Papers*, 308 (first quote); 1899 convention proceedings, in AFL, *Proceedings of the American Federation of Labor*, 11; executive council minutes, October 16–18, 1899, b. 1, RG4-08, Meany; charter application, Northern Mineral, November 30, 1895, f. 23, b. 4, RG28-03, Meany; Enyeart, *Quest*, 124–25; Taft, *A. F. of L.*, 152–55; Foner, *Policies and Practices*, 415–16.

36. Gompers to Weber, November 21, 1899 (first quote), and Gompers to Askew, November 23, 1899 (second and third quotes), both in reel 21; and Gompers to Weber, March 21, 1896, reel 9, all in *Samuel Gompers Letterbooks*.

37. *American Federationist*, February 1900, 44, 47 (all quotes); Charter 8055, Duenweg, December 1, 1899, 44–45; Charter 8057, Zincite, December 1, 1899, 44–45; and Charter 8082, Central City, December 26, 1899, 46–47, all in vol. 2, RG2-10, Meany.

38. 1899 convention proceedings, in AFL, *Proceedings of the American Federation of Labor*, 70 (first quote), 110 (second quote); *American Federationist*, February 1900, 47 (third quote).

39. *Miners Magazine*: February 1900, 8–9 (first quote); March 1900, 3 (second quote).

40. *Miners Magazine*, March 1900, 27 (first and fourth quotes); *Miners Magazine*, May 1900, 24–25 (second, third, and fifth quotes); "Solon P. Cress," in *Corbett and Ballenger's*, 88.

41. Ingalls, *Lead and Zinc*, 340, 345; *EMJ*, February 9, 1901, 188; Norris, "Missouri and Kansas," 330–31; MBM, *16th Annual Report*, 209–27.

42. *EMJ*, January 5, 1901, 23–24; *Miners Magazine*, March 1900, 27 (first quote); *American Federationist*, February 1900, 47 (second quote).

43. Gompers to Weber, January 1, 1900, reel 21, *Samuel Gompers Letterbooks*; Charter 8131, Galena, Kansas, January 20, 1900, vol. 2, RG2-10, Meany; *American Federationist*, May 1900, 145; "Miners Object to Less Pay," *Pittsburg Daily Headlight*, June 7, 1900; "The Oronogo Strike Ended," *Galena Daily Lever*, June 8, 1900; *American Federationist*, April 1901, 133 (quote); Foner, *Policies and Practices*, 26–28.

44. *Miners Magazine*, August 1900, 35 (quotes); *Miners Magazine*, September 1900, 47–49; MSS, *Official Manual, 1901–1902*, 106.

45. "Frisco Cons'd Mining Co.," December 12, 1900; "Gem, Idaho," February 15, 1901; and "Wardner, Idaho," February 20, 1901, all in f. 1, b. 1, Easton; Phipps, *Bull Pen to Bargaining Table*, 42–44; Wyman, *Hard Rock Epic*, 55–56; *EMJ*, February 23, 1901, 255 (quote).

46. "Burke, Idaho," February 19, 1901 (first quote); "Frisco Cons'd Mining Co.," December 12, 1900; "Gem, Idaho," February 15, 1901; "Gem, Idaho," February 25, 1901 (fourth quote); and "Gem, Idaho," April 10, 1901 (second and third quotes), all in f. 1, b. 1, Easton.

47. "Coeur D'Alenes Mine Owners' Association, Wallace, Id.," April 11, 1901 (first quote); and "Wallace, Id.," April 14, 1901 (second and third quotes), both in f. 1, b. 1, Easton.

48. *Miners Magazine*, May 1901, 6–7.

49. *EMJ*, July 20, 1901, 99; "Wallace, Idaho," February 23, 1901, f. 1, b. 1, Easton; "Le Roi Mining Company," *Economist*, February 22, 1902, 299–302; "Joplin Men Go to Northport," *JG*, September 12, 1901 (quote); Jensen, *Heritage of Conflict*, 87; Enyeart, *Quest*, 140.

50. "Rossland and Northport," *Nelson Tribune*, September 17, 1901; "Ran the Gauntlet," *Los Angeles Times*, September 18, 1901; "Disarming Workmen," *Los Angeles Times*, November 13, 1901 (first quote); "Mines and Mining," *Victoria Daily Colonist*, September 22, 1901 (second quote); "Bullets for the Chorus," *Los Angeles Times*, November 10, 1901 (third and fourth quotes); Boucher, "1901 Rossland Miners Strike."

51. "Gun Fighters from Joplin," *Nelson Tribune*, November 11, 1901 (first and second quotes); Cripple Creek WFM notices, 1901, reprinted in Carroll D. Wright, "Report on Labor Disturbances," 148–49 (final quotes p. 148).

52. *Miners Magazine*, November 1901, 5–6 (first four quotes); *Miners Magazine*, January 1902, 46 (remaining quotes).

53. *Miners Magazine*, March 1902, 39–40 (first three and sixth quotes); *Miners Magazine*, April 1902, 6–7 (fourth and fifth quotes); Foner, *Policies and Practices*, 415–24.

54. *American Federationist*, August 1901, 320 (first quote); Missouri General Assembly, *Laws of Missouri Passed at the Session*, 211; Missouri General Assembly, *Journal of the House*, 694–95; *American Federationist*, December 1901, 564; *American Federationist*, June 1902, 321 (second quote); MBLS, *Twenty-Third Annual Report*, 273, 283; Gompers to Dodson, June 16, 1903, reel 61, *Samuel Gompers Letterbooks*; *American Federationist*, June 1903, 502.

55. *Miners Magazine*, June 1903, 28 (first quote); *Miners Magazine*, May 1903, 38; "Miners Favor Organization," *JG*, April 30, 1903; WFM, *Eleventh Annual Convention*, 120–21 (second quote); Record Book of New and Defunct Unions, 43:4–7, WFM/IUMM&SW; *Miners Magazine*, August 27, 1903, 8; *Miners Magazine*, November 19, 1903, 9; *Miners Magazine*, February 18, 1904, 7.

56. "Men Wanted," *JG*, September 5, 1903; "Want Miners for Colorado," *JG*, September 10, 1903; "Wanted," *JG*, September 13, 1903; "Wanted," *JG*, September 22, 1903; "Joplin Miners in California," *JG*, September 23, 1903; "Mine Owners Send a Commission to Missouri to Hire Non-union Men," *Denver Post*, September 10, 1903, scrapbook 1, b. 1, Peabody Collection, UCL (quotes); Jensen, *Heritage of Conflict*, 121–53; Dubofsky, *We Shall Be All*, 49–56.

57. *Miners Magazine*, September 10, 1903, 11; *Miners Magazine*, December 17, 1903, 8;

Miners Magazine, January 14, 1904, 7 (first, second, and third quotes); WFM, *Twelfth Annual Convention*, 177–78 (fourth quote p. 178); *Miners Magazine*, February 18, 1904, 7 (last quote).

58. "Strike Breakers Have Reached Denver," *Denver Post*, September 18, 1903; "All Quiet in San Miguel," *Colorado News*, December 1903; and "Missouri Miners Coming From Idaho," *Colorado News*, October 13, 1903, all in scrapbook 1, b. 1, Peabody Collection, UCL; "P." to "Dear Sir," September 26, 1903, Pinkerton Detective Agency Reports, UCL; *EMJ*, September 12, 1903, 402; WFM, *Twelfth Annual Convention*, 177–78.

59. Dubofsky, *We Shall Be All*, 54–55; Jensen, *Heritage of Conflict*, 147–51; Jameson, *All That Glitters*, 202–25; "Assassination Intolerable," *WCR*, June 9, 1904 (first quote); "Unionism Must Go in Cripple Creek," *WCR*, June 10, 1904; WFM, *Thirteenth Annual Convention*, 259 (second quote).

60. WFM, *Thirteenth Annual Convention*, 257–66 (first three quotes p. 258, fourth quote p. 259).

61. WFM, *Thirteenth Annual Convention*, 257–66 (first quote p. 259 and second quote p. 262); Dubofsky, *We Shall Be All*, 54–55; Jensen, *Heritage of Conflict*, 147–51.

62. Harper, *White Man's Heaven*, 69–108 (*JG* quoted p. 93).

63. MSS, *Official Manual, 1905 and 1906*, 469; Kendi, *Stamped from the Beginning*, 295–322; Ingalls, *Lead and Zinc*, 235–36 (quotes).

CHAPTER 5

1. U.S. Bureau of the Census, *Historical Statistics*, 149–51; *Marine Engineering*, June 1899, 52.

2. *EMJ*, January 15, 1910, 199; A. J. Martin, *Summarized Statistics*, 16.

3. Norris, *AZn*, 13–15; Knerr, *Eagle-Picher Industries*, 65–67; Clarence A. Wright, *Mining and Treatment*, 6 (second quote); *EMJ*, October 15, 1910, 759–60 (first and third quotes).

4. Clarence A. Wright, *Mining and Treatment*, 6 (quote), 9–10; *Lead and Zinc News*, February 24, 1908, 5–8; *EMJ*, January 15, 1910, 199; *EMJ*, December 3, 1910, 1110.

5. Montgomery, *Fall of the House*, 333–34; Meyer, *Manhood on the Line*, 3–6; Kessler-Harris, "Treating the Male," 201; Kimmel, *Manhood in America*, 57–79.

6. Clarence A. Wright, *Mining and Treatment*, 36; "Who is the Champion Shoveler?," *JG*, April 14, 1907 (last quote).

7. "Galena, Kansas, Aug. 31st 1896," f. 4, b. 1, LSR (first two quotes); "A Day with the Shovellers," *JNH*, August 25, 1901 (remaining quotes); Graziosi, "Common Laborers," 517, 519.

8. "A Day with the Shovellers," *JNH*, August 25, 1901 (quotes); Nieberding, *History of Ottawa County*, 82.

9. "Miners Favor Organization," *JG*, April 30, 1903; *EMJ*, May 4, 1907, 120; "Shovelers Best Paid," *WCR*, August 10, 1907; WFM, *Thirteenth Annual Convention*, 258 (quote).

10. "A Sudden Death" and "An Old Citizen Gone," *WCR*, November 4, 1903; "Miner's Consumption," *WCR*, March 23, 1905 (first quote); "Ben Peppers Very Ill," *WCR*, April 26, 1907 (second quote); Rosner and Markowitz, *Deadly Dust*, 13–31; "Benjamin F. Peppers" and "Harrison Peppers," Webb City Ward 4, Jasper County, Missouri, 1900 federal manuscript census.

11. MBM, *17th Annual Report*, 26–27; MBM, *18th Annual Report*, 284–85, 293–94 (quote p. 284); MBM, *19th Annual Report*, 464–65; MBM, *20th Annual Report*, 112–15.

12. *Lead and Zinc News*, January 23, 1905, 80 (first two quotes); MBM, *20th Annual Report*, xiv–xv (third and fourth quotes), 456–62; MBM, *Twenty-Third Annual Report*, 8–9.

13. MBM, *19th Annual Report*, 461; "Death of George Wilson," *WCR*, October 9, 1906; MBM, *20th Annual Report*, 101–10; Jameson, *All That Glitters*, 90–93, 105–13; Levy, *Freaks of Fortune*, 60–88, 222; Roll, "Faith Powers," 82–88.

14. *Annotated Statutes*, 4079–80; State v. Cantwell, 179 Mo. 245 (1904); Cantwell v. Mis-

souri, 199 U.S. 602 (1905); State of New York, *Department of Labor Bulletin*, 121; William G. Ross, *Muted Fury*, 41–42; Freund, *Police Power*, 25; *Lead and Zinc News*, November 20, 1905, 6–7 (quote).

15. *Lead and Zinc News*, November 20, 1905, 6 (second quote); *EMJ*, December 16, 1905, 1121 (first quote); *EMJ*, February 3, 1906, 234–35; *EMJ*, January 4, 1908, 92.

16. *Lead and Zinc News*, February 24, 1908, 7; "Shovelers Best Paid," *WCR*, August 10, 1907; "The Spade Hand," *JG*, April 14, 1907; "Shoveler Wants to Know," *WCR*, August 13, 1907. On piece-rate mining elsewhere, see Mouat, *Roaring Days*, 91–93, 193; and Fishback, *Soft Coal, Hard Choices*, 66–72.

17. "Who is the Champion Shoveler?," *JG*, April 14, 1907 (quote); "Fox Beats the Record," *WCR*, April 24, 1907; "Shovelers Best Paid," *WCR*, August 10, 1907; "Shovels," *Mineral Museum News*, no. 1 (Fall 1987): 3, Rolla; "George Wright Breaks All Records for Work and Wages," *Galena Evening Times*, September 21, 1912; *Mines and Minerals*, November 1907, 156; "Jack Fox," Webb City Ward 1, and "Harvey E. McAuliffe," Webb City Ward 3, both in Jasper County, Missouri, 1910 federal manuscript census.

18. "Opportunities for Poor Men," *JG*, February 24, 1907 (first three quotes); "Fox Beats the Record," *WCR*, April 24, 1907; Frederic J. Haskin, "Zinc Boom Builds Cities Over Night," *Cleveland Plain Dealer*, October 31, 1916; Mills, "Joplin Zinc," 658, 661 (fourth quote).

19. Lanza and Higgins, *Pulmonary Disease*, 38 (first and third quotes); Frederic J. Haskin, "Zinc Boom Builds Cities Over Night," *Cleveland Plain Dealer*, October 31, 1916 (second and fourth quotes); Charles W. Landrum to H. S. Kimball, April 1, 1907, f. 4/1, b. 1, section 1, AZR; "Conference on Health and Working Conditions in the Tri-State District, Joplin Missouri, April 23, 1940," 10, b. 52, FPP ("fancy" quote).

20. "A Day with the Shovellers," *JNH*, August 25, 1901 (first and second quotes); Nieberding, *History of Ottawa County*, 82 (third quote); marriage statistics based on analysis of the 1910 federal manuscript census for Jasper, Newton, and Lawrence Counties; *Mines and Minerals*, November 1907, 156 (fourth quote).

21. "Three Hundred and 3 Cans of Dirt," *JG*, May 1, 1907 (first three quotes); "What Lem Smith, of Alba, Actually Did," *JG*, May 3, 1907 (fifth quote); "Shovelers May Contest," *WCR*, May 8, 1907 (fourth quote).

22. Taylor, *Principles of Scientific Management*, 39–72 (quote p. 46).

23. *EMJ*, November 9, 1907, 887.

24. "The Spade Hand," *JG*, April 14, 1907 (first and second quotes); "What Lem Smith, of Alba, Actually Did," *JG*, May 3, 1907 (remaining quotes).

25. "Character Study of the Miners Employed in Joplin," *JG*, February 24, 1907 (first three quotes); "Negroes Were Organized," *WCR*, April 6, 1908 (last quote).

26. "What Lem Smith, of Alba, Actually Did," *JG*, May 3, 1907 (third quote); *EMJ*, October 15, 1910, 761 (first two quotes); Bederman, *Manliness and Civilization*, 1–15, 171–87.

27. "Men Wanted for the Mines," *JG*, April 11, 1907; Landrum to Kimball, March 30 (first and second quotes), April 1 (third and fourth quotes), 1907, and Kimball to Landrum, April 4, 1907, all in f. 4/1, b. 1, section 1, AZR; *EMJ*, January 4, 1908, 92.

28. "Men Wanted for the Mines," *JG*, April 11, 1907 (second quote); *EMJ*, June 15, 1907, 1156; "Attack Led by Editor O'Neill," *Daily Arizona Silver Belt*, June 15, 1907; Landrum to Kimball, April 1, 1907, f. 4/1, b. 1, section 1, AZR (first quote); Jensen, *Heritage of Conflict*, 358–60.

29. Landrum to Kimball, April 1, 1907 (third quote), and Kimball to Landrum, April 4, 1907 (first and second quotes), both in f. 4/1, b. 1, section 1, AZR.

30. "Shovelers Best Paid," *WCR*, August 10, 1907 (first five quotes); [Kimball] to W. P. Rossman, April 30, 1908, f. 16/1, b. 5, section 1, AZR (final quote).

31. "'Shoveler' Wants to Know," *WCR*, August 13, 1907.

32. "'Shoveler' Wants to Know," *WCR*, August 13, 1907.

33. "'Shoveler' Wants to Know," *WCR*, August 13, 1907.

34. Charter 12041, United Lead and Zinc Miners Union, Joplin, February 12, 1906, and Charter 12042, Lead and Zinc Miners Union, Webb City, February 20, 1906, both in 3:76–77, RG2-10, Meany; *American Federationist*, April 1906, 262; *American Federationist*, June 1906, 426; *Miners Magazine*, March 8, 1906, 12; *Miners Magazine*, May 17, 1906, 10 (first quote); *Miners Magazine*, July 25, 1907, 6; *Miners Magazine*, June 4, 1908, 11; Shore, *Talkin' Socialism*, 172, 196–201; WFM, *Fifteenth Annual Convention*, stenographic report, 167 (second and third quotes). The *Appeal* was not totally immune to the region's promises. Publisher J. A. Wayland explained in 1902 that he moved the paper to Girard because "it is surrounded by the richest lands, rich in soil, rich in coal, rich in lead, rich in zinc." See "The Man with the Corner Lot," *Appeal to Reason*, March 29, 1902. Editor Fred Warren, meanwhile, offered to give away land he owned in a subscription contest to boost reader numbers ahead of the 1912 election. The land was "just east of the wonderfully rich mines in the Joplin district." While Warren could not promise that the winner would discover metal there, he noted that "valuable deposits of lead and zinc have been opened" nearby. See "A Bonifide Offer," *Appeal to Reason*, September 14, 1912.

35. All in *EMJ*: January 4, 1908, 92; November 30, 1907, 1029 (quote); July 13, 1907, 79; September 21, 1907, 555; January 7, 1911, 25; August 31, 1907, 428–29; October 5, 1907, 697.

36. *EMJ*, January 4, 1908, 15, 16 (first quote); Frank Starkweather to Charles Landrum, August 12, 1908, f. 4/1, b. 1, section 1, AZR; *American Federationist*, July 1908, 550 (second quote).

37. Frank Starkweather to Landrum, August 12, September 30 (final quote), 1908; Landrum to Kimball, August 31 (first quote), September 8, 17, 1908; and Kimball to Landrum, September 4, 9, 1908, all in f. 4/1, b. 1; and [Kimball] to W. P. Rossman, April 30, 1908 (second quote), f. 16/1, b. 5, all in section 1, AZR; *EMJ*, January 9, 1909, 70. See Rosenow, *Death and Dying*, 68–97.

38. WFM, *Sixteenth Annual Convention*, 306 (first quote); Charles H. Moyer report, 23, and William Jinkerson report, 321 (remaining quotes), both in WFM, *Seventeenth Annual Convention*; MSS, *Official Manual, 1909–1910*, 699; Jensen, *Heritage of Conflict*, 245–46.

39. Kimball to Landrum, February 4, 1909, and Landrum to Kimball, April 30 (first two quotes), June 11, 1909, all in f. 4/2; and Landrum to Kimball, [November 1909], f. 4/3, all in b. 1, section 1, AZR.

40. Landrum to Kimball, February 7, 1910, f. 4/4, b. 1, section 1, AZR.

41. *EMJ*, December 4, 1909, 1140; *EMJ*, February 12, 1910, 391; "Getting Good Wages," January 28, 1910, 107, MF 55 "Homestake Strike, 1910–1911," WFM/IUMM&SW; Landrum to Kimball, March 26, 1910 (first quote), f. 4/4, b. 1, section 1, AZR; "Labor Shortage in District Would Provide Work for 1200 Men," *JG*, April 17, 1910 (second quote); WFM, *Eighteenth Annual Convention*, 28 (last quote).

42. Missouri General Assembly, *Regular and Extra Sessions of the Forty-Fourth*, 251–52, 365–66; Kimball to Landrum, September 17, 1908, f. 4/1, and Landrum to Kimball, January 11, 1910 (quote), f. 4/4, both in b. 1, section 1, AZR; "Here, Too, Is a Chance for the Poor Man," *JG*, November 18, 1909; Anthony Bale, "America's First Compensation Crisis," in Rosner and Markowitz, *Dying for Work*, 35, 48–49.

43. "List of Casualties during May, 1910" (second quote); "List of Casualties from June, 1910"; "From May 1, 1910, to April 30, 1911"; and Landrum to Kimball, June 4, 1910 (first quote), all in f. 256/1, b. 14, section 1, AZR.

44. P. H. Polhemius to Landrum, March 4, 1910, and Landrum to Kimball, March 6, 7

(first quote), 17, 1910, all in f. 4/4; and Landrum to Kimball, May 28, July 16, 1910, f. 4/5, all in b. 1, section 1, AZR; "Webb City Miners Strike for Increase," *JG*, March 6, 1910; "Miners Appointed Committee" and "The Operators' View Point," *WCR*, May 25, 1910; "Wages Reduced, Miners Strike," *JG*, July 17, 1910; "300 Miners Join Strikers," *JG*, July 19, 1910; *EMJ*, June 4, 1910, 1144–45 (remaining quotes); "Miners American Zinc Strike Is Again Over," *JG*, July 23, 1910; MBM, *22nd Annual Report*, 5.

45. *Miners Magazine*, July 22, 1909, 3 (first quote); WFM, *Eighteenth Annual Convention*, 28 (second and third quotes); executive board report, July 18, 1910, 3:5, WFM/IUMM&SW (fourth and fifth quotes).

46. "Miners Organize," *Missouri Trades Unionist*, March 9, 1910; "Miners Organizing," *Missouri Trades Unionist*, March 16, 1910; "Miners Organizing in Joplin," *Missouri Trades Unionist*, April 13, 1910 (first quote); "Miners' Officers Elected," *Missouri Trades Unionist*, June 8, 1910; Kimball to Landrum, May 31, 1910 (second quote), f. 4/5, b. 1, section 1, AZR; *Miners Magazine*, June 16, 1910, 4; *Miners Magazine*, August 25, 1910, 10 (remaining quotes).

47. "Joplin Miners in the Black Hills," *Missouri Trades Unionist*, March 9, 1910; WFM, *Eighteenth Annual Convention*, 28 (fifth quote); executive board report, July 18, 1910, 3:5 (third and fourth quotes); "The Scab's Lament," *Central City Register*, February 19, 1910; "The Song of the Missourian," *Central City Register*, March 19, 1910; "The Man from Missouri," *Lead Daily Call*, March 10, 1910 (first quote); and "Why We are Here," *Lead Daily Call*, March 28, 1910 (second quote), all in MF 55 "Homestake Strike, 1910–1911," WFM/IUMM&SW.

48. Landrum to Kimball, April 22, May 28 (quotes), 1910, and Kimball to Landrum, May 31, 1910, all in f. 4/5, b. 1, section 1, AZR.

49. *Missouri Trades Unionist*, quoted in *Miners Magazine*, June 9, 1910, 4; "Wages Reduced," *JG*, July 17, 1910 (first two quotes); *EMJ*, December 9, 1911, 1134 (third quote).

50. WFM charter application, June 17, 1910, f. 19, b. 4, RG28-003, Meany; MBLS, *Thirty-Third Annual Report*, 102; Fear to Gompers, May 19, 1910, and Gompers to Fear, May 20, 1910 (first quote), both in reel 45, *American Federation of Labor Records*; *Miners Magazine*, September 29, 1910, 6–7 (second and third quotes). On the WFM and the AFL, see Jensen, *Heritage of Conflict*, 236–55.

51. *EMJ*, December 9, 1911, 1134 (first quote); Landrum to Kimball, March 26, 1910 (second quote), f. 256/1, b. 14, and J. W. Houser to J. H. Polhemus, June 1, 1912, f. 33/1, b. 7, both in section 1, AZR.

52. *Miners Magazine*, September 21, 1911, 14; "Register of Local Union Assessments, Etc., 1907–1922," vol. 51, WFM/IUMM&SW; *Miners Magazine*, April 6, 1911, 8 (Miller quotes); *Miners Magazine*, June 22, 1911, 9; *EMJ*, December 9, 1911, 1134 (Ruhl quote); MBLS, *Thirty-Third Annual Report*, 153, 200; MBLS, *Thirty-Fourth Annual Report*, 105–10, 114–32.

53. Lanza and Higgins, *Pulmonary Disease*, 10–11; Missouri Board of Health, *Twenty-Ninth Annual Report*, 75, 103; Clarence A. Wright, *Mining and Treatment*, 39–40; "Miner's Consumption," *WCR*, May 17, 1914; MBLS, *Thirty-Third Annual Report*, 36–39, 67, 151–54, 196–200 (second quote p. 37); *Miners Magazine*, April 6, 1911, 8 (Miller quote); "I. Jack Fox," November 1, 1911, Jasper County, Missouri Death Certificates, MSA.

54. MBLS, *Thirty-Fourth Annual Report*, 105–10, 114–32; J. W. Houser to J. H. Polhemus, June 1, 22, 1912, March 29, 1913, f. 33/1, b. 7, section 1, AZR; MSS, *Official Manual, 1913 and 1914*, 779; Selig Pearlman, "The Lead and Zinc Mines in the Joplin, Missouri, District," March 11, 1914, in reel 8, *U.S. Commission on Industrial Relations*; *Miners Magazine*, March 13, 1913, 5–6 (first quote); *Miners Magazine*, March 20, 1913, 4; *Miners Magazine*, March 27, 1913, 1; *Miners Magazine*, March 27, 1913, 8–9; *American Federationist*, December 1912, 1030; *Miners Magazine*, August 14, 1913, 10; *Miners Magazine*, April 17, 1913, 13 (Edens quotes); *Miners Magazine*, October 9, 1913, 4.

55. *Miners Magazine*, December 4, 1913, 3; "Leave Joplin"; "Rube Ferns Says It Is a Finish Fight" (first quote); and "Agree to Leave District," all in *Missouri Trades Unionist*, January 7, 1914; "Stay Away," *Missouri Trades Unionist*, January 21, 1914; *Miners Magazine*, November 13, 1913, 6; testimony of Robert J. Copeland, in U.S. Congress, *Conditions in the Coal Mines*, 1277–83; *Miners Magazine*, April 23, 1914, 3; Missouri State Federation of Labor, *Proceedings*, 24–42 (second quote p. 25); Andrews, *Killing for Coal*, 1–19; Jensen, *Heritage of Conflict*, 272–88. Ferns, known in the ring as the "Kansas Rube," lost his title in 1901 to "Barbados" Joe Walcott, the second black boxing world champion.

56. *EMJ*, February 21, 1914, 446; *EMJ*, December 20, 1913, 1190; *EMJ*, January 20, 1914, 75–76; Weidman, *Miami-Picher*, 46–49; "Miners on Strike," *Missouri Trades Unionist*, February 11, 1914; "Miami Miners Join Union," *Missouri Trades Unionist*, February 25, 1914; "With Metal Miners," *Missouri Trades Unionist*, March 4, 1914; "Oklahoma News Notes," *Missouri Trades Unionist*, April 1, 1914; *Miners Magazine*, February 19, 1914, 9; MBLS, *Thirty-Sixth Annual Report*, 69; Gibson, *Wilderness Bonanza*, 227–29.

57. Lanza and Higgins, *Pulmonary Disease*, 9–11, 32; "Bess Hackett," in *Joplin, Missouri, Directory*, 192; Missouri Board of Health, *Thirty-First Annual Report*, 120; Rupert Blue to A. J. Lanza, October 13, 1914, f. 50307, b. 146, entry 6, RG70, NARA-CP.

58. *EMJ*, January 10, 1914, 74–75; Livingston, *History of Jasper County*, 275–85.

59. Alice Hamilton, *Exploring the Dangerous Trades*, 145–49; Alice Hamilton, "Lead Poisoning," 59–60 (first quote p. 59); "Miner's Consumption," *WCR*, May 17, 1914 (second quote); Lanza and Higgins, *Pulmonary Disease*, 12; *EMJ*, March 7, 1914, 536; *EMJ*, May 2, 1914, 924.

60. "Fight on Dust in Mines One Method against Disease," *JG*, April 10, 1914 (first, second, and sixth quotes); "Miner's Consumption," *WCR*, May 17, 1914 (third, fourth, and fifth quotes); *EMJ*, March 7, 1914, 536; Blue to Lanza, October 13, 1914 (seventh quote), and Higgins to Director of the Bureau of Mines, November 26, 1914, both in f. 50307, b. 146, entry 6, RG70, NARA-CP; Derickson, "Federal Intervention," 237–42.

61. Lanza to Surgeon General, January 8, 1915, f. 5153 (January–April 1915), b. 558, central file, 1897–1923, RG90, NARA-CP; Lanza and Higgins, *Pulmonary Disease*, 12, 22–24, 36–43 (first three quotes p. 38, fourth quote p. 39, fifth quote p. 43); *EMJ*, February 13, 1915, 331–33; *EMJ*, December 5, 1914, 1016.

62. *EMJ*, December 5, 1914, 1016 (first quote); *EMJ*, December 12, 1914, 1062 (second quote); J. E. Wommack to Director, December 8, 1914; S. Y. Ramage and W. B. Shackleford to Director, December 5, 1914; Rupert Blue to Acting Director, December 18, 1914; Young to Director, December 22, 1914 (third quote); Higgins to Acting Director, January 16, 1915; and Higgins to George Pope, February 6, 1915, all in f. 50307; and Higgins to George Rice, February 15, 1915, f. 51072, all in b. 182, entry 6, RG70, NARA-CP; Higgins et al., *Siliceous Dust*, 54–57.

63. Executive board minutes, August 5, 1914, 3:178, WFM/IUMM&SW (first quote); *EMJ*, January 8, 1916, 64; Young to J. H. Polhemus, August 22, November 14, 1914, f. 33/3, b. 7; and Young to Kimball, February 7, 20, March 10 (second and third quotes), 1915, f. 33/4, b. 7, and May 29, 1915, f. 33/5, b. 8, all in section 1, AZR; J. H. Bradford, "Digest of Report of C. J. Stowell," April 4, 1914, in reel 10, *U.S. Commission on Industrial Relations*.

64. *Miners Magazine*, August 27, 1914, 1 (quote); Jensen, *Heritage of Conflict*, 288, 325–53, 369–71.

65. Young to Kimball, May 29 (first quote), June 5, 7, 1915, f. 33/5, b. 8, section 1, AZR; Don C. Grafton to William J. Cassidy, August 10, 1947 (second quote), and Young to Cassidy, October 20, 1947, both in f. 2; and "Research Notes," f. 3, all in WCF; "Miners Union Grows," *Missouri Trades Unionist*, June 23, 1915.

66. "Miners Organize," *Missouri Trades Unionist*, May 26, 1915 (first quote); "Demands

of Unions," *Missouri Trades Unionist*, June 30, 1915; "Fake Mine Workers' Leader Exposed at Sunday Meeting," *Worker's Chronicle*, April 20, 1917 (second and third quotes); "Striking Miners Parade in Joplin," *JNH*, June 29, 1915 (fourth and fifth quotes); "No Trouble Reported as Strike Continues," *WCR*, June 29, 1915 (sixth and seventh quotes); "Strikers March on Joplin Mines," *JG*, July 1, 1915 (eighth quote).

67. "Stand Pat," undated strike handbill, enclosed in Young to Kimball, July 1, 1915, f. 33/5, b. 8, section 1, AZR (quotes); "1,000 Miners Go on Strike at Webb City," *JG*, June 27, 1915; "Striking Miners Parade in Joplin," *JNH*, June 29, 1915; "No Trouble Reported as Strike Continues," *WCR*, June 29, 1915; "30 Mines Down by Strike," *JG*, June 30, 1915; *EMJ*, July, 10, 1915; "Many Miners Joining" and "Demands of Unions," *Missouri Trades Unionist*, June 30, 1915; "Strikers March on Joplin Mines," *JG*, July 1, 1915; "Strikes Close Lead Mines," *New York Times*, July 1, 1915, 12; "Strikers Arrange Wage Schedule They Will Ask Adopted," *JG*, July 2, 1915; "Oklahoma Field Will Be Included in Strike, According to Report," *JG*, July 3, 1915; "Committees Now to Visit Mines; More are Closed," *JG*, July 4, 1915; "Webb City Union" and "Joplin Miners' Union," *Missouri Trades Unionist*, July 7, 1915; Higgins to George Rice, July 6, 1915, f. 51072, b. 182, entry 6, RG70, NARA-CP.

68. Kimball to Young, June 29, 1915 (first quote), and Young to Kimball, July 1 (second quote), 2, 8, 9, 1915, all in f. 33/5, b. 8, section 1, AZR; "Striking Miners Cheer Speech by Joplin's Mayor," *JNH*, July 2, 1915 (fourth quote); "Operators Issue First Statement Regarding Strike," *JG*, July 7, 1915 (third quote).

69. "Mines to Resume Operations Early in Week, Is Plan," *JG*, July 9, 1915 (first quote); "Zinc Miners' Strike Failing," *Wall Street Journal*, July 10, 1915, 2; "Central District Mines Prepare to Start Operations," *JG*, July 10, 1915; "Expect Strike to be Broken Today," *JG*, July 11, 1915; "Strike of Two Weeks is Ended," *JG*, July 13, 1915 (second, third, and fourth quotes); Young to Kimball, July 10, 12 (last quote), 1915, and "200 Strike Breakers Said to Have Been Here," *Webb City Daily Sentinel*, July 13, 1915, all in f. 33/5, b. 8, section 1, AZR; *EMJ*, July 17, 1915, 122; *EMJ*, July 24, 1915, 164; *EMJ*, August 14, 1915, 288.

70. "George Wallace Chosen," *Missouri Trades Unionist*, July 21, 1915; "Moyer Visits Joplin," *Missouri Trades Unionist*, July 28, 1915; "Miners Adopt Scale," *Missouri Trades Unionist*, October 6, 1915; Gompers to Charles Moyer, July 12, 28, 30 (first quote), August 11, 1915, reel 196, RG1-02, Meany; Don C. Grafton to William J. Cassidy, August 10, 1947, f. 2, b. 1, WCF; *American Federationist*, August 1915, 614; *American Federationist*, September 1915, 769; *Miners Magazine*, August 5, 1915, 1, 3; *Miners Magazine*, September 21, 1915, 3; *Miners Magazine*, October 7, 1915, 2; registers of locals, vol. 43, WFM/IUMM&SW; Young to Kimball, July 31, 1915, f. 33/5 (second quote), and August 18, November 3, 1915, f. 33/6, all in b. 8, section 1, AZR; "Fake Mine Workers' Leader Exposed at Sunday Meeting," *Worker's Chronicle*, April 20, 1917.

71. Moyer to Miller, September 11, 1915, 4:24, WFM/IUMM&SW (first and second quotes); *Miners Magazine*, September 21, 1915, 3 (third quote); Jensen, *Heritage of Conflict*, 376–77.

CHAPTER 6

1. Frederic J. Haskin, "Zinc Boom Builds Cities Over Night," *Cleveland Plain Dealer*, October 31, 1916 (first and last quotes); "Report on the Miami Zinc and Lead District," 1917, 4, f. 36, b. 2, R677, Rolla; *EMJ*, March 17, 1917, 478; "Deposits of Miami Area Pronounced Richest in World," *Daily Oklahoman*, August 12, 1917, 1 (second quote); Blosser, *Prairie Jackpot*, 66–69.

2. Frederic J. Haskin, "Zinc Boom Builds Cities Over Night," *Cleveland Plain Dealer*, October 31, 1916; "Many Miners Being Employed," *Missouri Trades Unionist*, August 18, 1915;

EMJ, March 17, 1917, 478; *EMJ*, January 12, 1918, 70; A. J. Martin, *Summarized Statistics*, 22, 28, 40, 59, 60, 63; U.S. Bureau of the Census, *Fifteenth Census*, 1:405, 611, 891.

3. *EMJ*, September 26, 1914, 589; *EMJ*, January 9, 1915, 65; U.S. Bureau of Indian Affairs, *Report of the Commissioner*, 68; Weidman, *Miami-Picher*, 46–49; Hallam v. Commerce Mining and Royalty Co., 49 F.2d 103 (10th Cir. 1931).

4. *JG*, February 25, 1915, quoted in Knerr, *Eagle-Picher Industries*, 67–72, 75–77 (quote p. 70); *EMJ*, February 23, 1918, 362–63; Blosser, *Prairie Jackpot*, 15–69; U.S. Geological Survey, *Mineral Resources, 1916*, 144.

5. MBLS, *Thirty-Seventh Annual Report*, 197; "Defunct and New Unions," vol. 43, WFM/IUMM&SW; "Miners Adopt Scale," *Missouri Trades Unionist*, October 6, 1915; "Miners Discuss Unionism," *Missouri Trades Unionist*, February 9, 1916; "Miners' Unions Grow," *Missouri Trades Unionist*, February 16, 1916; *Miners Magazine*, November 11, 1915, 1; *Miners Magazine*, December 2, 1915, 1; *Miners Magazine*, February 3, 1916, 2 (fourth quote); *Miners Magazine*, June 1, 1916, 1 (first and third quotes); F. C. Smith to Surgeon General, August 10, 1915, f. 51072, b. 182, entry 6, RG70, NARA-CP (second quote); *EMJ*, April 22, 1916, 751; *EMJ*, June 17, 1916, 1091; *Miners Magazine*, July 6, 1916, 1; MBM, *Twenty-Ninth Annual Report*, 10–17; "Gaining on Tuberculosis in Missouri," 169–70; "The Bull is Now Loose in China Shop," *WCR*, November 6, 1915; Wyman, *Hard Rock Epic*, 171.

6. *EMJ*, April 22, 1916, 751; A. J. Martin, *Summarized Statistics*, 22; "Thin Ground Mines Close Here This Evening; Men Out of Employment Today," *WCR*, July 15, 1916; Howard Young to Harry Kimball, May 31, 1916, f. 33/8; and Kimball to William A. Ogg, December 4, 1917, and Young to J. W. Houser, December 29, 1917, both in f. 33/13, all in b. 8, section 1, AZR; *EMJ*, February 23, 1918, 362–63; *EMJ*, January 6, 1917, 27–28.

7. *Miners Magazine*, July 6, 1916, 3 (third quote); *EMJ*, June 10, 1916, 1048 (first quote); Howard Young to Harry Kimball, June 3, 10 (second quote), 1916, f. 33/9, b. 8, section 1, AZR; "Trolley Slips, I. W. W. Organizer Fails to 'Get Away' With an Attack," *JG*, June 6, 1916; "Two Found Guilty of Being in Riot," *JG*, June 10, 1916.

8. "Webb City Boys Leave for War With Mexico," *WCR*, June 19, 1916; "Postmaster Michie Advocates Buying a Big Silk Flag," *WCR*, June 21, 1916; Pryor, *Southwest Missouri Mining*, 81, 92; Locals 203, 205, 206, 207, 208, 209, 217, 219, and 221, all in vol. 49, and Moyer report, vol. 176, both in WFM/IUMM&SW; Wilson, "Proclamation 1364" (first quote); Don C. Grafton to William J. Cassidy, August 10, 1947, f. 2, b. 1, WCF (second quote); "St. Louis Militiamen Ordered to Flat River to Check Labor Riots," *St. Louis Post-Dispatch*, July 14, 1917; Jensen, *Heritage of Conflict*, 376–77.

9. A. J. Martin, *Summarized Statistics*, 22; Norris, *AZn*, 66–72; *EMJ*, January 12, 1918, 70–71; *EMJ*, January 11, 1919, 59; miscellaneous, f. "Sale of Missouri Lands," b. 40, section 3, AZR; MBLS, *Fortieth and Forty-First Annual Reports*, 133 (quote).

10. "Many Miners Being Employed," *Missouri Trades Unionist*, August 18, 1915 (first quote); Nieberding, *History of Ottawa County*, 32–33; *EMJ*, March 17, 1917, 478 (third quote); "Deposits of Miami Area Pronounced Richest in World," *Daily Oklahoman*, August, 12, 1917, 1; "Report on the Miami Zinc and Lead District of Oklahoma," 1917, 4–5, f. 36, b. 2, R677, Rolla (second quote); Harry C. Hood, interview by Judith Conboy et al., n.d. [1979], transcript, tape 1, side 2, p. 6, Hood Collection, Missouri Southern State University, Spiva Library Archives and Special Collections, Joplin, Missouri (fourth quote); "Joe Jake Nolan," September 12, 1918, in U.S. World War I Draft Registration Cards; "John F. Nolan," January 19, 1919, and "Otto McCollum," February 1, 1918, both in Jasper County, Missouri Death Certificates, MSA; "Scott McCollum," Quapaw Township, Ottawa County, Oklahoma, 1920 federal manuscript census.

11. MM, *Twenty-Third Consecutive*, 8–18, 92, 128–29; *Miners Magazine*, June 1, 1916, 6; *Miners Magazine*, December 1917, 3; *Miners Magazine*, February 1919, 2.

12. *EMJ*, July 10, 1917, 36; *American Mining Manual*, 292–98; *Compressed Air Magazine*, May 1920, 9635–43.

13. *EMJ*, January 12, 1918, 70 (first quote); "Deposits of Miami Area Pronounced Richest in World," *Daily Oklahoman*, August 12, 1917, 2 (second quote); "The Famous Rich Miami Mining District" and "Commerce, Tar River, Picher, Century," *Miami Record-Herald*, April 27, 1917; Tri-State Survey Committee, *Preliminary Report*, 19–27; Harry Kimball to William A. Ogg, June 27, 1918, f. 33/15, b. 8, section 1, AZR (third quote); Mills, "Joplin Zinc," 662–63 (fifth and sixth quotes p. 662); First Annual Report, 10–19, RG70, NARA-FW (fourth quote p. 19); Charles Day to Bureau of Mines, December 3, 1915, f. 53461, b. 169, entry 6, RG70, NARA-CP; "Overheard on the Interurban," *Anti-hijacker*, August 17, 1918 (seventh quote); "Magic Progress Made by Century," *Independent*, September 27, 1917; Robertson, *Hard as the Rock Itself*, 126–27.

14. Harrington et al., "Dust-Ventilation Investigation," appendix C, 2, RG70, NARA-FW; Martinez, *Hard Rock Legacy*, 8–18; Lanza and Higgins, *Pulmonary Disease*, 40; Mills, "Joplin Zinc," 662; Harry C. Hood, interview by Judith Conboy et al., n.d. [1979], transcript, tape 1, side 1, pp. 1, 3; tape 1, side 2, p. 9; and tape 2, side 1, p. 1, Hood Collection, Missouri Southern State University, Spiva Library Archives and Special Collections, Joplin, Missouri; Pryor, *Southwest Missouri Mining*, 63–125; *American Zinc and Lead Journal*, November 1917, 8–10.

15. *EMJ*, March 2, 1918, 415; "Report on the Miami Zinc and Lead District of Oklahoma," 1917, 4, f. 36, b. 2, R677, Rolla; Weed, *Mines Handbook*, 1274 (second quote); Richard Ageton to George Rice, March 26, 1923, f. 112306 no. 3, b. 1059, entry 6, RG70, NARA-CP; Mills, "Joplin Zinc," 660–61 (first quote p. 661); Oklahoma State Election Board, *Directory, State of Oklahoma*, 27; record of insurance board hearing, January 1925, 55, f. 460, b. 15, RG70, NARA-FW; Nieberding, *History of Ottawa County*, 33; Evelyn Maurer, draft article, enclosed in Frances Perkins to Verne Zimmer, October 25, 1939, f. 7-0-6-13 Tri-State Survey, b. 29, Classified Central Files, 1937–41, RG100, NARA-CP (third quote); Robertson, *Hard as the Rock Itself*, 128.

16. Frederic J. Haskin, "Zinc Boom Builds Cities Over Night," *Cleveland Plain Dealer*, October 31, 1916 (first two quotes); Mills, "Joplin Zinc," 661 (third quote).

17. Lawrence Barr and Theo Sisco, interview by Joe Todd, 1985, H1985.13, Oral History Collection, OHC; Nieberding, *History of Ottawa County*, 33 (quote); "Anthony McTeer," September 12, 1918, in U.S. World War I Draft Registration Cards.

18. "Charles C. Chesnut," f. 2, R754, Rolla (first quote); "Intensive Mining Here," *King Jack*, February 26, 1920 (second quote); Mills, "Joplin Zinc," 657–61 (remaining quotes); draft report, May 1920, 20, and "The Speakers' Manual," 42, both in f. 2, b. 1, Interchurch World Movement Records, Union Theological Seminary, New York, New York; "Heroes Unsung," *King Jack*, January 31, 1926; Knerr, *Eagle-Picher Industries*, 86; Mary Murphy, *Mining Cultures*, 107–9; Jameson, *All That Glitters*, 38–41.

19. *EMJ*, May 21, 1921, 860 (third quote); Mills, "Joplin Zinc," 657 (first, fifth, and sixth quotes); *Compressed Air Magazine*, May 1920, 9635–43 (second quote p. 9639); Cassidy, "Tri-State Zinc-Lead," 142–43 (fourth quote p. 143); masthead, *King Jack*, December 11, 1919 (seventh quote); U.S. Bureau of the Census, *Fourteenth Census*, 822, 826; U.S. Bureau of the Census, *Fifteenth Census*, 3:562, 572; Suggs, *Union Busting*, 10–22.

20. Mills, "Joplin Zinc," 657 (first two quotes), 658 (eighth quote); "Coal Strike Supposed to Be Settled" and "Refuse to Haul Coal Miners," *King Jack*, December 11, 1919 (third and fourth quotes); "Labor Unions," *King Jack*, December 18, 1919 (fifth and sixth quotes); *EMJ*, May 21, 1921, 860 (seventh quote).

21. Martinez, *Hard Rock Legacy*, 17–26; Pub. L. No. 66-82, 41 Stat. 355 (1919); Carl Mayer

to W. C. Hale, October 30, 1920; Premier Zinc Co. to Mayer, June 9, 1920; and Mayer to Commissioner of Indian Affairs, May 27, 1920, all in f. 1, b. 1, entry 89, RG75, NARA-FW.

22. A. C. Wallace to Committee in Charge, August 21, 1920; Carl Mayer to Commissioner of Indian Affairs, October 5, 1920, February 19, 1921; and Wainwright, Hale, and Hocker to Commissioner of Indian Affairs, August 23, 1920 (quotes), all in f. 2, b. 1, entry 89, RG75, NARA-FW; "Lead Firm's Bid Accepted for Indian Land Leases," *Baltimore Sun*, July 28, 1922.

23. Klein, *Grappling with Demon Rum*, 154–69; Lawrence Barr and Theo Sisco, interview by Joe Todd, 1985, H1985.13, Oral History Collection, OHC (first quote); Nieberding, *History of Ottawa County*, 75; Evelyn Maurer, draft article, enclosed in Frances Perkins to Verne Zimmer, October 25, 1939, f. 7-0-6-13 Tri-State Survey, b. 29, Classified Central Files, 1937–41, RG100, NARA-CP (second quote).

24. Nieberding, *History of Ottawa County*, 74–76, 92 (second quote p. 74); First Annual Report, 7, RG70, NARA-FW (first quote); "Iva O. Simpson, Rural Quapaw," f. 2, R754, Rolla (last three quotes).

25. "Of Work (Mrs. B.)," appendix B, in Tri-State Survey Committee, *Preliminary Report*, 15–16.

26. *EMJ*, May 13, 1922, 819–20 (quote p. 820); *Bulletin of the American Zinc Institute*, May 1919, 1; Norris, "Missouri and Kansas," 333; Gordon, *New Deals*, 134–35. Yeatman earned his engineering degree at Washington University in 1883 alongside Joseph Gazzam and Seeley Mudd. See Rickard, *Interviews with Mining Engineers*, 391.

27. *EMJ*, November 17, 1917, 899; *Mining Congress Journal*, March 1919, 98; *Mining and Scientific Press*, September 17, 1921, 414; *EMJ*, January 18, 1919, 168; *Bulletin of the American Zinc Institute*, May 1919, 1.

28. *EMJ*, October 23, 1920, 831; "Joplin-Miami District," *EMJ*, March 19, 1921, 526; *Mining Congress Journal*, April 1921, 157; *Bulletin of the American Zinc Institute*, January–March 1921, 100–101 (quote p. 101); *Mining Congress Journal*, June 1921, 222; *EMJ*, November 12, 1921, 793; "Situation of Needy in Picher Relieved," *JG*, February 24, 1921; *Bulletin of the American Zinc Institute*, September–October 1921, 4–8.

29. *Bulletin of the American Zinc Institute*, April–May 1921, 98; *EMJ*, April 22, 1922, 690; A. J. Martin, *Summarized Statistics*, 22, 25, 28.

30. Oklahoma State Department of Health, *Fifth Annual Report*, 22, 92–93, 242–45 (first and second quotes p. 242); Mills, "Joplin Zinc," 662–66 (third quote p. 666, fourth and fifth quotes p. 662); *Bulletin of the American Zinc Institute*, January–March 1921, 101–2 (final quote); *American Zinc and Lead Journal*, November 1917, 8–10; *Bulletin of the American Zinc Institute*, September 1922, 21.

31. "Baxter a Part of Sanitary District," *Baxter Daily Citizen*, January 28, 1919; "Sanitary Men in Baxter Springs," *Baxter Daily Citizen*, May 6, 1919; "The Best Showing," *Columbus Daily Advocate*, August 29, 1919; Oklahoma State Department of Health, *Fifth Annual Report*, 242; U.S. Department of the Treasury, *Annual Report of the Surgeon General*, 51–63; Laas, *Bridging Two Eras*, 198; "Reminiscences of Dr. Thomas Parran," interview by Harlan Phillips, July 16, 1962, 49–52 (quote p. 50), Health Sciences Project, Columbia University, New York, New York.

32. *EMJ*, February 24, 1923, 374; Sayers et al., *Silicosis and Tuberculosis*, 1–2; Parran to Sayers, October 9, 1922 (quote); Matthew van Siclen to Ageton et al., February 10, 1923; and Ageton to George Rice, March 2, 1923, all in f. 112306 1922, b. 1059, entry 6, RG70, NARA-CP; Harrington et al., "Dust-Ventilation Investigation," 4, RG70, NARA-FW; Derickson, "'On the Dump Heap,'" 660–62.

33. Harrington et al., "Dust-Ventilation Investigation," 16–17, 19–21, 32 (quote), RG70, NARA-FW; Derickson, "'On the Dump Heap,'" 660–62.

34. Harrington et al., "Dust-Ventilation Investigation," 19–21, RG70, NARA-FW (quote p. 20); "Reminiscences of Dr. Thomas Parran," interview by Harlan Phillips, July 16, 1962, 49–52, Health Sciences Project, Columbia University, New York, New York.

35. Harrington et al., "Dust-Ventilation Investigation," 30–33, RG70, NARA-FW (first quote p. 30, second and third quotes p. 33); Derickson, "'On the Dump Heap,'" 660–63.

36. Ageton to Rice, March 26, 1923, f. 416.1 112306 no. 3, b. 1059, entry 6, RG70, NARA-CP (quote); Harrington et al., "Dust-Ventilation Investigation," 34, RG70, NARA-FW.

37. Ageton, "Workman's Compensation Insurance," f. 104518, b. 969 (quote), entry 6, RG70, NARA-CP; Harrington et al., "Dust-Ventilation Investigation," 34, RG70, NARA-FW; Sayers, *Silicosis among Miners*, 20–22; Derickson, "'On the Dump Heap,'" 664.

38. Ageton, "Workman's Compensation Insurance," f. 104518, b. 969, entry 6, RG70, NARA-CP (quote); Burke, "Evolution of Workers' Compensation Law," 344–47; "Workmen's Compensation Insurance for Metal Mines"; record of insurance board hearing, January 1925, 8–9, 47–56; and "Comparative Digest of Compensation Laws," August 1927, all in f. 460, b. 15, RG70, NARA-FW; "Magazine Section," *JG*, January 1929, 37; Derickson, "'On the Dump Heap,'" 659; Anthony Bale, "America's First Compensation Crisis," in Rosner and Markowitz, *Dying for Work*, 35–49.

39. Orten, "Present Economic Condition," 424–25; record of insurance board hearing, January 1925, 1, 21–45 (second quote p. 29), 48 (first quote), 56, 64, 83–84, f. 460, b. 15, RG70, NARA-FW; Mine Safety Rules Analysis, [1925], f. 256–2, b. 14, section 1, AZR; minutes of the board of directors, OPA, July 16, 1924, 2:4, b. 165, PMC (third quote); "Siebenthal to Complete Geology," *JG*, July 23, 1925; Knerr, *Eagle-Picher Industries*, 92–94; Derickson, "'On the Dump Heap,'" 664–65.

40. *EMJ*, May 30, 1925, 894; A. J. Martin, *Summarized Statistics*, 16, 22, 25, 28; Ageton, "Workman's Compensation Insurance," f. 104518, b. 969, entry 6, RG70, NARA-CP; Knerr, *Eagle-Picher Industries*, 89–91; "Accident Prevention Statistics," August 1, 1924, f. 256–2, b. 14, section 1, AZR; Derickson, "'On the Dump Heap,'" 662.

41. "Protecting the District," *JG*, September 30, 1923 (quote); *EMJ*, September 29, 1923, 560; *EMJ*, April 26, 1924, 698; *EMJ*, June 21, 1924, 1018; "General Wage Schedule," f. 21, b. 100, PMC; Oklahoma State Health Commissioner, *Ninth Annual Report*, 8–9, 60–62.

42. Harlow, *Makers of Government*, 636; McGinnis, *Oklahoma's Depression Radicals*, 6–9; Scales and Goble, *Oklahoma Politics*, 108–26; "Realm of Missouri Knights of the Ku Klux Klan," *JG*, January 23, 1923; Martinez, *Happy Birthday Picher*, 179; *Homer Wear v. State of Oklahoma* transcript, March 5, 1923, Criminal Court of Appeals Records, Oklahoma State Archives, Oklahoma City, Oklahoma; Sellars, *Oil, Wheat, and Wobblies*, 172–74.

43. "Lead and Zinc Mine Workers Flocking into Unions," *Oklahoma Federationist*, April 11, 1924 (quote); McGinnis, *Oklahoma's Depression Radicals*, 9–11; Locals 130, 134, 136, 138, and 139, Book of Defunct and New Unions, 43:57, WFM/IUMM&SW; Zieger, Minchin, and Gall, *American Workers, American Unions*, 62; Jensen, *Heritage of Conflict*, 460–63; UMW, *Proceedings*, 285–86; "Welcome to Oklahoma Mine, Mill and Smelter Workers," *Oklahoma Federationist*, May 2, 1924.

44. "Welcome to Oklahoma Mine, Mill and Smelter Workers," *Oklahoma Federationist*, May 2, 1924 (first quote); "Lead and Zinc Mine Workers Flocking into Unions," *Oklahoma Federationist*, April 11, 1924 (second quote).

45. "Lead and Zinc Mine Workers Flocking into Unions" (Finley quotes) and "Let Congress Know!," *Oklahoma Federationist*, April 11, 1924; Gompers to "All Organized Labor," March 24, 1924, f. 1, b. 1, OFLMC; "Zinc Miner Gives Answer to Operator," *Oklahoma Federationist*, July 18, 1924 (Beggs quote).

46. All in *Oklahoma Federationist*: "Zinc Miner Gives Answer to Operators," July 18, 1924

(second quote); "Welcome to Oklahoma Mine, Mill and Smelter Workers," May 2, 1924 (first and third quotes); "Zinc Miner Dies" and "Zinc Miners Have Large Free Dinner," July 18, 1924. See Levine, "Workers' Wives," 45–52.

47. All in *Oklahoma Federationist*: "Lead and Zinc Mine Workers Flocking into Unions," April 11, 1924 (third and fourth Finley quotes); "Monster Crowd of Miners Greets Finley," April 25, 1924 (first and second Finley quotes); "Word from the Battle Front," April 11, 1924 (Beggs quotes).

48. All in *Oklahoma Federationist*: "Lead and Zinc Mine Workers Flocking into Unions," April 11, 1924 (first and fourth quotes); "Monster Crowd of Miners Greets Finley," April 25, 1924 (second quote); "Welcome to Oklahoma Mine, Mill and Smelter Workers," May 2, 1924 (fifth quote). See Gordon, *New Deals*, 87 (third quote).

49. "Organization of Zinc Miners Growing Rapidly," *Oklahoma Federationist*, June 6, 1924; "Zinc Miners Have Large Free Dinner," *Oklahoma Federationist*, July 18, 1924; local union ledger summaries, vol. 53, WFM/IUMM&SW; *EMJ*, June 21, 1924, 1016 (quote); "Oklahoma," *American Federationist*, July 1924, 593.

50. All in *Oklahoma Federationist*: "Monster Crowd of Miners Greets Finley," April 25, 1924 (Finley quotes); "Zinc Miner Gives Answer to Operator," July 18, 1924 (Beggs quotes); "Zinc Miners to Enjoy Labor Day," August 22, 1924 (Dunivin quotes). See Levine, "Workers' Wives," 45–46, 51–52.

51. "Organization of Zinc Miners Growing Rapidly," *Oklahoma Federationist*, June 6, 1924; "Lead and Zinc Mine Workers Flocking into Unions," *Oklahoma Federationist*, April 11, 1924; Resolution 3, Twenty-First Annual Convention, Muskogee, September 15–17, 1924, f. 5, b. 1, OFLMC (quotes).

52. "Mines to Stay on Open Shop Basis," *MNR*, July 7, 1924 (quotes); *EMJ*, June 21, 1924, 1016.

53. American Mining Congress Western Division, *Tri-State Mining District*, 31, 36–37; A. J. Martin, *Summarized Statistics*, 20–21; *EMJ*, November 29, 1924, 863; *EMJ*, January 3, 1925, 26; *EMJ*, June 28, 1924, 1051; *EMJ*, January 24, 1925, 175; *EMJ*, February 28, 1925, 375; "Sliding Wage Scale is Profit Sharing," *JG*, April 8, 1926 (quote); "General Wage Schedule," f. 21, b. 100, PMC.

54. "Zinc Miners Gaining," *Oklahoma Federationist*, December 5, 1924 (quotes); Locals 143 and 155, Book of Defunct and New Unions, 43:57, WFM/IUMM&SW.

55. Federal Writers' Project of Oklahoma, *Labor History of Oklahoma*, 60, 72; Phelan, *William Green*, 29–31; McGinnis, *Oklahoma's Depression Radicals*, 9–11; Jensen, *Heritage of Conflict*, 463–64; "Secretary Treasurer's Report," Twenty-Third Annual Convention, f. 6, b. 1, OFLMC.

56. *EMJ*, November 7, 1925, 740; "Drainage Committee to Meet at Webb City," *JG*, February 15, 1925; A. J. Martin, *Summarized Statistics*, 20–22, 25, 28; *EMJ*, November 20, 1926, 824; "General Wage Schedule," f. 21, b. 100; and exhibits 3 and 4, f. "AI, Kansas, Minutes, 1930," b. 65, both in PMC; Nieberding, *History of Ottawa County*, 67 (quote).

57. Martinez, *Hard Rock Legacy*, 22–26, 30; Oklahoma State Health Commissioner, *Tenth Annual Report*, 94.

58. Lawrence Barr and Theo Sisco, interview by Joe Todd, 1985, H1985.13, Oral History Collection, OHC (first quote); Nieberding, *History of Ottawa County*, 74–75 (last two quotes); *EMJ*, December 18, 1926, 961 (second quote); "Grandpa Was the Meanest Man in Town," *Tri-State Tribune*, September 15, 1988; Robertson, *Hard as the Rock Itself*, 121–83.

59. Twelve Anniversary Program, May 6, 1927, f. 1, R754, Rolla (first two quotes); "Picher Party Was a Distinct Success" and "Fiery Crosses Burn in Picher Chat Piles" (last quote), *MNR*, May 8, 1927.

60. *EMJ*, April 30, 1927, 739; *EMJ*, June 18, 1927, 995; *EMJ*, May 18, 1929, 796–97; *EMJ*, October 19, 1929, 640; *EMJ*, December 7, 1929, 899; Knerr, *Eagle-Picher Industries*, 90–91.

61. Record of insurance board hearing, January 1925, 1, 21–45, 56–57, 64 (first quote), 83–84 (second quote), f. 460, b. 15, RG70, NARA-FW.

62. Mine Safety Rules Analysis, [1925], f. 256–2, b. 14, section 1, AZR; *EMJ*, February 28, 1925, 381; *EMJ*, June 20, 1925, 1018; *EMJ*, June 25, 1927, 1057; "Here's More About District's Ore," *MNR*, January 2, 1927; "Magazine Section," *JG*, January 1929, 37; Ageton speech, January 30, 1929, f. "OK Mining Law," b. 51, PMC; Knerr, *Eagle-Picher Industries*, 92–93.

63. "Physical Exams Are Needed in District," *JG*, November 4, 1926; Lanza to Sayers, November 12, 1926; Sayers to Lanza, December 1, 1926; and Wade Wright to Sayers, February 24, 1927, all in f. 437 4143 Metro Ins., b. 1219, entry 6, RG70, NARA-CP; Ageton to E. C. Rogers, October 14, 1927, f. 021, b. 1 (first quote), and First Annual Report, 31–41, both in RG70, NARA-FW; "Sanitation and Health in Mines," *MNR*, May 13, 1927; *EMJ*, May 14, 1927, 816; "Several Cases of Trachoma in the Mines," *JG*, May 5, 1927; Sayers to Richard Jenkins, April 28, 1926, f. B7902, b. 40; Ageton to S. N. Davis, May 29, 1926, f. 886, b. 79 (second and third quotes); OPA Silicosis Committee Report, March 21, 1927, f. "Clinic, General," b. 88; OPA directors meeting minutes, June 5, 1925, vol. 2, b. 165; and OPA directors meeting minutes, September 21, 1927, vol. 3, b. 165, all in PMC; Derickson, "'On the Dump Heap,'" 664–67; Derickson, *Black Lung*, 74–102.

64. "Sanitation and Health in Mines," *MNR*, May 13, 1927 (first, second, third, and final quotes); "Bulletin Explains Ratings of Clinic," *JG*, July 24, 1927; First Annual Report, 39–59, RG70, NARA-FW (ratings quotes p. 51); Sayers et al., *Silicosis and Tuberculosis*, 2–4.

65. First Annual Report, 39–59, RG70, NARA-FW; Sayers et al., *Silicosis and Tuberculosis*, 6, 12–13, 15–16, 26 (quotes); Derickson, "'On the Dump Heap,'" 669–70.

66. Clinic report, October 1, 1927, f. 91, b. 5, and First Annual Report, 27–28 (first quote), 48–49, both in RG70, NARA-FW; "Expects to Stamp Out Silicosis Here," *JG*, July 28, 1927; "Producers' Association to Meet in Annual Session Tomorrow," *MNR*, October 21, 1928; "Let's Go!" and "Notice" (second quote), both in f. 1790, b. 83; and OPA directors meeting minutes, September 21, 1927, vol. 3, b. 165, all in PMC; *EMJ*, June 9, 1928, 941.

67. Welfare nurse report, March 1927, f. B7902, b. 40; memorandum to Honorable Joseph J. Marlowe, [1929], and Meriwether to M. D. Harbaugh, October 3, 1929, both in f. "Clinic, General," b. 87; Royd Sayers to J. D. Conover, January 9, 1929, and Third Annual Report, 1929–30, 3, both in f. 1682, b. 88 (quote); and Parran, "Syphilis is Bad Business," 1937, f. 570, b. 62, all in PMC.

68. Second Annual Report, 1928–29 (first, second, and third quotes); Sayers to Meriwether, October 10, 1929; and Meriwether to M. D. Harbaugh, October 3, 1929 (fourth and fifth quotes), all in f. 91, b. 5; and Meriwether to Sayers, December 15, 1928, f. 700, b. 17, all in RG70, NARA-FW; "Increase in the Base Rate," October 1929, 6, f. "Adjuster's Meetings 1930," b. 65, PMC.

CHAPTER 7

1. A. J. Martin, *Summarized Statistics*, 16, 20–21; Unemployment Association, f. 705, b. 51, PMC; "Magazine Section," *JG*, January 1929, 37; Harbaugh, *Story*, 25; Weidman, *Miami-Picher*, 127–28; Oklahoma State Department of Mines and Mining, *Twenty-Fifth Annual Report*, 57, 59.

2. Fronczak, "Fascist Game," 563–88.

3. "Picher's Chest Fund Depleted," *MNR*, February 9, 1930; Welfare Department Report, 1930, f. 705, b. 51, PMC; Knerr, *Eagle-Picher Industries*, 97–99 (quote p. 98); "Free Coal Given Out to Needy at Picher," *MNR*, December 12, 1930.

4. "Mine Labor Petitions Out," news clipping, and "Kansas Miners Air Troubles in Mass Meeting," *Baxter Springs Citizen*, April 7, 1930, both in scrapbook 3, b. 163, PMC; "Of Rackets (Mr. Z.)," in Tri-State Survey Committee, *Preliminary Report*, appendix B, 17; F. V. Meriwether to Royd Sayers, May 5, 1930, f. 93, b. 6, RG70, NARA-FW.

5. "First Arrest is Made in Thrust at 'Buddy Cars,'" *MNR*, April 3, 1930 (first two quotes, fourth quote); "Miners to Hold Mass Meeting," *MNR*, April 6, 1930 (third quote); "Unemployed Meet in Mass at Theatre," *King Jack*, April 3, 1930.

6. "First Arrest is Made in Thrust at 'Buddy Cars,'" *MNR*, April 3, 1930 (second quote); "Miners to Hold Mass Meeting," *MNR*, April 6, 1930 (first quote).

7. Meriwether to Sayers, May 5, 1930, f. 93 (first quote), and April 11, 1930, f. 102, both in b. 6, RG70, NARA-FW; Derickson, "'On the Dump Heap,'" 673 (second quote).

8. Clinic meeting minutes, March 8, 1932, f. "Minutes of Meetings of Board," b. 88, PMC (quote); Meriwether to Sayers, December 4, 1930, January 5, February 3, 1931, f. 93, b. 6, RG70, NARA-FW; Derickson, "'On the Dump Heap,'" 674–75.

9. Meriwether to Sayers, March 4, 1931, f. 93, April 27, 1931, f. 102 (first quote), and January 4, 1932, f. 93 (fourth quote), all in b. 6, RG70, NARA-FW; Meriwether to Sayers, April 28, May 1, 1931, f. 4374143, and Harbaugh to Sayers, June 16, 1931, f. 4374143 no. 4 (second and third quotes), all in b. 1219, entry 6, RG70, NARA-FW; "Clinic Cards are Explained," *MNR*, November 1, 1931.

10. Clinic meeting minutes, March 8, 1932 (third quote); clinic memo, May 7, 1932 (fourth quote); OPA Executive Secretary to OPA members, July 22, 1932 (second quote); and "Minutes of Meeting of Members," August 1, 1934, all in f. "Minutes of Meetings of Board," b. 88, PMC; M. D. Harbaugh, "Tri-State Labor Relations," draft, September 1935, f. 85, b. 100, PMC (first, fifth, and sixth quotes); "Mr. U," in Tri-State Survey Committee, *Preliminary Report*, appendix B, 27 (last quote); Derickson, "'On the Dump Heap,'" 673–74.

11. Mortality statistics, 1927–32, f. 43981384, and "Venereal Diseases among the Mining Population," f. 4391932x, both in b. 1586, RG70, NARA-CP; Richard Herman Legg, "A Personal Family History," f. 2, R754, Rolla (first quote); "Edward R. Legg," Galena Township, Jasper County, Missouri, 1930 federal manuscript census; "Mrs. I," in Tri-State Survey Committee, *Preliminary Report*, appendix B, 34 (second quote).

12. "Relief Work to Be Centralized" and "Eagle-Picher Co. Devises Novel Scheme to Keep Pot Boiling for Many Unemployed Miners," both in *MNR*, November 1, 1931; Ottawa County Red Cross activity report, 1933–34, f. "Red Cross Public Assistance," b. 38, PMC; Knerr, *Eagle-Picher Industries*, 99–101.

13. "Vote for Your Homes, Business and Community," *MNR*, October 30, 1932 (first quote); "Let's Get Back What We Had!," *MNR*, November 6, 1932; Oklahoma State Election Board, *Directory of the State of Oklahoma*, 166; MSS, *Official Manual, 1933–1934*, 288.

14. "1690 Tons of Zinc Concentrates Sold," *MNR*, February 12, 1933; "Like Old Times," *MNR*, July 23, 1933 (quote); "Oklahoma City, July 12," *MNR*, July 12, 1933; "General Wage Schedule," f. 21, b. 100, PMC; "Civil Work Jobs Open in County Tomorrow," *MNR*, November 21, 1933.

15. Editorial, *MNR*, July 23, 1933 (quote); Gordon, *New Deals*, 170–73.

16. "Like Old Times," *MNR*, July 23, 1933; M. D. Harbaugh to Hugh Johnson, July 26, 1933, f. "President's Re-Employment Agreement," and Harbaugh to the Mining Companies, June 16, 1933, f. 1752, both in b. 115; and "Basic Wage Scale for 8 hours," August 10, 1933, f. 91, b. 142, all in PMC; "Zinc Code in Effect," *MNR*, August 3, 1933; "Mine Operators Sign Own Code," *MNR*, August 6, 1933; "Personal Confidential," September 1, 1933, f. "Executive Committee Letters (1932–1934)," b. 110, section 3, AZR; Roosevelt, "President's Reemployment Agreement."

17. National Industrial Recovery Act, Pub. L. No. 73-67, 48 Stat. 195 (1933) (quote); "Miners Forming Organization in District," *Tri-State Tribune*, August 31, 1933; "Mine Union Meeting Held at Galena," *JG*, September 8, 1933; Anderson, "'Such Contented Workers,'" 4–7; Jensen, *Nonferrous Metals Industry*, 4–12.

18. "Thousands of Visitors Join Miamians in Blue Eagle Day Celebration" (first quote); "Mines Shovelers Vie, Bands Toot, Fat Men Race, Dancers Whirl—All in Celebration Under Blue Eagle" (last quote); and "Notes of Parade" (second quote), all in *MNR*, October 8, 1933.

19. Henry Bower, M. R. Corwin, Ralph Crabtree, Roy Hill, George House, Jack Jones, and W. P. McGinnis, f. "Questionnaires," b. 835, case file 616, RG25, NARA-CP (first quote); editorial, *MNR*, July 23, 1933 (second quote); NLRB, *Decisions*, 802 (third quote); "Mine Union Meeting Held at Galena," *JG*, September 8, 1933 (fourth quote); "Zinc Code Meets with Opposition from Labor," *Reno Gazette*, December 9, 1933 (fifth quote); "Hearing on Code of Fair Competition for the Zinc Industry," December 8, 1933, f. 1737, b. 115, PMC (sixth quote); Suggs, *Union Busting*, 33.

20. Tri-State Examining Bureau, 1935, f. 1691, b. 88 (first two quotes); memorandum, "Mine, Mill," April 12, 1935, f. 85, b. 100; "Minutes of Meeting," August 1, 1934, f. "Minutes, Tri-State Ind. Exam. Bureau," b. 108; and M. D. Harbaugh, "Labor Relations in the Tri-State District," September 1935, 8–14, f. 92, b. 142 (last quote p. 9), all in PMC; Cassidy, "Tri-State Zinc-Lead," 266–67; Resolution 26F, f. 30, b. 16, OFLMC (third quote); "Recollections of Glenn A. Hickman," December 1981, f. 6, b. 3997, SP (fourth quote); Wadleigh, "Strike," 5–6, Pittsburg State University, Axe Library Special Collections and Archives, Pittsburg, Kansas; Suggs, *Union Busting*, 33–36; Rosner and Markowitz, *Deadly Dust*, 146–47.

21. Resolutions 26A, 26B, 26C, 26D, 26E, 26F, and 26G, f. 30, b. 16, OFLMC.

22. Zinc code hearing, December 8, 1933, 21–23 (second quote p. 22); J. D. Conover call memo, September 4, 17, 1934; and J. D. Conover to M. D. Harbaugh, January 31, 1934, all in f. 1757, b. 115; Harbaugh to Conover, February 3, 1934, f. 21, b. 100; and M. D. Harbaugh, "Labor Relations in the Tri-State District," September 1935, 15, f. 92, b. 142, all in PMC; NLRB, *Decisions*, 741 (third quote), 802 (first quote).

23. J. F. Callbreath to OPA, May 1, 1934, f. "American Mining Congress, 1934," b. 80 (second quote), and memo in support of exceptions, April 7, 1934, f. 1757, b. 115 (first and third quotes), both in PMC; NRA, *Code of Fair*, 41–45.

24. "Official Report of Proceedings," C-73, Joplin, December 1937, 375, b. 832, case file 616, RG25, NARA-CP; Dearing, *ABC of the NRA*, 69–76; Ross, *Death of a Yale Man*, 189–90; "FERA Work Relief at Standstill in County Owing to Lack of Funds," *MNR*, October 14, 1934; "Broth is Given Out at Picher," *MNR*, January 2, 1935; "FERA Call is Delayed Here," *MNR*, January 13, 1935; "Summary by Files Shows County Received $263,272 During Nine Months in Three Relief Phases," *MNR*, January 20, 1935.

25. Exhibit B14, b. 833, case file 616, RG25, NARA-CP; *Report of the Proceedings of the Thirty-First Convention of the International*, 6–7, 14 (quote), f. 3, b. 1, WFM/IUMM&SW; MM, *Thirty-Second Convention*, 33–35; Anderson, "'Such Contented Workers,'" 6–7; Bernstein, *Turbulent Years*, 106–9; Kelley, *Hammer and Hoe*, 144–46.

26. "Ted Schastein," Lyon Township, Cherokee County, Kansas 1938 County Census; Glenn Hickman to George Suggs, April 28, 1982, typescript 5, f. 6, b. 3997, SP (quote); "Official Report of Proceedings," C-73, Joplin, December 1937, 49–58, b. 832, case file 616, RG25, NARA-CP.

27. Memorandum, "Mine, Mill," April 12, 1935, f. 85, b. 100; NRA labor compliance officer letter, March 16, 1935, and "Miners Make Wage Protest," *Daily Oklahoman*, April 28, 1935, both in f. 1740, b. 115; and OPA secretary annual report, November 11, 1935, f. "Thirteenth

Annual Meeting," b. 123, all in PMC; "Tri-State Mines Start Closing in a General Strike," *JG,* May 9, 1935; Suggs, *Union Busting,* 37–42.

28. "Tri-State Mines Start Closing in a General Strike," *JG,* May 9, 1935 (all other quotes); Robinson to Lewis, April 10, 1935, reel 8, pt. 1, *CIO Files of John L. Lewis* (third quote); "Ore Mills Idle as Miners Quit Jobs in Strike," *MNR,* May 9, 1935; "Official Report of Proceedings," C-73, Joplin, December 1937, 49–70, b. 832, case file 616, RG25, NARA-CP; NLRB, *Decisions,* 741; Suggs, *Union Busting,* 42–44.

29. "Tri-State Mines Start Closing in a General Strike," *JG,* May 9, 1935; "Walkout Closes All but 3 Small Tri-State Mines," *JG,* May 10, 1935; "No Violence Reported in Strike Area," *Tri-State Tribune,* May 16, 1935; "Ore Mills Idle as Miners Quit Jobs in Strike," *MNR,* May 9, 1935; Press Committee, District Four, to James Robinson, May 15, 1935, f. 7-0-6-13, b. 29, Classified General Files, 1934–49, RG100, NARA-CP (quote); Suggs, *Union Busting,* 47–49.

30. Suggs, *Union Busting,* 50–54 (Campbell quoted p. 53); "No Funds Available for Strikers Here," *JG,* May 10, 1935; "No Settlement in Miners Strike Indicated," *Tri-State Tribune,* May 23, 1935; Knerr, *Eagle-Picher Industries,* 105.

31. "Miners Discuss Return to Jobs," *JG,* May 18, 1935 (first quote); "1,150 Miners Ask for Their Jobs Back," *JG,* May 19, 1935 (second quote); "Thomas L. Armer," Augusta City, Butler County, Kansas, 1930 federal manuscript census.

32. "Efforts to Hold Meetings Fail," *JG,* May 21, 1935 (quote); Suggs, *Union Busting,* 54–57.

33. "Efforts to Hold Meetings Fail," *JG,* May 21, 1935; "3,000 at Union Miners' Meeting" (second and third quotes) and "Large Crowd Attends a Meeting at Picher," both in *JG,* May 22, 1935; Wadleigh, "Strike," 6 (first quote), Pittsburg State University, Axe Library Special Collections and Archives, Pittsburg, Kansas; McGinnis, *Oklahoma's Depression Radicals,* 20–22, 31–33, 49–60; Suggs, *Union Busting,* 57–58.

34. "Recollections of Glenn A. Hickman," 1–3, f. 6, b. 3997, SP (second quote p. 1); "Zinc Fields Figures Recalled," *Tulsa Tribune,* October 19, 1976, f. 1, Cuddeback Collection, University of Oklahoma, Western History Collections, Norman, Oklahoma; "Fred W. Evans," Quapaw Township, Ottawa County, Oklahoma, 1930 federal manuscript census; "The Greatest Risk Taker," news clipping, May 17, 1990, f. 3, R754, Rolla (first quote); NLRB, *Decisions,* 744–46 (third quote p. 745); Suggs, *Union Busting,* 59–69.

35. "3,150 in New Union," *JG,* May 30, 1935; "Mr. M.," appendix B, 20, in Tri-State Survey Committee, *Preliminary Report* (quote); Suggs, *Union Busting,* 69–72.

36. NLRB, *Decisions,* 744 (first quote), 760–64; "Militia Reaches Field after Day of Rioting," *JG,* May 28, 1935 (second, third, and fourth quotes); "Quiet Prevails in Mine Strike," *MNR,* May 29, 1935; Suggs, *Union Busting,* 75–82.

37. NLRB, *Decisions,* 746–53, 761 (first quote p. 747); "Willingness to Reopen Mines in Tri-State Shown," *JG,* May 30, 1935 (second quote); "Close to 1,000 Miners Back to Work This Week," *JG,* June 5, 1935; Suggs, *Union Busting,* 79–85.

38. "Mines Gradually Returning to Work," *JG,* June 16, 1935; Merle Chambers (first quote) and Tom Hood (second quote), questionnaires, b. 835, case file 616, RG25, NARA-CP; Suggs, *Union Busting,* 117–19.

39. "Kansas Militia Is En Route to Baxter Springs," *JG,* June 8, 1935; "3 Under Arrest after Disorder in Kansas Field," *JG,* June 9, 1935; Dick Helman and C. E. Shouse to Landon, June 10, 1935, f. 9; and Harry Burr telegram to George Blakeley, June 8, 1935, and C. E. Shouse telegram to adjutant general of Kansas, June 7, 1935, f. 12, all in b. 1, Baxter; "Martial Law to be Invoked at Galena," *JG,* June 29, 1935; "Col. Head, Back in Picher, Orders Field to Disarm," *JG,* June 30, 1935; "Working Miners Seek Protection," *JG,* July 3, 1935; "Operation of District Mines Now Normal," *Tri-State Tribune,* August 1, 1935; Suggs, *Union Busting,* 85–100.

40. Memorandum, operators meeting, Baxter Springs, Kansas, June 9, 1935 (first quote), and memorandum, operators meeting, Picher, Oklahoma, June 10, 1935 (second quote), both in f. 85, b. 100, PMC; Ross, *Death of a Yale Man*, 196.

41. NLRB, *Decisions*, 756–62 (first quote p. 759); "Lest We Forget," *Metal Mine and Smelter Worker*, August 10, 1935, f. 85, b. 100, PMC (second quote); Suggs, *Union Busting*, 119–31.

42. James Robinson to Lewis, April 10, 1935, reel 8, pt. 1, *CIO Files of John L. Lewis*; MM, *Thirty-Second Convention*, 22–25, 32–35, 46–47, 140–41 (Brown quote p. 35); Jensen, *Nonferrous Metals Industry Unionism*, 26–35; Suggs, *Union Busting*, 52.

43. AFL, *Fifty-Fifth Annual Convention*, 294–95, 426–28 (quote p. 427); Tomlins, *State and the Unions*, 141–45.

44. Locals 17, 106, 107, 108, 110, and 111, vol. 55, WFM/IUMM&SW; McTeer to O. L. Crain, December 4, 1935, f. 32, b. 18 (first quote); Local 15 monthly reports, 1935–36, f. 8, b. 20; and president's annual report, September 1935, 6, b. 65 (second quote), all in OFLMC; "Hard Rock Miners Ask Aid of Labor in Controversy," *Oklahoma Federationist*, November 1935 (third quote); "To All Central Labor Bodies and Crafts," [1935], f. 3, b. 3, Baxter; *American Federationist*, November 1917, 1001.

45. Tony McTeer to O. L. Crain, December 4, 1935, f. 32, b. 18, and Local 15 monthly reports, 1935–36, f. 8, b. 20, both in OFLMC; Pratt to NLRB, June 24, 1935; Nels Anderson to NLRB, June 27, 1935; Roy Wisdom and Richard Murray, Local 17, to Roosevelt, October 9, 1935; Russell Hasp, Local 111, to Roosevelt, November 30, 1935; and Millner to Roosevelt, December 24, 1935 (quotes), all in case no. XVII-C-19, b. 115, entry 155, RG25, NARA-CP; "Hard Rock Miners Ask Aid of Labor in Controversy," *Oklahoma Federationist*, November 1935; "To All Central Labor Bodies and Crafts," [1935], f. 3, b. 3, Baxter; Suggs, *Union Busting*, 177–78.

46. Millner to Roosevelt, December 24, 1935, case no. XVII-C-19, b. 115, entry 155, RG25, NARA-CP (first quote); "Hard Rock Miners Ask Aid of Labor in Controversy," *Oklahoma Federationist*, November 1935 (second quote); "To All Central Labor Bodies and Crafts," [1935], f. 3, b. 3, Baxter.

47. "Hard Rock Miners Ask Aid of Labor in Controversy," *Oklahoma Federationist*, November 1935 (quote); "To All Central Labor Bodies and Crafts," [1935], f. 3, b. 3, Baxter.

48. Brophy to Benedict Wolf, March 26, 1936; Wolf to Brophy, March 27, 1936; and Pratt to NLRB, March 2, 1936 (quote), all in case no. XVII-C-19, b. 115, entry 155, RG25, NARA-CP; Lewis to Logan Beckman, March 12, 1936, and John Sherwood to Lewis, February 5, 1936, both in reel 8, pt. 1, *CIO Files of John L. Lewis*.

49. "Order Prevents Hearing against Eagle-Picher Co.," *MNR*, May 25, 1936; Ross, *Death of a Yale Man*, 207–9.

50. Ross, *Death of a Yale Man*, 194–201 (first quote p. 198, second quote p. 199, and fifth quote p. 201); "Hearing Prevented by Court Restraining Order," *Metal Mine and Smelter Worker*, May 30, 1936 (third and fourth quotes); NLRB, *Decisions*, 758 (sixth and seventh quotes); Suggs, *Union Busting*, 127–40.

51. NLRB, *Decisions*, 760–62 (first quote p. 758, second quote p. 762, fourth quote p. 760n57); "Back To Work," [1935], f. 10, b. 2, Baxter (third quote); Ross, *Death of a Yale Man*, 197–200; Suggs, *Union Busting*, 119–31.

52. "Record Smashing Crowd Celebrates" (first quote) and "Why Shouldn't We Celebrate?" (second quote), *Metal Mine and Smelter Worker*, May 30, 1936; Ross, *Death of a Yale Man*, 200 (third quote); M. D. Harbaugh, "Labor Relations in the Tri-State Mining District," June 1936, 21 (fourth and eighth quotes), and M. D. Harbaugh, "Tri-State Labor Relations," draft article, 1–3 (fifth and sixth quotes p. 1, seventh quote p. 3), both in f. 85, b. 100, PMC; Wad-

leigh, "Strike," 4–6, (ninth and tenth quotes p. 5, eleventh quote p. 4), Pittsburg State University, Axe Library Special Collections and Archives, Pittsburg, Kansas.

53. "'Wild Red' Berry Wins Picher Duel," *MNR*, November 1, 1936 (first quote); "Blue Card Union Dance Tonight," *MNR*, April 22, 1936; "Blue Card Wrestling!," *MNR*, April 22, 1936; "Record Smashing Crowd Celebrates," *Metal Mine and Smelter Worker*, May 30, 1936 (remaining quotes); Suggs, *Union Busting*, 137–40.

54. "Industrial Union Issue Becomes a Powder Keg within Oklahoma Labor Ranks," *Daily Oklahoman*, September 13, 1936; "State Labor Gives Support to Roosevelt," *Daily Oklahoman*, September 17, 1936; Resolutions 1–5, f. 12, b. 18a; AFL to the State Federations of Labor, September 5, 1936, f. 16, b. 19; and OFL Executive Board meeting minutes, September 15, 1936, f. 16, b. 19, all in OFLMC; MM, *Thirty-Third Annual Convention*, 22–23; Local 111 to All Organized Labor, [August 1936], f. 201, b. 16, Classified General Files, 1934–49, RG100, NARA-CP (quotes).

55. OPA secretary annual report, November 11, 1935, f. "Thirteenth Annual Meeting," b. 123, PMC (first quote); NLRB, *Decisions*, 756–57, 792–94; "Mr. M.," in Tri-State Survey Committee, *Preliminary Report*, appendix B, 20–21 (second quote); Nieberding, *History of Ottawa County*, 79 (third quote); "CIOers Would Show Up Officials of Blue Card Union with Letter," *Blue Card Record*, April 9, 1937 (fourth quote); Suggs, *Union Busting*, 119–26.

56. A. J. Martin, *Summarized Statistics*, 18; Tri-State Survey Committee, *Preliminary Report*, 37; "Wages Hiked by Mining Firms," December 21, 1936; "Wages Raised 25 Cents a Day," January 17, 1937; "Miners' Wages Up Another Peg," February 21, 1937; "Tri-State Miners' Wage Scale Now at the Highest Level in Its History," March 7, 1937; and "Tri-State Miners' Pay to New Highs," March 14, 1937, all in news clipping file, f. 19, b. 100, PMC; "E. C. Mantle," Commerce, Ottawa County, Oklahoma, 1940 federal manuscript census; Leavy, *Last Boy*, 39–45.

57. Robinson report to CIO Executive Session, October 14, 1937, Atlantic City, reel 2, pt. 2, *CIO Files of John L. Lewis*; Schasteen to Roosevelt, December 10, 1936, f. 201, b. 16, Classified General Files, 1934–49, RG100, NARA-CP (second quote); Schasteen to John L. Lewis, January 14, 1937, reel 8, pt. 1, *CIO Files of John L. Lewis*; MM, *Official Proceedings of the Thirty-Fourth Convention*, 36–37 (first quote, p. 36).

58. *Blue Card Record*, May 7, 1937, quoted in Suggs, *Union Busting*, 203 (first quote); *Blue Card Record*, March 5, 1937, quoted in "Official Report of Proceedings," C–73, December 1937, 1122–24, b. 832, case file 616, RG25, NARA-CP (second and fourth quotes p. 1122, third and fifth quotes p. 1124); MM, *Official Proceedings of the Thirty-Fourth Convention*, Denver, August 1937, 187:36–37, WFM/IUMM&SW (sixth quote p. 36).

59. MM, *Official Proceedings of the Thirty-Fourth Convention*, Denver, August 1937, 187: 36–37, WFM/IUMM&SW; Robinson to Lewis, March 30, 1937, and Lewis to Robinson, March 31, 1937 (second quote), both in reel 8, pt. 1, *CIO Files of John L. Lewis*; Suggs, *Union Busting*, 160–61.

60. "Rival Charges of Blame Made in Riot Probe," *JG*, April 13, 1937 (second quote); "Peace Returns to Mine Field after Gunplay," *MNR*, April 12, 1937 (first quote); NLRB, *Decisions*, 763–65; Suggs, *Union Busting*, 161–64.

61. "Rival Charges of Blame Made in Riot Probe," *JG*, April 13, 1937 (third quote); "Peace Returns to Mine Field after Gunplay," *MNR*, April 12, 1937; Suggs, *Union Busting*, 165–68 (first quote p. 165, second quote p. 167).

62. "Rival Charges of Blame Made in Riot Probe," *JG*, April 13, 1937; Suggs, *Union Busting*, 168–71.

63. "Peace Returns to Mine Field after Gunplay," *MNR*, April 12, 1937 (second quote); "4 Decades Since Mine Area Labor War," *MNR*, April 12, 1975 (first quote); Suggs, *Union Busting*, 168–71.

64. "A. F. of L. and Blue Card Leaders Silent," *MNR*, April 15, 1937 (third quote); "A. F. of L., Granting Charter to Blue Carders, Will Push Rigid Ban on C. I. O. in District," *MNR*, April 16, 1937; "Blue Card Union Joins A. F. of L. with Setup Unchanged," *JG*, April 16, 1937 (first and second quotes); Green to Warren, April 13, 1937, reel 23, RG1-017, Meany.

65. "Blue Card Mass Meeting Called," *Blue Card Record*, April 16, 1937 (first quote); "Blue Carders to Celebrate Sunday at Fairgrounds," *MNR*, April 16, 1937; "5,000 Upraised Arms Gesture Indorsement by Blue Carders of Affiliation with the A. F. of L. and Rigid Defiance of C. I. O.," *MNR*, April 19, 1937 (remaining quotes).

66. "5,000 Upraised Arms Gesture Indorsement by Blue Carders of Affiliation with the A. F. of L. and Rigid Defiance of C. I. O.," *MNR*, April 19, 1937 (all other quotes); "Green Welcomes Blue Card Union," *JG*, April 24, 1937; NLRB, *Decisions*, 769-7 (third quote p. 768-69).

67. Executive council meeting minutes, May 23, 26-30, 1937, 18-19, b. 26, RG4-008, Meany (first quote p. 18, second quote p. 19); Hickman to L. T. Johnson, June 38, 1937, f. 14, b. 20, OFLMC.

68. "Green Closes A. F. of L. Door to Insurgents," *MNR*, April 23, 1937; Phelan, *William Green*, 144-51.

69. "20,000 Celebrate at Miners' Picnic; State Chief Speaks," *MNR*, May 27, 1937; "Green Closes A. F. of L. Door to Insurgents," *MNR*, April 23, 1937; Green to J. A. Canfield, May 7 (first quote), 10 (second quote), 1937, reel 23, RG1-017, Meany; "The President's Column," *Oklahoma Federationist*, April 1937; Resolution 11, f. 6, b. 22 (third quote), and Resolution of IAM Great Falls Lodge, May 17, 1937, f. 19, b. 19, both in OFLMC; AFL, *Fifty-Seventh Annual Convention*, 112-13, 374 (fourth quote p. 113).

70. NLRB, *Decisions*, 770-74 (first quote p. 771); "20,000 Celebrate at Miners' Picnic; State Chief Speaks," *MNR*, May 27, 1937 (Warren quotes); Green to Blue Card Union, May 26, 1937, f. 18, b. 4, RG28-003, Meany (Green quotes).

71. Galena Defense Committee to All Organized Labor, [April 1937], and Cassell to Lewis, April 21, 1937 (second quote), both in reel 8, pt. 1, *CIO Files of John L. Lewis*; "To Enforce A. F. of L. Rule against the C. I. O.," *JG*, April 17, 1937 (first quote).

72. Reid Robinson report to CIO executive session, October 14, 1937, reel 2; UMW Executive Board to Lewis, June 2, 1938, reel 9; and Fowler to Lewis, April 20, 1937, reel 12, all in *CIO Files of John L. Lewis*, pt. 2; Fowler to Lewis, April 21, 1937; Fowler to Green, April 21, 1937 (first quote); and "Mass Meeting of C. I. O. Unions is Planned Here," *Muskogee Times*, April 20, 1937 (second and third quotes), all in f. 30, b. 153, UMW-P; MM, *Official Proceedings of the Thirty-Fourth Convention*, Denver, August 1937, 187:38, WFM/IUMM&SW (fourth quote).

73. Reid Robinson report to CIO executive session, October 14, 1937, reel 2, pt. 2, *CIO Files of John L. Lewis*; "Cards Stacked against Blue Card in Labor Disputes Board Hearing," *Blue Card Record*, [December 1937] (first quote), and "Rules A. F. of L. not Involved in Labor Suit," *JNH*, December 8, 1937, both in f. 85, b. 100; and "NLRB Aide Admits Taking Union Side," *New York Times*, [January 1940], f. 11, b. 1 (second quote), all in PMC; Ogburn to Frank Morrison, October 11, 1937, f. 18, b. 4, RG28-003, Meany; William J. Avrutis oral history, November 29, 1969, 6, f. 10, b. 6, Kheel Collection, Cornell University, Kheel Center for Labor-Management Documentation and Archives, Ithaca, New York (final quotes); Suggs, *Union Busting*, 194-96.

74. "Evans Resigns," *JG*, November 12, 1937; AFL Executive Council minutes, April 25-May 3, 1938, 79, b. 26, RG4-008, Meany; "Blue Card Union's Attorney Labels NLRB Examiner's Report a Defeat," *MNR*, September 16, 1938 (quote); "An Organizing Program for the Non-ferrous Metal Industry," [1939], f. 25, b. 17, Brophy Papers, CUA; Phelan, *William Green*, 148-52; Bernstein, *Turbulent Years*, 635-714.

75. Charge C-73, November 3, 1937, b. 834, case file 616, RG25 NARA-CP; NLRB, *Decisions*, 867–82; Suggs, *Union Busting*, 193–98.

76. AFL Executive Council minutes, April 25–May 5, 1938, 79, b. 26, RG4-008, Meany; Zieger, *CIO*, 95–96; "Looking Ahead" (quote) and "AFL Company Union Hit as Tool of Employers," both in *CIO News*, September 17, 1938; "Heat Wave Hits AFL Leaders," *CIO News*, August 13, 1938; "Looking Ahead," *CIO News*, April 24, 1939; "Southwest CIO Meets Called One of the Best," *CIO News*, December 26, 1938; Arkansas-Oklahoma Industrial Union Council first convention minutes, December 1938, f. 33, b. 186, UMW-P; "NLRB Orders Ford to Rehire Workers," *Bradford Evening Star and Daily Record*, January 18, 1941.

77. Cassell to Verne Zimmer, November 23, 1938 (second quote); Zimmer to Cassell, December 5, 1938; Cassell to Eleanor Roosevelt, September 4, 1938 (first and third quotes); Cassell to Perkins, August 21, 1938; Perkins to Zimmer, October 25, 1939; and "The Tri-State Mining Area," [1939], all in f. 7-0-6-13; and Sheldon Dick to Zimmer, October 10, 1939, and J. Raymond Walsh to Perkins, May 8, 1939, both in f. 7-0-6-13, 1941, all in b. 29, Classified Central Files, 1937–41, RG100, NARA-CP; Tri-State Survey Committee, *Preliminary Report*, 1 (fourth quote), 82; Davidson, *South of Joplin*; "Zinc Stink," *Time*, December 4, 1939, 63; "American Plague Spot," *New Republic*, January 1, 1940.

78. "New Charter in Tri-State as Eagle-Picher Appeals," *CIO News: Mine, Mill and Smelter Workers Edition*, November 27, 1939; Suggs, *Union Busting*, 193–95; Cassidy, "Tri-State Zinc-Lead," 271; Ferns to John L. Lewis, December 15, 17, 1939, f. 5, and Oklahoma Industrial Union Council annual convention report, November 1939, f. 20, both in b. 136, UMW-P.

79. "Silicosis and Tuberculosis," *CIO News: Mine, Mill and Smelter Workers Edition*, December 11, 1939 (first quote); Sheldon Dick and Dick, *Men and Dust*, 6:56–7:19 (McTeer quotes); Tri-State Survey Committee, *Preliminary Report*, appendix A, 14.

80. Perkins to Payne Ratner, March 28, 1940, f. 7-0-4(3) "Tri-State Area Mining Conference," b. 26, Classified Central Files, 1937–41, RG100, NARA-CP; Sheldon Dick and Dick, *Men and Dust*; William Alexander, *Films on the Left*, 287–93; Eleanor Roosevelt, My Day, syndicated news column, March 11, 1940, https://www2.gwu.edu/~erpapers/myday/display doc.cfm?_y=1940&_f=md055524 (quote). Geer played Zebulon Walton in the 1970s series *The Waltons*.

81. "Uniform Working Regulations for Tri-State Urged" and "Miss Perkins Speaks at Picher Meeting" (first quote), *JG*, April 24, 1940; "20 Reinstated," *CIO News: Mine, Mill and Smelter Workers Edition*, June 17, 1940 (other quotes); Jim Ferns to John L. Lewis, August 21, 1940, reel 11, pt. 2, *CIO Files of John L. Lewis*; "Convention Next Year in Joplin," *CIO News: Mine, Mill and Smelter Workers Edition*, August 19, 1940; Suggs, *Union Busting*, 219.

82. "Conference on Health and Working Conditions in the Tri-State District, Joplin, Missouri, April 23, 1940," 25–26, b. 52, FPP (McTeer quotes); Tri-State Survey Committee, *Preliminary Report*, appendix B, 20 (second quote).

83. "Conference on Health and Working Conditions in the Tri-State District, Joplin, Missouri, April 23, 1940," 9–10 (Titus quote), 36 (Just quote), b. 52, FPP; Nieberding, *History of Ottawa County*, 81 (1980s quotes); "4 Decades Since Mine Area Labor War," *MNR*, April 12, 1975 (last quote).

84. "Eagle Picher Buys Commerce Properties," *Wall Street Journal*, December 31, 1938; "Tri-State Industrial Examining Bureau, 1932–1938," f. 1732, b. 88, PMC; Alice Hamilton, "Mid-American Tragedy," 436 (quote); "Joe Nolan," *MNR*, January 14, 1940; "8,750-Ton Zinc Concentrate Turnover is District's Largest in Seven Weeks," *MNR*, April 20, 1941; "Elven C. Mantle of Commerce Dies," *JG*, May 7, 1952; Mantle, *Mick*, 7; Suggs, *Union Busting*, 216.

85. Tri-State Survey Committee, *Preliminary Report*, 36–40 (quote p. 39); "Tri-State Zinc & Lead Production," f. 34, b. 100, PMC; "Historical Slide Sequence," f. 1, Cuddeback Collec-

tion, University of Oklahoma, Western History Collections, Norman, Oklahoma; "Ted Scha-steen," Bakersfield, Kern County, California, 1940 federal manuscript census; "Lead and Zinc Mining as a Factor in the Labor Market," 1950, Kansas Collection, Pittsburg State University, Axe Library Special Collections and Archives, Pittsburg, Kansas.

86. Just statement, June 1939, f. "Silicosis Scandal," b. 69 (second quote), and Evan Just, "Living and Working Conditions in the Tri-State Mining District," 2, f. 963, b. 76 (first quote), both in PMC; Lawrence Barr and Theo Sisco, interview by Joe Todd, 1985, H1985.13, Oral History Collection, OHC (third quote); Orville Ray, 1999 interview, and John Mott, 1999 interview, both quoted in Robertson, *Hard as the Rock Itself*, 144–45 (fourth quote p. 144, fifth quote p. 145).

87. Report of welfare nurse, October 31, 1942, f. "Annual Meeting, 1941–42," b. 126, PMC; Tri-State Survey Committee, *Preliminary Report*, appendix B, 23–25, 33–34 (D. quotes p. 23, R. quotes p. 24, P. quotes p. 25, I. quotes p. 33–34); "Conference on Health and Working Conditions in the Tri-State District, Joplin, Missouri, April 23, 1940," 2, b. 52, FPP (Perkins quote).

88. Hannon to Perkins, May 6, 1940 (Hannon quotes), and Carey to Perkins, June 2, 1939 (Carey quotes), both in f. 7-0-6-13, b. 29, Classified Central Files, 1937–41, RG100, NARA-CP; "Evelyn Hannon," Treece, Cherokee County, Kansas, 1940 federal manuscript census.

89. U.S. Department of Labor press release, October 27, 1940, and memorandum on housing (first quote), both in f. 70613 "Continuing Committee on Housing," b. 30, Classified General Files, 1934–49, RG100, NARA-CP; "Union Asks Quick Action on Tri-State Problems," *CIO News: Mine, Mill and Smelter Workers Edition*, February 17, 1941 (second quote); William Cassidy to Secretary of Labor, August 2, 1947, f. 7-0-6 "Safety and Health Surveys," b. 27, Classified Central Files, 1945–1949, RG100, NARA-CP.

EPILOGUE

1. Norris, *AZn*, 166–69; A. J. Martin, *Summarized Statistics*, 20–21.

2. A. J. Martin, *Summarized Statistics*, 16; Lawrence Barr and Theo Sisco, interview by Joe Todd, 1985, H1985.13, Oral History Collection, OHC (quote); "Average Recoveries of Zinc and Lead," f. "Tri-State Zinc," b. 89, and "Lead and Zinc Concentrates," f. 1231, b. 105, both in PMC.

3. Frances Perkins to Verne Zimmer, February 3, 1942, f. "Committees on the Tri-State Area," b. 127, General Subject File, RG174, NARA-CP; Suggs, *Union Busting*, 218.

4. "St. Louis Smelting NLRB Election Victory," *CIO News: Mine, Mill and Smelter Workers Edition*, May 19, 1941; Alex Cashin to Reid Robinson, February 26, 1942, f. 21, b. 34; Gobel Cravens to Ben Riskin, f. 29, b. 36; St. Louis Smelting wage schedule, f. "Local 514," b. 131; and Hain to Riskin, July 27, 1942 (quote), f. "E. Hain," b. 166, all in WFM/IUMM&SW; "Hearings Held at Joplin on Union's Pay Hike Request," *MNR*, January 31, 1943; "Seven New Charters Is Two-Week Record," *Union*, April 20, 1942; "$134,000 in Back Pay," *Union*, March 15, 1943; "Nine Locals Win Raises," *Union*, April 26, 1943; U.S. Bureau of Labor Statistics, *Wages*, 2–13; Suggs, *Union Busting*, 221.

5. All in *Union*: "$250,000 in Back Pay" and "Organize Arkansas Smelter," October 17, 1945; "District 4 Roundup" and "District 4," February 6, 1946; "Henryetta Local to Strike," April 3, 1946; "Organize Eagle-Picher," April 17, 1946; "Ask Election at Eagle Picher," May 1, 1946.

6. "Fight for OPA in Oklahoma," *Union*, May 13, 1946; "Tri-State Unions in Save OPA Rally," *Union*, May 27, 1946; "Tri-State Council Says," *Union*, July 22, 1946; "Surface Victory at Eagle Picher," *Union*, September 2, 1946; "EP Miners Give MMSW 2-to-1 Vote," *Union*, November 25, 1946; "Urges Premiums Re-instated," *JG*, July 12, 1946; "Miners to Discuss

Future Wage Plans," *JG*, July 25, 1946; "Tri-State Area's Mines to Reopen," *JG*, July 26, 1946; Local 861, December 1946, f. "IUMM&SW, Local 861," b. 26, PE1, Tamiment; Elwood Hain to Leonard Douglas, October 20, 1948, f. 72, b. 44, WFM/IUMM&SW (quote); Cassidy, "Tri-State Zinc-Lead," 275–80; Jacobs, "How About Some Meat?," 910–41.

7. "Truman Vetoes Zinc, Lead Subsidy," *JG*, August 9, 1947; "Dear Union Family," September 13, 1947, f. 45, b. 79, and Howard Lee to Maurice Travis, July 13, 1947, f. 4, b. 80 (quote), both in WFM/IUMM&SW; "Eagle-Picher Mine Strike is Called," *JG*, December 31, 1947; "Eagle-Picher Pay Offer Rejected," *JG*, June 29, 1948; "Lead and Zinc Concentrates," f. 1231, b. 105, PMC.

8. "2,000 Tri-State Miners are Made Idle," *JG*, July 2, 1948; "Who are Robinson and Travis?," *JG*, August 8, 1948; "Union Proposal Rejected," *JG*, August 14, 1948; "Why 2,000 Miners and Millmen are Unemployed," f. "Local 861," b. 44, WFM/IUMM&SW; Jensen, *Nonferrous Metals Industry Unionism*, 215–40; Storch, *Working Hard*, 114–20.

9. "Mine Strikers Meet in Picher," *JG*, August 31, 1948; "Back-to-Work Meeting," *JG*, September 1, 1948; "Striking Miners Turn Down Offer," *JG*, September 2, 1948; "Eagle-Picher Walkout Ends," *JG*, September 4, 1948; "New Miners' Group Changes Its Name," *JG*, September 10, 1948 (second quote); "New Mine Union to Ask Election," *JG*, November 18, 1948; "Independent Union Formed," *JNH*, September 2, 1948; "Predicts Mine Union Will Vote to End Strike," *JNH*, September 3, 1948 (first quote); "Walter Cherry," Quapaw Township, Ottawa County, Oklahoma, 1920 federal manuscript census; "Walter L. Cherry," Galena, Cherokee County, Kansas 1943 County Census.

10. "Insulation Workers Quit the C. I. O," *JG*, November 20, 1948; "AFL Expects Election within Three Weeks," *MNR*, April 28, 1949; Hain to Leonard Douglas, October 20, 1948 (quote), f. 72, b. 44, and Hain report, April 30, 1949, f. "Elwood Hain," b. 95, both in WFM/IUMM&SW.

11. "Mining Cutback," *Wall Street Journal*, April 27, 1949; "Gypsum Union Branching Out," *MNR*, June 26, 1949; "Tri-State Locals Fight Bosses' Wage Cut," *Union*, May 9, 1949.

12. "To All Members of Eagle-Picher Local 861," April 23, 1949 (first quote), and "Text for Rebroadcast," [May 1949] (remaining quotes), both in f. "Oklahoma Secession," b. 293, WFM/IUMM&SW; *Tri-State Miner*, May 31, 1949, f. 16; "Hey Joe, Remember?," [May 1949], f. 8; and "Let's Set the Record Straight," [May 1949], f. 8, all in b. 1, TAM 63, Tamiment.

13. "A. F. L. Union Wins," *JG*, June 3, 1949; "Eagle-Picher Shutdown," *MNR*, June 29, 1949; "Back-to-Work Movement," *MNR*, July 26, 1949; "Sign Pact Cutting Wages," *New York Times*, August 7, 1949, 49; "Lead and Zinc Demand Good," *MNR*, August 14, 1949; U.S. Bureau of the Census, "Oklahoma," 69.

14. Gibson, *Wilderness Bonanza*, 267–72 (first quote, from *Daily Oklahoman*, p. 268); U.S. Tariff Commission, *Lead and Zinc*, table 4; "Saga of a Dying Town," *Washington Post*, April 15, 1951, R9 (last quote); "Lead and Zinc Concentrates," f. 1231, b. 105, PMC.

15. Martin Burns to David McDonald, June 3, 1953, and Burns to I. W. Abel, September 10, 1953, both in f. 23, b. 74; James Dickerson to McDonald, June 7, 1950, f. 5, b. 100; and Burns to McDonald strike telegrams, 1955, 1956, f. 33, b. 102, all in USW, Pennsylvania State University, Eberly Family Special Collections Library, State College, Pennsylvania; "Miners at Picher Leave Their Jobs," *JG*, June 21, 1953; "Thousands of Workers of Eagle-Picher Go On Wage Strike," *Ada Weekly News*, June 25, 1953; "Mine Union Claims Contract Violation," *JG*, January 17, 1954; "Union Accepts 10-Cent Boost," *MNR*, July 11, 1955; "Eagle-Picher Strike Put Off," *JG*, December 23, 1955; "Miners Accept E-P Pay Offer," *MNR*, September 2, 1956; Gibson, *Wilderness Bonanza*, 246–48.

16. "Eagle-Picher to Lay Off 1,100," *New York Times*, April 24, 1957; "Eagle-Picher Suspends Mine, Mill Operations," *Wall Street Journal*, July 31, 1957; "Miners to Study E-P Pro-

posal," *MNR*, November 15, 1957; "Union Planning Party at Picher," *MNR*, December 21, 1958 (quote).

17. Storch, *Working Hard*, 117–26; Zieger, *CIO*, 279–351 (quote p. 290); Phelan, *William Green*, 168–72; Rosswurm, *CIO's Left-Led Unions*, 1–16, 86–94, 149–57; Stromquist, *Labor's Cold War*; Bernstein, *Turbulent Years*, 773–74; Cowie, *Great Exception*, 162–63; Norrell, "Caste in Steel," 684–85; Goldfield, "Race and the CIO," 14–15.

18. Basso, *Meet Joe Copper*, 4–14, 159–93, 277; Cowie, *Great Exception*, 129–46.

19. Lawrence Barr and Theo Sisco, interview by Joe Todd, 1985, H1985.13, Oral History Collection, OHC.

20. Genevieve Stovall Craig, "Picher, Oklahoma: Churndrill to Chat Pile," *Ford Times*, September 1955, 27–31 (quotes p. 31); "Picher Story," *MNR*, August 21, 1955; "John T. Stovall," March 23, 1916, Jasper County, Missouri Death Certificates, MSA.

21. Genevieve Stovall Craig, "Picher, Oklahoma: Churndrill to Chat Pile," *Ford Times*, September 1955, 27–31 (quotes p. 29).

22. "Mickey's a Hero, But He's No Cut Out to the Hero Pattern," *San Mateo Times*, June 1, 1956 (first two quotes); Leavy, *Last Boy*, 151–65, 417–28 (third quote p. 419).

23. Mantle, *Mick*, 12 (first four quotes); Mantle, *Quality of Courage*, 16 (last quote); Leavy, *Last Boy*, 79–82, 198–200; Elven C. Mantle and Joseph J. Nolan gravesites, Grand Army of the Republic Cemetery, Miami, Oklahoma.

24. Nieberding, *History of Ottawa County*, 82.

BIBLIOGRAPHY

PRIMARY SOURCES

ARCHIVAL COLLECTIONS

Colorado
  History Colorado, Denver
    Leadville Strike Reports Collection, 1896–1899, MS334
  University of Colorado Boulder Libraries, Special Collections and Archives, Boulder
    James H. Peabody Collection
    Pinkerton Detective Agency Reports
    Western Federation of Miners/International Union of Mine, Mill and Smelter
      Workers Collection
Idaho
  University of Idaho, Special Collections and Archives, Moscow
    Stanley Easton (1873–1961) Papers, 1900–1916, Manuscript Group 5
Kansas
  Baxter Springs Heritage Center and Museum, Baxter Springs
    Small Manuscripts Collection
  Kansas State Historical Society, Topeka
    J. J. Luck Mining Company Records, Unit ID 220708 (donated by David and Michael
      Hayes)
    Joseph W. Ridge Papers, 1858–1906, Manuscripts Collection 487
  Pittsburg State University, Axe Library Special Collections and Archives, Pittsburg
    Kansas Collection
    Picher Mining Collection
    Wadleigh, James. "The Strike of Mine, Mill and Smelter Workers of the Tri-State
      District, 1935."
Maryland
  National Archives and Records Administration, College Park
    RG25, National Labor Relations Board Records
    RG70, Records of the U.S. Bureau of Mines
    RG90, Records of the U.S. Public Health Service
    RG100, Records of the Division of Labor Standards, U.S. Department of Labor
    RG174, General Records of the Department of Labor
  University of Maryland Libraries, Special Collections and University Archives,
    College Park
    George Meany Memorial AFL-CIO Archives
      RG1-02, Gompers Copy Books
      RG1-17, Office of the President Copy Books, 1925–1960
      RG2-10, AFL Secretary-Treasurer's Office, Charter Books, 1891–1966
      RG4-08, Executive Council Minutes, 1893–1955

RG28-003, Organization and Field Services Department, International and
National Union Charter Files, 1886–1989
Massachusetts
Harvard University Business School, Baker Library, Cambridge
R. G. Dun & Co. Credit Report Volumes
Missouri
Jasper County Public Library, Local History Room, Joplin
Haase, Dixie, comp. "Granby, Mo., 'The Oldest Mining Town in the Southwest.'"
Self-published, 1984.
Kansas City, Fort Scott & Memphis Railroad Company. *The Klondike of Missouri.*
No publisher, 1898.
Joplin History and Mineral Museum
Historical Mining Photographs Collection
Missouri History Museum Library and Research Center, St. Louis
Blow Family Papers, 1837–1960
James B. Eads Collection, 1776–1974
Joseph Parker Gazzam Papers, 1788–1953
Missouri Southern State University, Spiva Library Archives and Special Collections,
Joplin
Arrell Gibson Collection
Harry Hood Collection, 1893–1980
Missouri State Archives, Jefferson City
Missouri Death Certificates, 1910–1968
Missouri Online Business Filings
Supreme Court of Missouri Historical Records, http://www.sos.mo.gov/records
/archives/archivesdb/supremecourt
Southeast Missouri State University, Special Collections and Archives, Cape Girardeau
George R. Suggs Papers
State Historical Society of Missouri Research Center, Columbia
C216, Missouri State Labor Council, AFL-CIO, Records, 1891–1975
Missouri Union Provost Marshal Papers, 1861–1866
State Historical Society of Missouri Research Center, Rolla
Mineral Museum News
R10, American Zinc, Lead, and Smelting Company Records, 1901–1965
R57, Frank Carmany Wallower Collection, 1882–1966
R369, Blow & Kennett Mining Company Record Book, 1858–1861
R371, Center Valley Lead and Zinc Company, 1892–1900
R677, Harry J. Cantwell Collection, 1898–1950
R754, Picher (Okla.) Collection, c. 1975–1995
R873, *Southwestern Miner*, 1892
New York
Columbia University Rare Book and Manuscript Library, New York
Health Sciences Project
Frances Perkins Papers, 1895–1965
Cornell University, Kheel Center for Labor-Management Documentation and Archives,
Ithaca
Kheel Collection of Oral Histories, 1965–1975, Collection 6058
National War Labor Board Non-ferrous Metals Commission Records, Collection
5314

Western Federation of Miners/International Union of Mine, Mill and Smelter
Workers Records, Collection 5268
New-York Historical Society Museum and Library, New York
Gustavus Vasa Fox Collection, 1823–1919
New York University, Tamiment Library/Robert F. Wagner Archives, New York
William Cassidy Files on the Tri-State Mining District of Missouri, Kansas, and
Oklahoma, TAM 063
Printed Ephemera Collection on Trade Unions, PE1
"Record of the Proceedings of the General Assembly of the Knights of Labor," Film
R-7522
Union Theological Seminary, Burke Library Special Collections, New York
Interchurch World Movement Records, 1919–1921
Oklahoma
Grand Army of the Republic Cemetery, Miami
Oklahoma History Center, Oklahoma City
Federal Writers' Project Records
Oral History Collection
Oklahoma State Archives, Oklahoma City
Criminal Court of Appeals Records
University of Oklahoma, Western History Collections, Norman
Frank J. Cuddeback Collection
Eagle-Picher Mining and Smelting Company Collection
Oklahoma State Federation of Labor Manuscript Collection, 1907–1958
Pennsylvania
Pennsylvania State University, Eberly Family Special Collections Library, State College
Oral History Transcripts
United Mine Workers of America
President's Office Correspondence with Districts, 1894–1983
President's Office Records, 1898–2010
United Steel Workers of America, President's Office Records, 1916–1980
University of Pittsburgh, Hillman Library, Archives and Special Collections,
Pittsburgh
Thomas Parran Papers, 1892–1968
Texas
National Archives and Records Administration, Fort Worth
RG70, Records of the U.S. Bureau of Mines, Records of the Picher Clinic
First Annual Report, 1927–1928, Folder 23, Box 2
Daniel Harrington, Richard V. Ageton, F. Flinn, and W. M. Myers, "Dust-
Ventilation Investigation in the Mines of the Picher, Oklahoma District,"
Folder 82, Box 2
RG75, Records of the Bureau of Indian Affairs
Washington
Northwest Museum of Arts and Culture, Spokane
Amasa B. Campbell Papers, MS38
Washington, D.C.
Catholic University of America, American Catholic History Research Center and
University Archives
Papers of John Brophy
John W. Hayes Papers

Philip Murray Papers
Records of the Congress of Industrial Organizations
Library of Congress
Abraham Lincoln Papers, Series I: General Correspondence, 1833–1916

NEWSPAPERS AND PERIODICALS

Ada (Okla.) Weekly News
American Federationist
American Zinc and Lead Journal
Anti-hijacker (Tar River, Okla.)
Appeal to Reason
Aspen (Colo.) Evening Chronicle
Aspen (Colo.) Tribune
Atchison (Kans.) Daily Globe
Baltimore Sun
Bates County Record (Butler, Mo.)
Baxter (Kans.) Daily Citizen
Blue Card Record
Boston Evening Journal
Boulder County (Colo.) Courier
Bradford (Pa.) Evening Star and Daily Record
Bulletin of the American Zinc Institute
Butte (Mont.) Weekly Miner
Carthage (Mo.) Banner
Carthage (Mo.) Press
Central City (Colo.) Register
Cherokee (Kans.) Index
CIO News
CIO News: Mine, Mill and Smelter Workers Edition
Cleveland Plain Dealer
Colorado News (Pueblo)
Colorado Weekly Chieftain (Pueblo)
Columbus (Kans.) Daily Advocate
Compressed Air Magazine
Crisis
Daily Arizona Silver Belt (Globe)
Daily Missouri Republican (St. Louis)
Daily National Intelligencer
Daily Oklahoman (Oklahoma City)
Denver Daily News
Denver Evening Post
Denver Post
Denver Republican
Economist
Emporia (Kans.) Daily Gazette
Emporia (Kans.) News

Engineering and Contracting
Engineering and Mining Journal
Ford Times
Frank Leslie's Illustrated Newspaper (New York)
Galena (Kans.) Daily Lever
Galena (Kans.) Evening Times
Galena (Kans.) Miner
Independent (Douthat, Okla.)
Joplin (Mo.) Daily Globe
Joplin (Mo.) Daily Herald
Joplin (Mo.) Daily News
Joplin (Mo.) Globe
Joplin (Mo.) Morning Herald
Joplin (Mo.) News Herald
Journal of the Knights of Labor
Kansas City (Mo.) Times
King Jack (Commerce and Picher, Okla.)
Kinsley (Kans.) Graphic
Lead and Zinc News (Joplin, Mo.)
Lead (S.D.) Daily Call
Leadville (Colo.) Herald Democrat
Los Angeles Times
Marine Engineering
Metal Mine and Smelter Worker
Miami (Okla.) Daily News-Record
Miami (Okla.) News-Record
Miami (Okla.) Record-Herald
Miners Magazine
Mines and Minerals
Mining and Scientific Press
Mining Congress Journal
Missouri Cash-Book (Jackson)
Missouri Patriot (Springfield)
Missouri Trades Unionist (Joplin)
Missouri Weekly Patriot (Springfield)
Nelson (BC) Tribune
Neosho (Mo.) Times
Nevada (Mo.) Daily Mail
New Republic
New York Times
Oklahoma Federationist

People's Tribune (Jefferson City, Mo.)
Pittsburg (Kans.) Daily Headlight
Reno (Nev.) Gazette
Rocky Mountain News (Denver)
San Francisco Chronicle
San Mateo (Calif.) Times
Spokesman-Review (Spokane, Wash.)
St. Joseph (Mo.) Gazette-Chronicle
St. Louis Globe-Democrat
St. Louis Post-Dispatch
State Line Herald (Joplin, Mo., and Galena, Kans.)

Time
Tri-State Tribune (Picher, Okla.)
Union (Denver)
Victoria (BC) Daily Colonist
Wall Street Journal
Washington Post
Washington Times
Webb City (Mo.) Daily Register
Webb City (Mo.) Register
Weekly Raleigh (N.C.) Register
Worker's Chronicle (Pittsburg, Kans.)

GOVERNMENT PUBLICATIONS

The Annotated Statutes of the State of Missouri, 1906. Vol. 4. St. Paul, Minn.: West, 1906.

Broadhead, Garland C. Report of the Geological Survey of the State of Missouri, Including Field Work of 1873–1874. Jefferson City, Mo.: Regan, 1874.

Buckley, E. R., and H. A. Buehler. The Geology of the Granby Area. Jefferson City: Missouri Bureau of Geology and Mines, 1906.

Federal Writers' Project of Oklahoma. Labor History of Oklahoma. Oklahoma City: A. M. Van Horn, 1939.

"First Annual Report of the State Mine Inspector." In Missouri Bureau of Labor Statistics, Ninth Annual Report of the Bureau of Labor Statistics of the State of Missouri. Jefferson City, Mo.: Tribune, 1887.

Hamilton, Alice. Lead Poisoning in the Smelting and Refining of Lead. Bulletin of the United States Department of Labor Statistics, no. 141. Washington, D.C.: Government Printing Office, 1914.

Haworth, Erasmus. The University Geological Survey of Kansas. Vol. 8, Special Report on Lead and Zinc. Topeka: State Printing Office, 1904.

Higgins, Edwin, A. J. Lanza, F. B. Laney, and George S. Rice. Siliceous Dust in Relation to Pulmonary Disease among Miners in the Joplin District, Missouri. Bureau of Mines Bulletin 132. Washington, D.C.: Government Printing Office, 1917.

Jackson, Andrew. State of the Union Address, December 8, 1829. Teaching American History. https://teachingamericanhistory.org/library/document/state-of-the-union -address-39/.

Lanza, A. J., and Edwin Higgins. Pulmonary Disease among Miners in the Joplin District, Missouri: A Preliminary Report. Washington, D.C.: Government Printing Office, 1915.

Martin, A. J. Summarized Statistics of Production of Lead and Zinc in the Tri-State (Missouri-Kansas-Oklahoma) Mining District. Bureau of Mines Information Circular 7383. Washington, D.C.: Government Printing Office, 1946.

Missouri Board of Health. Twenty-Ninth Annual Report of the State Board of Health of Missouri, 1911. Jefferson City, Mo.: Hugh Stephens, 1912.

———. Thirty-First Annual Report of the State Board of Health. Jefferson City, Mo.: Hugh Stephens, 1914.

Missouri Bureau of Labor Statistics. First Annual Report of the Bureau of Labor Statistics for the Year Ending January 1, 1880. Jefferson City, Mo.: Carter and Regan, 1880.

———. Second Annual Report of the Bureau of Labor Statistics of the State of Missouri. Jefferson City, Mo.: Tribune, 1881.

————. *Ninth Annual Report of the Bureau of Labor Statistics of the State of Missouri.* Jefferson City, Mo.: Tribune, 1887.

————. *Tenth Annual Report of the Bureau of Labor Statistics of the State of Missouri.* Jefferson City, Mo.: Tribune, 1888.

————. *Eleventh Annual Report of the Bureau of Labor Statistics of the State of Missouri.* Jefferson City, Mo.: Tribune, 1889.

————. *Twenty-Third Annual Report of the Bureau of Labor Statistics and Inspection of the State of Missouri for the Year Ending November 5th, 1901.* Jefferson City, Mo.: Tribune, 1901.

————. *Thirty-Third Annual Report of the Bureau of Labor Statistics of the State of Missouri.* Jefferson City, Mo.: Hugh Stephens, 1912.

————. *Thirty-Fourth Annual Report of the Bureau of Labor Statistics of the State of Missouri.* Jefferson City, Mo.: Hugh Stephens, 1913.

————. *Thirty-Sixth Annual Report of the Bureau of Labor Statistics of the State of Missouri.* Jefferson City, Mo.: Hugh Stephens, 1915.

————. *Thirty-Seventh Annual Report of the Bureau of Labor Statistics of the State of Missouri.* Jefferson City, Mo.: Allied, 1916.

————. *Fortieth and Forty-First Annual Reports of the Bureau of Labor Statistics of the State of Missouri.* Jefferson City, Mo.: Allied, 1921.

Missouri Bureau of Mines. *Report of the State Mine Inspector of the State of Missouri for the Year Ending November 5, 1890.* Jefferson City, Mo.: Tribune, 1890.

————. *Fifth Annual Report of the State Mine Inspector of the State of Missouri for the Year Ending June 30, 1891.* Jefferson City, Mo.: Tribune, 1891.

————. *Seventh Annual Report of the State Mine Inspectors of the State of Missouri for the Year Ending June 30, 1893.* Jefferson City, Mo.: Tribune, 1893.

————. *Eighth Annual Report of the State Mine Inspectors of the State of Missouri for the Year Ending June 30, 1894.* Jefferson City, Mo.: Tribune, 1894.

————. *Ninth Annual Report of the State Mine Inspectors of the State of Missouri for the Year Ending June 30, 1895.* Jefferson City, Mo.: Tribune, 1896.

————. *Tenth Annual Report of the State Mine Inspectors of the State of Missouri for the Year Ending June 30, 1896.* Jefferson City, Mo.: Tribune, 1896.

————. *Eleventh Annual Report of the State Mine Inspectors of the State of Missouri for the Year Ending June 30, 1897.* Jefferson City, Mo.: Tribune, 1898.

————. *13th Annual Report of the State Mine Inspectors of the State of Missouri for the Year Ending June 30, 1899.* Jefferson City, Mo.: Tribune, 1900.

————. *16th Annual Report of the State Lead and Zinc Mine Inspector of the State of Missouri for the Year Ending December 31, 1902.* Jefferson City, Mo.: Tribune, 1903.

————. *17th Annual Report of the State Lead and Zinc Mine Inspectors of the State of Missouri.* Jefferson City, Mo.: Hugh Stephens, 1904.

————. *18th Annual Report of the Bureau of Mines and Mine Inspection of the State of Missouri.* Jefferson City, Mo.: Hugh Stephens, 1905.

————. *19th Annual Report of the Bureau of Mines and Mine Inspection of the State of Missouri.* Jefferson City, Mo.: Hugh Stephens, 1906.

————. *20th Annual Report of the Bureau of Mines and Mine Inspection of the State of Missouri.* Jefferson City, Mo.: Hugh Stephens, 1907.

————. *22nd Annual Report of the Bureau of Mines, Mining and Mine Inspection of the State of Missouri.* Jefferson City, Mo.: Hugh Stephens, 1909.

————. *Twenty-Third Annual Report of the Bureau of Mines, Mining and Mine Inspection of the State of Missouri.* Jefferson City, Mo.: Hugh Stephens, 1910.

———. *Twenty-Ninth Annual Report of the Bureau of Mines, Mining and Mine Inspection of the State of Missouri*. Jefferson City, Mo.: Hugh Stephens, 1916.

Missouri General Assembly. *Journal of the House of Representatives of the 41st General Assembly of the State of Missouri*. Jefferson City, Mo.: Tribune, 1901.

———. *Journal of the Senate of the State of Missouri at the Adjourned Session of the Twenty-Third General Assembly*. Jefferson City, Mo.: Emerson S. Foster, 1865–66.

———. *Laws of Missouri Passed at the Regular and Extra Sessions of the Forty-Fourth General Assembly*. Jefferson City, Mo.: Hugh Stephens, 1907.

———. *Laws of Missouri Passed at the Session of the Forty-First General Assembly*. Jefferson City, Mo.: Tribune, 1901.

Missouri Secretary of State. *State Almanac and Official Directory of Missouri for 1879*. St. Louis: John J. Daly, 1879.

———. *Official Directory of Missouri for 1881*. St. Louis: John J. Daly, 1881.

———. *Official Directory of Missouri for 1883*. St. Louis: John J. Daly Stationery, 1883.

———. *Official Directory of Missouri for 1885*. St. Louis: John J. Daly Stationery, 1885.

———. *Official Manual of the State of Missouri for the Years 1895–1896*. Jefferson City, Mo.: Tribune, 1895.

———. *Official Manual of the State of Missouri for the Years 1897–1898*. Jefferson City, Mo.: Tribune, 1897.

———. *Official Manual of the State of Missouri for the Years 1899–1900*. Jefferson City, Mo.: Tribune, 1899.

———. *Official Manual of the State of Missouri for the Years 1901–1902*. Jefferson City, Mo.: Tribune, 1901.

———. *Official Manual of the State of Missouri for the Years 1905 and 1906*. Jefferson City, Mo.: Hugh Stephens, 1905.

———. *Official Manual of the State of Missouri for the Years 1909–1910*. Jefferson City, Mo.: Hugh Stephens, 1910.

———. *Official Manual of the State of Missouri for the Years 1913 and 1914*. Jefferson City, Mo.: Hugh Stephens, 1913.

———. *Official Manual of the State of Missouri, 1933–1934*. Jefferson City, Mo.: Midland, 1934.

Missouri Writers' Project. *Missouri: A Guide to the "Show-Me" State*. New York: Duell, Sloan, and Pearce, 1941.

National Labor Relations Board. *Decisions of National Labor Relations Board*. Vol. 16. Case No. C-73. Washington, D.C.: Government Printing Office, 1940.

National Recovery Administration. *Code of Fair Competition for the Zinc Industry*. Washington, D.C.: Government Printing Office, 1935.

Oklahoma State Department of Health. *Fifth Annual Report of the State Department of Health of Oklahoma*. Oklahoma City: Printery, 1921.

Oklahoma State Department of Mines and Mining. *Twenty-Fifth Annual Report of the Department of Mines and Mining for the Fiscal Year Ending June 30, 1932*. Oklahoma City: Oklahoma Printing Company, 1932.

Oklahoma State Election Board. *Directory of the State of Oklahoma, 1935*. Oklahoma City: State Election Board, 1935.

———. *Directory, State of Oklahoma*. Oklahoma City: Walker-Wilson-Tyler, 1927.

Oklahoma State Health Commissioner. *Ninth Annual Report of the State Department of Health of Oklahoma*. Guthrie: Oklahoma Printing Company, 1928.

———. *Tenth Annual Report of the State Department of Public Health*. Oklahoma City: Oklahoma Printing Company, 1930.

Roosevelt, Franklin D. "The President's Reemployment Agreement." July 27, 1933. American Presidency Project. https://www.presidency.ucsb.edu/documents/the -presidents-reemployment-agreement.

Sayers, R. R. *Silicosis among Miners*. Bureau of Mines Technical Paper 372. Washington, D.C.: Government Printing Office, 1925.

Sayers, R. R., F. V. Meriwether, A. J. Lanza, and W. W. Adams. *Silicosis and Tuberculosis among Miners of the Tri-State District of Oklahoma, Kansas and Missouri-I*. Bureau of Mines Technical Paper 545. Washington, D.C.: Government Printing Office, 1933.

State of New York. *Department of Labor Bulletin, 1911*. Vol. 13. Albany: E. Lyon, 1912.

Swallow, G. C. *The First and Second Annual Reports of the Geological Survey of Missouri*. Jefferson City, Mo.: James Lusk, 1855.

U.S. Bureau of Indian Affairs. *Report of the Commissioner of Indian Affairs, 1907*. Washington, D.C.: Government Printing Office, 1907.

U.S. Bureau of Labor Statistics. *Wages in the Nonferrous-Metals Industry, June 1943*. Bulletin 765. Washington, D.C.: Government Printing Office, 1944.

U.S. Bureau of the Census. *Fourteenth Census of the United States, 1920*. Vol. 3, *Population*. Washington, D.C.: Government Printing Office, 1922.

———. *Fifteenth Census of the United States, 1930*. 6 vols. Washington, D.C.: Government Printing Office, 1931–33.

———. *Historical Statistics of the United States, 1789–1945: A Supplement to the Statistical Abstract of the United States*. Washington, D.C.: Government Printing Office, 1949.

———. "Oklahoma." Pt. 36 of *Characteristics of the Population*, vol. 2 of *Census of Population: 1950*. Washington, D.C.: Government Printing Office, 1953.

———. *Statistics for Missouri: Thirteenth Census of the United States, 1910*. Washington, D.C.: Government Printing Office, 1913.

U.S. Census Office. *Census Reports*. Vol. 1, *Population, Part 1*. Washington, D.C.: Government Printing Office, 1901.

———. *Report on Population of the United States at the Eleventh Census: 1890*. Pt. 1. Washington, D.C.: Government Printing Office, 1895.

———. *Report on the Mining Industries of the United States*. Washington, D.C.: Government Printing Office, 1886.

———. *Seventh Census of the United States, 1850*. Vol. 1. Washington, D.C.: Robert Armstrong, 1853.

———. *Statistics of the Population of the United States at the Tenth Census (June 1, 1880)*. Washington, D.C.: Government Printing Office, 1883.

U.S. Centennial Commission. *International Exhibition: 1876 Official Catalogue*. Pt. 3, *Machinery Hall, Annexes, and Special Buildings*. Philadelphia: John R. Nagle, 1876.

U.S. Congress. *Conditions in the Coal Mines of Colorado: Hearings before a Subcommittee of the Committee on Mines and Mining*. Pt. 4. Washington, D.C.: Government Printing Office, 1914.

U.S. Department of the Treasury. *Annual Report of the Surgeon General of the Public Health Service of the United States for the Fiscal Year 1920*. Washington, D.C.: Government Printing Office, 1920.

U.S. Geological Survey. *The Mineral Resources of the United States*. Washington, D.C.: Government Printing Office, 1883.

———. *Mineral Resources of the United States, 1883 and 1884*. Washington, D.C.: Government Printing Office, 1885.

———. *Mineral Resources of the United States, 1886*. Washington, D.C.: Government Printing Office, 1887.

———. *Mineral Resources of the United States, 1916*. Pt. 1. Washington, D.C.: Government Printing Office, 1917.

U.S. Tariff Commission. *Lead and Zinc*. Washington, D.C.: Government Printing Office, 1961.

The War of the Rebellion: A Compilation of the Official Records of the Union and Confederate Armies. 128 vols. Washington, D.C.: Government Printing Office, 1880–1901.

Wilson, Woodrow. "Proclamation 1364: Declaring that a State of War Exists between the United States and Germany," April 2, 1917. The American Presidency Project. https://www.presidency.ucsb.edu/documents/proclamation-1364-declaring-that-state-war-exists-between-the-united-states-and-germany.

Winslow, Arthur. *Missouri Geological Survey*. Vol. 6, *Lead and Zinc Deposits: Section I*. Jefferson City, Mo.: Tribune, 1894.

———. *Missouri Geological Survey*. Vol. 7, *Lead and Zinc Deposits: Section II*. Jefferson City, Mo.: Tribune, 1894.

Wright, Carroll D. "A Report on Labor Disturbances in the State of Colorado, from 1880 to 1904." S. Doc. No. 58-122 (1905).

Wright, Clarence A. *Mining and Treatment of Lead and Zinc Ores in the Joplin District, Missouri: A Preliminary Report*. Bureau of the Mines Technical Paper 41. Washington, D.C.: Government Printing Office, 1913.

UNPUBLISHED GOVERNMENT MATERIAL

All censuses were accessed via www.ancestry.com.

1850 federal manuscript census and slave schedules
1860 federal manuscript census and slave schedules
1870 federal manuscript census
1880 federal manuscript census
1900 federal manuscript census
1910 federal manuscript census
1920 federal manuscript census
1930 federal manuscript census
1940 federal manuscript census
Kansas 1938 County Census
Kansas 1943 County Census
U.S. General Land Office Records, 1796–1907
U.S. World War I Draft Registration Cards, 1917–1918

MICROFILM

American Federation of Labor Records: Gompers Era, 1878–1949. 144 Reels. Pt. 1, "Records Held by the American Federation of Labor." Sanford, N.C.: Microfilming Corporation of America, 1981.

CIO Files of John L. Lewis. 25 Reels. Pt. 1, "Correspondence with CIO Unions, 1929–1962." Frederick, Md.: University Publications of America, 1988.

CIO Files of John L. Lewis. 20 Reels. Pt. 2, "General Files on the CIO and AFL, 1929–1955." Frederick, Md.: University Publications of America, 1988.

Samuel Gompers Letterbooks, 1883–1924. 340 Reels. Washington, D.C.: Letter of Congress Photoduplication Service, 1967.

Terence V. Powderly Papers, 1864–1937. 94 Reels. Glen Rock, N.J.: Microfilming Corporation of America, 1974.

U.S. Commission on Industrial Relations, 1912–1915. 15 Reels. Frederick, Md.: University Publications of America, 1985.

OTHER PUBLISHED PRIMARY SOURCES

American Dialect Society. *Dialect Notes*. Vol. 5. Pt. 1. New Haven, Conn.: Tuttle, Morehouse, and Taylor, 1918.

American Federation of Labor. *Proceedings of the American Federation of Labor: 1899, 1900, 1901*. Bloomington, Ill.: Pantagraph, n.d.

———. *Report of Proceedings of the Fifty-Fifth Annual Convention of the American Federation of Labor*. Washington, D.C.: Judd and Detweiler, 1935.

———. *Report of Proceedings of the Fifty-Seventh Annual Convention of the American Federation of Labor*. Washington, D.C.: Judd and Detweiler, 1937.

American Mining Congress Western Division. *The Story of the Tri-State Mining District*. Joplin, Mo.: Joplin, 1931.

American Mining Manual, 1920. Chicago: Mining Manual Company, 1920.

Aurora Centennial Committee. *Aurora Centennial, 1870–1970: Yesterday and Today*. Aurora, Mo.: MWM Color, 1970.

Bishop, A. W. *Loyalty on the Frontier or Sketches of Union Men of the South-West*. St. Louis: E. P. Studley, 1863.

Blosser, Howard W. *Prairie Jackpot: Beginning of the Rich Picher Oklahoma Mining Field*. Webb City, Mo.: Sentinel, 1973.

Bryan, William Jennings, and Mary Baird Bryan. *The Memoirs of William Jennings Bryan*. Chicago: John C. Winston, 1925.

Carter, Susan B., Scott Sigmund Gartner, Michael R. Haines, Alan L. Olmstead, Richard Sutch, and Gavin Wright, eds. *Economic Sectors*. Vol. 4 of *Historical Statistics of the United States: Earliest Times to the Present*. New York: Cambridge University Press, 2006.

Centennial History of Newton County, Missouri. 1876. Reprint, Newton County Historical Society, 1996.

Clark, Root, and Co.'s City Directory of Leadville, and Business Directory of Carbonateville, Kokomo, and Malta, for 1879. Denver: Daily Times Steam Printing, 1879.

Conrad, Howard, ed. *Encyclopedia of the History of Missouri: A Compendium*. Vols. 2–3. New York: Southern History, 1901.

Corbett and Ballenger's Second Annual Leadville City Directory. Leadville, Colo.: Corbett and Ballenger, 1881.

Dacus, Joseph A., and James William Buel. *A Tour of St. Louis*. St. Louis: Western, 1878.

Davidson, L. S. *South of Joplin: Story of a Tri-State Diggin's*. New York: W. W. Norton, 1939.

Dearing, Charles Lee. *The ABC of the NRA*. Washington, D.C.: Brookings Institution, 1934.

Dick, Sheldon, and Lee Dick. *Men and Dust*. 16 mm film. New York: Lee Dick Inc. and Tri-State Survey Committee, 1940.

Draper, William R., and Mable Draper. *Old Grubstake Days in Joplin: The Story of the Pioneers Who Discovered the Largest and Richest Lead and Zinc Mining Field in the World*. Girard, Kans.: E. Haldeman-Julius, 1946.

Freund, Ernst. *The Police Power, Public Policy, and Constitutional Rights*. Chicago: Callaghan, 1904.

"Gaining on Tuberculosis in Missouri." *Survey* 34, no. 8 (May 22, 1915), 169–70.

Gazzam, Joseph P. "The Leadville Strike of 1896." *Bulletin of the Missouri Historical Society* 7, no. 1 (October 1950): 89–94.

Hamilton, Alice. *Exploring the Dangerous Trades: The Autobiography of Alice Hamilton, M.D.* 1943. Reprint, Boston: Northeastern University Press, 1985.

———. "A Mid-American Tragedy." *Survey Graphic* 29 (August 1940): 434–37.

Harlow, Rex, comp. *Makers of Government in Oklahoma*. Oklahoma City: Harlow, 1930.

Holibaugh, John R. *The Lead and Zinc Mining Industry of Southwest Missouri and Southeast Kansas*. New York: Scientific Publishing Company, 1895.

Holley, Clifford. *The Lead and Zinc Pigments*. New York: John Wiley and Sons, 1909.

An Illustrated Historical Atlas Map of Jasper County, Mo. Philadelphia: Brink, McDonough, 1876.

Ingalls, Walter Renton. *Lead and Zinc in the United States*. New York: Hill, 1908.

———. *Lead Smelting and Refining, with Some Notes on Lead Mining*. New York: Engineering and Mining Journal, 1906.

International Union of Mine, Mill and Smelter Workers. *Official Proceedings of the Twenty-Third Consecutive and Third Biennial Convention, International Union of Mine, Mill and Smelter Workers*. Denver: Publishers Press, 1918.

———. *Official Proceedings of the Thirty-Second Convention of the International Union of Mine, Mill and Smelter Workers*. Salt Lake City: Allied Printing, 1935.

———. *Official Proceedings of the Thirty-Fourth Convention of the International Union of Mine, Mill and Smelter Workers*. Denver: Allied Printing, 1937.

———. *Official Proceedings of the 38th Convention of the International Union of Mine, Mill and Smelter Workers*. Joplin, Mo.: Allied Printing, 1941.

Jobe, Sybil Shipley, ed. *Reprint of Centennial History of Newton County, Missouri*. Neosho, Mo.: Newton County Historical Society, 1976.

Joplin, Missouri, Directory: Fifth Annual Issue, 1913. Joplin, Mo.: Joplin, 1913.

"Journal of Mining and Mines: Lead Mines in South-West Missouri." *Western Journal* 3, no. 6 (September 1850): 411–12.

Kaufman, Stuart B., Peter J. Albert, and Elizabeth A. Fones-Wolf, eds. *The Samuel Gompers Papers*. Vol. 2, *The Early Years of the American Federation of Labor, 1887–1890*. Urbana: University of Illinois Press, 1987.

Kaufman, Stuart B., Peter J. Albert, and Grace Paladino, eds. *The Samuel Gompers Papers*. Vol. 4, *A National Labor Movement Takes Shape, 1895–98*. Urbana: University of Illinois Press, 1991.

Knights of Labor. "Preamble and Declaration of Principles of the Knights of Labor of America." 1878. Duke University Libraries Digital Repository. https://repository.duke.edu/dc/broadsides/bdspa032486.

Laas, Virginia Jeans, ed. *Bridging Two Eras: The Autobiography of Emily Newell Blair, 1877–1951*. Columbia: University of Missouri Press, 1999.

"Lead in Southwest Missouri—Moseley's Mines." *Western Journal and Civilian* 13, no. 2 (January 1855): 119–22.

"Lead Mining in Southern Missouri." *Debow's Review* 18, no. 3 (March 1855): 389–91.

Leonard, J. W. *Industries of St. Louis: Her Relations as a Center of Trade*. St. Louis: J. M. Elstner, 1887.

Livingston, Joel T. *A History of Jasper County, Missouri, and Its People*. Vol. 1. Chicago: Lewis, 1912.

Lloyd, E., and D. Baumann. *The Mineral Wealth of Southwest Missouri*. Joplin, Mo.: Lloyd and Baumann, 1874.

Mantle, Mickey. *The Mick: An American Hero; The Legend and the Glory*. With Herb Gluck. New York: Doubleday, 1985.

———. *The Quality of Courage: Heroes in and out of Baseball*. Lincoln: University of Nebraska Press, 1999.

McGregor, Malcolm G. *The Biographical History of Jasper County, Missouri*. Chicago: Lewis, 1901.

Mills, Charles Morris. "Joplin Zinc: Industrial Conditions in the World's Greatest Zinc Center." *Survey* 45 (1921): 657–66.

"The Missouri Lead Mines." *United States Mining Journal* 10, no. 2 (1857): 2.

"Missouri Lead Mines in the Hands of the Rebels." *Journal of Mining and Manufactures* 45, no. 6 (December 1861): 626–27.

Missouri State Federation of Labor. *Proceedings of the Twenty-Third Annual Convention of the Missouri State Federation of Labor.* Hannibal, Mo.: Journal Printing Company, 1914.

North, F. A., ed. *The History of Jasper County, Missouri.* Des Moines, Iowa: Mills, 1883.

Orten, M. D. "The Present Economic Condition of the Tri-State Zinc District." *Journal of Business of the University of Chicago* 1, no. 4 (October 1928): 417–28.

Parker, Nathan H. *Missouri as It Is in 1867: An Illustrated Historical Gazetteer of Missouri.* Philadelphia: J. B. Lippincott, 1867.

Pryor, Jerry. *Southwest Missouri Mining.* Charleston, S.C.: Arcadia, 2000.

Reavis, L. U. *St. Louis: The Future Great City of the World.* St. Louis: C. R. Barns, 1876.

Richardson, Albert D. *Beyond the Mississippi: From the Great River to the Great Ocean.* Hartford, Conn.: American Publishing Company, 1867.

Rickard, Thomas A. *Interviews with Mining Engineers.* San Francisco: Mining and Scientific Press, 1922.

Ross, Malcolm. *Death of a Yale Man.* New York: Farrar and Rinehart, 1939.

Rothwell, Richard P., ed. *The Mineral Industry, Its Statistics, Technology and Trade in the United States and Other Countries to the End of 1899.* New York: Scientific Publishing Company, 1900.

Sang, Alfred. "The Corrosion of Iron and Steel." *Proceedings of the Engineers' Society of Western Pennsylvania* 24, no. 10 (January 1909): 493–542.

Scharf, John Thomas. *History of St. Louis City and County.* Vol. 1. Philadelphia: Louis H. Everts, 1883.

Schoolcraft, Henry Rowe. *A View of the Lead Mines of Missouri.* New York: C. Wiley, 1819.

Shaner, Dolph. *The Story of Joplin.* New York: Stratford House, 1948.

Siebenthal, C. E. "The Early Days of the Zinc Industry in the United States." *Steel and Metal Digest* 9, no. 6 (July 1919): 341–44.

Swallow, G. C. *Geological Report of the Country along the Line of the Southwestern Branch of the Missouri Pacific Railroad.* St. Louis: George Knapp, 1859.

Taylor, Frederick Winslow. *The Principles of Scientific Management.* New York: Harper and Brothers, 1915.

Tri-State Survey Committee. *A Preliminary Report on Living, Working and Health Conditions in the Tri-State Mining Area (Missouri, Oklahoma and Kansas).* New York: Rotograph, 1939.

Tunnard, William H. *A Southern Record: The History of the Third Regiment, Louisiana Infantry.* Baton Rouge, La.: Printed by the author, 1866.

United Mine Workers. *Proceedings of the Twenty-Seventh Consecutive and Fourth Biennial Convention of the United Mine Workers of America.* Vol. 1. Indianapolis: Bookwalter-Ball, 1919.

Washington University. *A Catalogue of the Officers and Students in Washington University, with the Courses of Study for the Academic Year, 1888–89.* St. Louis: Nixon-Jones, 1889.

Weed, Walter Harvey. *The Mines Handbook.* New York: W. M. Weed, 1920.

Weidman, Samuel. *Miami-Picher Zinc-Lead District.* Norman: University of Oklahoma Press, 1932.

Western Federation of Miners. *Official Proceedings of the Eleventh Annual Convention.* Denver: Western Newspaper Union, 1903.

———. *Official Proceedings of the Twelfth Annual Convention.* Denver: Western Newspaper Union, 1904.

———. *Official Proceedings of the Thirteenth Annual Convention.* Denver: Reed, 1905.

———. *Official Proceedings of the Fifteenth Annual Convention.* Stenographic report edition. Denver: Allied, 1907.

———. *Official Proceedings of the Sixteenth Annual Convention.* Denver: W. H. Kistler, 1908.

———. *Official Proceedings of the Seventeenth Annual Convention.* Denver: W. H. Kistler, 1909.

———. *Official Proceedings of the Eighteenth Annual Convention of the Western Federation of Miners.* Denver: Great Western, 1910.

———. *Official Proceedings of the Twenty-Second Consecutive and Second Biennial Convention, Western Federation of Miners.* Denver: Allied, 1916.

Year Book of the Commercial, Banking, and Manufacturing Interests of St. Louis. St. Louis: Nixon-Jones, 1882.

SECONDARY SOURCES

Alexander, W., and A. Street. *Metals in the Service of Man.* London: Penguin Books, 1956.

Alexander, William. *Films on the Left: American Documentary Film from 1931 to 1942.* Princeton, N.J.: Princeton University Press, 1981.

Anderson, David M. "'Such Contented Workers': Mine-Mill Organizers in the Ely, Nevada Copper District, 1920–1943." Master's thesis, University of Nevada, Las Vegas, 1994.

Andrews, Thomas G. *Killing for Coal: America's Deadliest Labor War.* Cambridge, Mass.: Harvard University Press, 2008.

Appleby, Joyce. *The Relentless Revolution: A History of Capitalism.* New York: W. W. Norton, 2010.

Arnesen, Eric. *Waterfront Workers of New Orleans: Race, Class, and Politics, 1863–1923.* Urbana: University of Illinois Press, 1994.

Arnold, Andrew B. *Fueling the Gilded Age: Railroads, Miners, and Disorder in Pennsylvania Coal.* New York: New York University Press, 2014.

Aron, Stephen. *American Confluence: The Missouri Frontier from Borderland to Border State.* Bloomington: Indiana University Press, 2006.

Astor, Aaron. *Rebels on the Border: Civil War, Emancipation, and the Reconstruction of Kentucky and Missouri.* Baton Rouge: Louisiana State University Press, 2012.

Baron, Ava. "Masculinity, the Embodied Male Worker, and the Historian's Gaze." *International Labor and Working-Class History,* no. 69 (Spring 2006): 143–60.

Basso, Matthew L. *Meet Joe Copper: Masculinity and Race on Montana's World War II Home Front.* Chicago: University of Chicago Press, 2013.

Bederman, Gail. *Manliness and Civilization: A Cultural History of Gender and Race in the United States, 1880–1917.* Chicago: University of Chicago Press, 1996.

Bernstein, Irving. *Turbulent Years: A History of the American Worker, 1933–1941.* Boston: Houghton Mifflin, 1970.

Bezis-Selfa, John. *Forging America: Ironworkers, Adventurers, and the Industrial Revolution.* Ithaca, N.Y.: Cornell University Press, 2004.

Blevins, Brooks. *A History of the Ozarks.* Vol. 1, *The Old Ozarks.* Urbana: University of Illinois Press, 2018.

Boucher, Gerald R. "The 1901 Rossland Miners Strike: The Western Federation of Miners Responds to Individual Capitalism." Bachelor's thesis, University of Victoria, 1986.

Brattain, Michelle. *The Politics of Whiteness: Race, Workers, and Culture in the Modern South.* Princeton, N.J.: Princeton University Press, 2001.

Brody, David. *Steelworkers in America: The Nonunion Era*. Cambridge, Mass.: Harvard University Press, 1960.

Burke, Bob. "The Evolution of Workers' Compensation Law in Oklahoma: Is the Grand Bargain Still Alive?" *Oklahoma City University Law Review* 41, no. 3 (Winter 2016): 337–423.

Buttrick, John. "The Inside Contract System." *Journal of Economic History* 12, no. 3 (Summer 1952): 205–21.

Caldemeyer, Dana M. "Run of the Mine: Miners, Farmers, and the Non-union Spirit of the Gilded Age, 1886–1896." PhD diss., University of Kentucky, 2016.

———. *Acting Against the Order: Capitalism, Miners, and the Problem of Unionism in the Gilded Age*. Urbana: University of Illinois Press, forthcoming.

Callahan, Richard J., Jr. *Work and Faith in the Kentucky Coal Fields: Subject to Dust*. Bloomington: Indiana University Press, 2009.

Carter, Margaret S. *New Diggings on the Fever River*. Madison: University of Wisconsin Press, 1959.

Case, Theresa Ann. "Losing the Middle Ground: Strikebreakers and Labor Protest on the Southwestern Railroads." In *Rethinking U.S. Labor History: Essays on the Working-Class Experience, 1756–2009*, edited by Donna T. Haverty-Stacke and Daniel J. Walkowitz, 54–81. New York: Continuum International, 2010.

Cassidy, William James. "The Tri-State Zinc-Lead Mining Region: Growth, Problems and Prospects." PhD diss., University of Pittsburgh, 1955.

Christensen, Lawrence O., Gary Kremer, William Foley, and Kenneth Winn, eds. *Dictionary of Missouri Biography*. Columbia: University of Missouri Press, 1999.

Churchwell, Sarah. *Behold, America: The Entangled History of "America First" and "The American Dream."* New York: Basic Books, 2018.

Clawson, Dan. *Bureaucracy and the Labor Process: The Transformation of U.S. Industry, 1860–1920*. New York: Monthly Review Press, 1980.

Cohen, Lizabeth. *Making a New Deal: Industrial Workers in the Chicago, 1919–1939*. New York: Cambridge University Press, 1990.

Corbin, David Alan. *Life, Work, and Rebellion in the Coal Fields*. Urbana: University of Illinois Press, 1981.

Cowie, Jefferson. *The Great Exception: The New Deal and the Limits of American Politics*. Princeton, N.J.: Princeton University Press, 2016.

Cowie, Jefferson, and Nick Salvatore. "The Long Exception: Rethinking the Place of the New Deal in American History." *International Journal of Labor and Working-Class History* 74, no. 1 (Fall 2008): 3–32.

Currarino, Rosanne. *The Labor Question in America: Economic Democracy in the Gilded Age*. Urbana: University of Illinois Press, 2011.

Davis, Mike. *Prisoners of the American Dream: Politics and Economy in the History of the U.S. Working Class*. London: Verso, 1986.

Dawley, Alan. *Struggles for Justice: Social Responsibility and the Liberal State*. Cambridge, Mass.: Harvard University Press, 1993.

DeMoss, Matt. "A Missed Opportunity: The Failure to Unionize Little Balkan Miners during the Strikes of 1893." Pittsburg State University Digital Commons. December 10, 2009. https://digitalcommons.pittstate.edu/cgi/viewcontent.cgi?article=1034&context=hist.

Derickson, Alan. *Black Lung: Anatomy of a Public Health Disaster*. Ithaca, N.Y.: Cornell University Press, 1998.

———. *Dangerously Sleepy: Overworked Americans and the Cult of Manly Wakefulness*. Philadelphia: University of Pennsylvania Press, 2014.

————. "Federal Intervention in the Joplin Silicosis Epidemic, 1911–1916." *Bulletin of the History of Medicine* 62, no. 2 (Summer 1988): 236–51.

————. "'On the Dump Heap': Employee Medical Screening in the Tri-State Zinc-Lead Industry, 1924–1932." *Business History Review* 62, no. 4 (Winter 1988): 656–77.

————. *Workers' Health, Workers' Democracy: The Western Miners' Struggle, 1891–1925.* Ithaca, N.Y.: Cornell University Press, 1988.

Dew, Charles B. "Disciplining Slave Ironworkers in the Antebellum South: Coercion, Conciliation, and Accommodation." *American Historical Review* 79, no. 2 (April 1974): 393–418.

Dick, Everett. *The Lure of the Land: A Social History of the Public Lands from the Articles of Confederation to the New Deal.* Lincoln: University of Nebraska Press, 1970.

Douglas, Richard L. "A History of Manufactures in the Kansas District." *Collections of the Kansas State Historical Society, 1909–1910,* no. 11 (1910): 81–215.

Dubofsky, Melvyn. *We Shall Be All: A History of the Industrial Workers of the World.* Chicago: Quadrangle Books, 1969.

Dubofsky, Melvyn, and Joseph McCartin. *Labor in America: A History.* Hoboken, N.J.: John Wiley and Sons, 2017.

Du Bois, W. E. B. *Black Reconstruction.* New York: Russell and Russell, 1935.

Durr, Kenneth D. *Behind the Backlash: White Working-Class Politics in Baltimore, 1940–1980.* Chapel Hill: University of North Carolina Press, 2003.

Emmons, David M. *The Butte Irish: Class and Ethnicity in an American Mining Town, 1875–1925.* Urbana: University of Illinois Press, 1989.

Enyeart, John P. *The Quest for "Just and Pure Law": Rocky Mountain Workers and American Social Democracy, 1870–1924.* Stanford, Calif.: Stanford University Press, 2009.

Ervin, Keona. *Gateway to Equality: Black Women and the Struggle for Economic Justice in St. Louis.* Lexington: University Press of Kentucky, 2017.

Feurer, Rosemary, and Chad Pearson, eds. *Against Labor: How U.S. Employers Organized to Defeat Union Activism.* Urbana: University of Illinois Press, 2017.

Fink, Leon. "Culture's Last Stand: Gender and the Search for Synthesis in American Labor History." *Labor History* 34, no. 2–3 (1993): 183–87.

————. *Workingmen's Democracy: The Knights of Labor and American Politics.* Urbana: University of Illinois Press, 1983.

Fishback, Price V. *Soft Coal, Hard Choices: The Economic Welfare of Bituminous Coal Miners, 1890–1930.* New York: Oxford University Press, 1992.

Foley, William E. *A History of Missouri.* Vol. 1, *1673 to 1820.* Columbia: University of Missouri Press, 1971.

Foner, Philip S. *From Colonial Times to the Founding of the American Federation of Labor.* Vol. 1 of *History of the Labor Movement in the United States.* 1947. Reprint, New York: International Publishers, 1972.

————. *From the Founding of the A. F. of L. to the Emergence of American Imperialism.* Vol. 2 of *History of the Labor Movement in the United States.* New York: International Publishers, 1975.

————. *The Policies and Practices of the American Federation of Labor, 1900–1909.* Vol. 3 of *History of the Labor Movement in the United States.* 1964. Reprint, New York: International Publishers, 1973.

Fones-Wolf, Ken, and Elizabeth Fones-Wolf. *Struggle for the Soul of the Postwar South: White Evangelical Protestants and Operation Dixie.* Urbana: University of Illinois Press, 2015.

Freeman, Joshua B. "Hard Hats: Construction Workers, Manliness, and the 1970 Pro-war Demonstrations." *Journal of Social History* 26, no. 4 (1993): 725–45.

Fronczak, Joseph. "The Fascist Game: Transnational Political Transmission and the Genesis of the U.S. Modern Right." *Journal of American History* 105, no. 3 (December 2018): 563–88.

Garlock, Jonathan. *Guide to the Local Assemblies of the Knights of Labor.* Westport, Conn.: Greenwood, 1982.

Gates, William B., Jr. *Michigan Copper and Boston Dollars: An Economic History of the Michigan Copper Mining Industry.* Cambridge, Mass.: Harvard University Press, 1951.

Gaventa, John. *Power and Powerlessness: Quiescence and Rebellion in an Appalachian Valley.* Urbana: University of Illinois Press, 1980.

Gellman, Erik S., and Jarod Roll. *The Gospel of the Working Class: Labor's Southern Prophets in New Deal America.* Urbana: University of Illinois Press, 2011.

Gerstle, Gary. *American Crucible: Race and Nation in the Twentieth Century.* Princeton, N.J.: Princeton University Press, 2001.

Gibson, Arrell M. *Wilderness Bonanza: The Tri-State District of Missouri, Kansas, and Oklahoma.* Norman: University of Oklahoma Press, 1972.

Gier, Jaclyn J., and Laurie Mercier, eds. *Mining Women: Gender in the Development of a Global Industry, 1670 to 2005.* New York: Palgrave Macmillan, 2006.

Glenn, Evelyn Nakano. *Unequal Freedom: How Race and Gender Shaped American Citizenship and Labor.* Cambridge, Mass.: Harvard University Press, 2002.

Glickman, Lawrence. "Inventing the 'American Standard of Living': Gender, Race and Working-Class Identity, 1880–1925." *Labor History* 34, nos. 2–3 (1993): 221–35.

Goldfield, Michael. "Race and the CIO: The Possibilities for Racial Egalitarianism during the 1930s and 1940s." *International Journal of Labor and Working-Class History* 44 (Fall 1993): 1–32.

Gordon, Colin. *New Deals: Business, Labor and Politics in America, 1920–1935.* New York: Cambridge University Press, 1994.

Gracy, David B., II. *Moses Austin: His Life.* San Antonio, Tex.: Trinity University Press, 1987.

Graziosi, Andrea. "Common Laborers, Unskilled Workers, 1880–1915." *Labor History* 22, no. 4 (Fall 1981): 512–44.

Green, James. *The Devil Is Here in These Hills: West Virginia's Coal Miners and Their Battle for Freedom.* New York: Atlantic Monthly Press, 2015.

Greene, Julie. *Pure and Simple Politics: The American Federation of Labor and Political Activism, 1881–1917.* New York: Cambridge University Press, 1998.

Griswold, Don L., and Jean Harvey Griswold. *History of Leadville and Lake County, Colorado: From Mountain Solitude to Metropolis.* Vol. 2. Boulder: University Press of Colorado, 1996.

Grob, Gerald. *Workers and Utopia: A Study of Ideological Conflict in the American Labor Movement, 1865–1900.* Evanston, Ill.: Northwestern University Press, 1961.

Gutman, Herbert. *Work, Culture, and Society in Industrializing America.* New York: Vintage Books, 1973.

Hahn, Steven. *A Nation Without Borders: The United States and Its World in an Age of Civil Wars, 1830–1910.* New York: Viking, 2016.

Hamilton, Shane. *Trucking Country: The Road to America's Wal-Mart Economy.* Princeton, N.J.: Princeton University Press, 2008.

Harper, Kimberly. *White Man's Heaven: The Lynching and Expulsion of Blacks in the Southern Ozarks, 1894–1909.* Fayetteville: University of Arkansas Press, 2010.

Harris, Howell. *Bloodless Victories: The Rise and Fall of the Open Shop in the Philadelphia Metal Trades, 1890–1940*. New York: Cambridge University Press, 2000.

Hild, Matthew. *Greenbackers, Knights of Labor, and Populists: Farmer-Labor Insurgency in the Late-Nineteenth-Century South*. Athens: University of Georgia Press, 2007.

Hoganson, Kristin L. *The Heartland: An American History*. New York: Penguin, 2019.

Honey, Michael K. *Southern Labor and Black Civil Rights: Organizing Memphis Workers*. Urbana: University of Illinois Press, 1993.

Hunter, Tera W. *To 'Joy My Freedom: Southern Black Women's Lives and Labors after the Civil War*. Cambridge, Mass.: Harvard University Press, 1997.

Jackson, Robert W. *Rails across the Mississippi: A History of the St. Louis Bridge*. Urbana: University of Illinois Press, 2001.

Jacobs, Meg. "'How About Some Meat?': The Office of Price Administration, Consumption Politics, and State Building from the Bottom Up, 1941–1946." *Journal of American History* 84, no. 3 (December 1997): 910–41.

James, Larry A., comp. *"Earth's Hidden Treasures": Mining in Newton County, Missouri*. Pt. 1. Cassville, Mo.: Shoal Creek Heritage Preservation, 2012.

———, comp. *"Earth's Hidden Treasures": Mining in Newton County, Missouri*. Pt. 2. Cassville, Mo.: Litho, 2013.

Jameson, Elizabeth. *All That Glitters: Class, Conflict, and Community in Cripple Creek*. Urbana: University of Illinois Press, 1998.

Jensen, Vernon H. *Heritage of Conflict: Labor Relations in the Nonferrous Metals Industry up to 1930*. Ithaca, N.Y.: Cornell University Press, 1950.

———. *Nonferrous Metals Industry Unionism, 1932–1954: A Story of Leadership Controversy*. Ithaca, N.Y.: Cornell University Press, 1954.

Jolly, James H. *U.S. Zinc Industry: A History, Statistics, and Glossary*. Baltimore: American Literary Press, 1997.

Kelley, Robin D. G. *Hammer and Hoe: Alabama Communists and the Great Depression*. Chapel Hill: University of North Carolina Press, 1990.

Kelly, Brian. *Race, Class, and Power in the Alabama Coal Fields, 1908–1921*. Urbana: University of Illinois Press, 2001.

Kendi, Ibram X. *Stamped from the Beginning: The Definitive History of Racist Ideas in America*. New York: Nation Books, 2016.

Kessler-Harris, Alice. "Treating the Male as 'Other': Redefining the Parameters of Labor History." *Labor History* 34, nos. 2–3 (1993): 190–204.

———. *Women Have Always Worked: A Historical Overview*. New York: Feminist Press, 1981.

Kimeldorf, Howard. *Battling for American Labor: Wobblies, Craft Workers, and the Making of the Union Movement*. Berkeley: University of California Press, 1999.

Kimmel, Michael S. *The History of Men: Essays in the History of American and British Masculinities*. Albany: State University of New York Press, 2005.

———. *Manhood in America: A Cultural History*. New York: Oxford University Press, 2006.

Klein, James E. *Grappling with Demon Rum: The Cultural Struggle over Liquor in Early Oklahoma*. Norman: University of Oklahoma Press, 2008.

Knerr, Douglas. *Eagle-Picher Industries: Strategies for Survival in the Industrial Marketplace, 1840–1980*. Columbus: Ohio State University Press, 1992.

Krause, David J. *The Making of a Mining District: Keweenaw Native Copper, 1500–1870*. Detroit: Wayne State University Press, 1992.

Laas, Virginia Jeans. "Nineteenth-Century Rugged Individualism: Lead and Zinc Mining in Joplin, Missouri, 1870–1914." Paper presented at the Southern Historical Association annual meeting, New Orleans, November 14, 1987.

Larson, John Lauritz. *The Market Revolution in America: Liberty, Ambition, and the Eclipse of the Common Good*. New York: Cambridge University Press, 2009.

Laurie, Bruce. *Artisans into Workers: Labor in Nineteenth Century America*. 1989. Reprint, Urbana: University of Illinois Press, Illinois version, 1997.

Leavy, Jane. *The Last Boy: Mickey Mantle and the End of America's Childhood*. New York: Harper Perennial, 2010.

Letwin, Daniel. *Challenge of Interracial Unionism: Alabama Coal Miners, 1878–1921*. Chapel Hill: University of North Carolina Press, 1998.

Levine, Susan. "Workers' Wives: Gender, Class and Consumerism in the 1920s United States." *Gender and History* 3, no. 1 (Spring 1991): 45–64.

Levy, Jonathan. *Freaks of Fortune: The Emerging World of Capitalism and Risk in America*. Cambridge, Mass.: Harvard University Press, 2012.

Lichtenstein, Nelson. *State of the Union: A Century of American Labor*. Princeton, N.J.: Princeton University Press, 2013.

Lingenfelter, Richard E. *The Hardrock Miners: A History of the Mining Labor Movement in the American West, 1863–1893*. Berkeley: University of California Press, 1974.

Lombardo, Timothy J. *Blue-Collar Conservatism: Frank Rizzo's Philadelphia and Populist Politics*. Philadelphia: University of Pennsylvania Press, 2018.

Long, Clarence D. *Wages and Earnings in the United States, 1860–1890*. Princeton, N.J.: Princeton University Press, 1960.

Lynn, Joshua A. *Preserving the White Man's Republic: Jacksonian Democracy, Race, and the Transformation of American Conservatism*. Charlottesville: University of Virginia Press, 2019.

MacLean, Nancy. *Behind the Mask of Chivalry: The Making of the Second Ku Klux Klan*. New York: Oxford University Press, 1995.

———. *Freedom is Not Enough: The Opening of the American Workplace*. Cambridge: Harvard University Press, 2006.

Martin, Lou. *Smokestacks in the Hills: Rural-Industrial Workers in West Virginia*. Urbana: University of Illinois Press, 2015.

Martinez, Lynda R., ed. and comp. *Happy Birthday Picher: 90 Years of Memories*. Self-published, 2008.

———, comp. *Hard Rock Legacy: Memories and History of Picher Oklahoma*. N.p.: Sunset, 2007.

Matthews, Norval M. *The Promise Land: A Story about the Ozark Mountains and the Early Settlers of Southwest Missouri*. Point Lookout, Mo.: School of the Ozarks Press, 1974.

McGinnis, Patrick E. *Oklahoma's Depression Radicals: Ira M. Finley and the Veterans of Industry of America*. New York: Peter Lang, 1991.

McKillen, Elizabeth. *Making the World Safe for Workers: Labor, the Left, and Wilsonian Internationalism*. Urbana: University of Illinois Press, 2013.

Mellinger, Philip J. *Race and Labor in Western Copper: The Fight for Equality, 1896–1918*. Tucson: University of Arizona Press, 1995.

Merritt, Keri Leigh. *Masterless Men: Poor Whites and Slavery in the Antebellum South*. New York: Cambridge University Press, 2017.

Metzl, Jonathan M. *Dying of Whiteness: How the Politics of Racial Resentment is Killing America's Heartland*. New York: Basic Books, 2019.

Meyer, Stephen. *Manhood on the Line: Working-Class Masculinities in the American Heartland*. Urbana: University of Illinois Press, 2016.

Minchin, Timothy. *What Do We Need a Union For? The TWUA in the South, 1945–1955.* Chapel Hill: University of North Carolina Press, 1997.

Miner, H. Craig. *The St. Louis–San Francisco Transcontinental Railroad: The Thirty-Fifth Parallel Project, 1853–1890.* Lawrence: University Press of Kansas, 1972.

Mink, Gwendolyn. *Old Labor and New Immigrants in American Political Development: Union, Party, and State, 1875–1920.* Ithaca, N.Y.: Cornell University Press, 1986.

Montgomery, David. *The Fall of the House of Labor: The Workplace, the State, and American Labor Activism, 1865–1925.* New York: Cambridge University Press, 1987.

———. "The Mythical Man." *International Journal of Labor and Working-Class History* 74, no. 1 (Fall 2008): 56–62.

Moody, Kim. *An Injury to All: The Decline of American Unionism.* London: Verso, 1988.

Moreton, Bethany. *To Serve God and Wal-Mart: The Making of Christian Free Enterprise.* Cambridge, Mass.: Harvard University Press, 2010.

Morgan, David. "Materiality, Social Analysis, and the Study of Religion." In *Religion and Material Culture: The Matter of Belief,* edited by David Morgan, 55–74. London: Routledge, 2005.

Mouat, Jeremy. *Roaring Days: Rossland's Mines and the History of British Columbia.* Vancouver: University of British Columbia Press, 1995.

Murphy, Lucy Eldersveld. *Gathering of Rivers: Indians, Metis, and Mining in the Western Great Lakes, 1737–1832.* Lincoln: University of Nebraska Press, 2000.

Murphy, Mary. *Mining Cultures: Men, Women, and Leisure in Butte, 1914–1941.* Urbana: University of Illinois Press, 1997.

Mutti Burke, Diane. *On Slavery's Border: Missouri's Small Slaveholding Households, 1815–1865.* Athens: University of Georgia Press, 2010.

Nelson, Bruce. *Divided We Stand: American Workers and the Struggle for Black Equality.* Princeton, N.J.: Princeton University Press, 2001.

———. "Working-Class Agency and Racial Inequality." *International Review of Social History* 41, no. 3 (December 1996): 407–20.

Nieberding, Velma. *The History of Ottawa County.* Miami, Okla.: Walsworth, 1983.

Norrell, Robert J. "Caste in Steel: Jim Crow Careers in Birmingham, Alabama." *Journal of American History* 73, no. 3 (December 1986): 669–94.

Norris, James D. *AZn: A History of the American Zinc Company.* Madison: State Historical Society of Wisconsin, 1968.

———. "The Missouri and Kansas Zinc Miners' Association, 1899–1905." *Business History Review* 40, no. 3 (Autumn 1966): 321–34.

Norwood, Stephen H. *Strike-Breaking and Intimidation: Mercenaries and Masculinity in Twentieth-Century America.* Chapel Hill: University of North Carolina Press, 2002.

Painter, Nell Irvin. *The History of White People.* New York: W. W. Norton, 2010.

Paul, Rodman Wilson. *Mining Frontiers of the Far West, 1848–1880.* Revised and expanded by Elliott West. 1963. Reprint, Albuquerque: University of New Mexico Press, 2001.

Pearson, Chad. *Reform or Repression: Organizing America's Anti-union Movement.* Philadelphia: University of Pennsylvania Press, 2015.

Perkins, Blake. *Hillbilly Hellraisers: Federal Power and Populist Defiance in the Ozarks.* Urbana: University of Illinois Press, 2017.

Perlman, Selig. *A History of Trade Unionism in the United States.* New York: Macmillan, 1922.

Phelan, Craig. *William Green: Biography of a Labor Leader.* Albany: State University of New York Press, 1989.

Philpott, William. *The Lessons of Leadville.* Denver: Colorado Historical Society, 1994.

Phipps, Stanley S. *From Bull Pen to Bargaining Table: The Tumultuous Struggle of the Coeur d'Alenes Miners for the Rights to Organize, 1887–1942*. New York: Garland, 1988.

Postel, Charles. *The Populist Vision*. New York: Oxford University Press, 2009.

Primm, James Neal. *Lion of the Valley: St. Louis, Missouri, 1764–1980*. 1981. Reprint, St. Louis: Missouri Historical Society Press, 1998.

Renner, G. K. *Joplin: From Mining Town to Urban Center*. Northridge, Mo.: Windsor, 1985.

Richards, Lawrence. *Union-Free America: Workers and Antiunion Culture*. Urbana: University of Illinois Press, 2008.

Robertson, David. *Hard as the Rock Itself: Place and Identity in the American Mining Town*. Boulder: University of Colorado Press, 2006.

Rodgers, Daniel T. *The Work Ethic in Industrial America, 1850–1920*. Chicago: University of Chicago Press, 1974.

Rohrbough, Malcolm. *Trans-Appalachian Frontier: People, Societies, and Institutions, 1775–1850*. Bloomington: Indiana University Press, 2008.

Roll, Jarod. "Faith Powers and Gambling Spirits in Late Gilded Age Metal Mining." In *The Pew and the Picket Line: Christianity and the American Working Class*, edited by Christopher D. Cantwell, Heather W. Carter, and Janine Giordano Drake, 74–95. Urbana: University of Illinois Press, 2016.

———. "Sympathy for the Devil: The Notorious Career of Missouri's Strikebreaking Metal Miners, 1896–1910." *Labor* 11, no. 4 (Winter 2014): 11–37.

Rosenow, Michael K. *Death and Dying in the Working Class, 1865–1920*. Urbana: University of Illinois Press, 2015.

Rosner, David, and Gerald Markowitz. *Deadly Dust: Silicosis and the Politics of Occupational Disease in Twentieth-Century America*. Princeton, N.J.: Princeton University Press, 1991.

———, eds. *Dying for Work: Workers' Safety and Health in Twentieth-Century America*. Bloomington: Indiana University Press, 1987.

Ross, William G. *A Muted Fury: Populists, Progressives, and Labor Unions Confront the Courts, 1890–1937*. Princeton, N.J.: Princeton University Press, 1994.

Rosswurm, Steve, ed. *The CIO's Left-Led Unions*. New Brunswick, N.J.: Rutgers University Press, 1994.

Rowe, John. *The Hard-Rock Men: Cornish Immigrants and the North American Mining Frontier*. New York: Barnes and Noble Books, 1974.

Rutherford, Danilyn. "The Enchantments of Secular Belief." Paper presented to the Martin Marty Seminar for the Advanced Study of Religion, University of Chicago Divinity School, October 2008.

Salvatore, Nick. *Eugene V. Debs: Citizen and Socialist*. Urbana: University of Illinois Press, 1982.

Sanders, Elizabeth. *Roots of Reform: Farmers, Workers, and the American State, 1877–1917*. Chicago: University of Chicago Press, 1999.

Saxton, Alexander. *The Indispensable Enemy: Labor and the Anti-Chinese Movement in California*. Berkeley: University of California Press, 1971.

Scales, James R., and Danney Goble. *Oklahoma Politics: A History*. Norman: University of Oklahoma Press, 1982.

Scranton, Philip. *Figured Tapestry: Production, Markets and Power in Philadelphia Textiles, 1855–1941*. New York: Cambridge University Press, 1989.

Self, Robert O. *All in the Family: The Realignment of American Democracy since the 1960s*. New York: Hill and Wang, 2012.

Sellars, Nigel Anthony. *Oil, Wheat, and Wobblies: The Industrial Workers of the World in Oklahoma, 1905–1930*. Norman: University of Oklahoma Press, 1998.

Shore, Elliott. *Talkin' Socialism: J. A. Wayland and the Radical Press*. Lawrence: University of Kansas Press, 1988.

Simon, Bryant. "The Appeal of Cole Blease of South Carolina: Race, Class, and Sex in the New South." *Journal of Southern History* 62, no. 1 (February 1996): 57–86.

———. *Fabric of Defeat: The Politics of South Carolina Millhands, 1910–1948*. Chapel Hill: University of North Carolina Press, 1998.

———. "Rethinking Why There Are So Few Unions in the South." *Georgia Historical Quarterly* 81, no. 2 (Summer 1997): 465–84.

Steele, Phillip W., and Steve Cottrell. *Civil War in the Ozarks*. Gretna, La.: Pelican, 2000.

Stein, Judith. *Running Steel, Running America: Race, Economic Policy, and the Decline of Liberalism*. Chapel Hill: University of North Carolina Press, 1998.

Storch, Randi. *Working Hard for the American Dream: Workers and Their Unions, World War I to the Present*. Chichester, UK: Wiley-Blackwell, 2013.

Stromquist, Shelton, ed. *Labor's Cold War: Local Politics in a Global Context*. Urbana: University of Illinois Press, 2008.

Suggs, George, Jr. *Union Busting in the Tri-State: The Oklahoma, Kansas, and Missouri Metal Workers' Strike of 1935*. Norman: University of Oklahoma Press, 1986.

Taft, Philip. *The A. F. of L. in the Time of Gompers*. New York: Octagon Books, 1970.

Taillon, Paul Michael. *Good, Reliable, White Men: Railroad Brotherhoods, 1877–1917*. Urbana: University of Illinois Press, 2009.

Todd, Arthur Cecil. *The Cornish Miner in America: The Contribution to the Mining History of the United States by Emigrant Cornish Miners—the Men Called Cousin Jacks*. Glendale, Calif.: Arthur H. Clark, 1967.

Tomlins, Christopher L. *The State and the Unions: Labor Relations, Law, and the Organized Labor Movement in America, 1880–1960*. New York: Cambridge University Press, 1985.

Twitty, Anne. *Before* Dred Scott*: Slavery and Legal Culture in the American Confluence, 1787–1857*. New York: Cambridge University Press, 2016.

Violette, Eugene Morrow. *A History of Missouri*. Boston: D. C. Heath, 1918.

Weir, Robert E. *Beyond Labor's Veil: The Culture of the Knights of Labor*. University Park: Pennsylvania State University Press, 1996.

Wilentz, Sean. *Chants Democratic: New York City and the Rise of the American Working Class, 1788–1850*. New York: Oxford University Press, 1984.

Wilkerson, Jessica. *To Live Here, You Have to Fight: How Women Led Appalachian Movements for Social Justice*. Urbana: University of Illinois Press, 2019.

Windham, Lane. *Knocking on Labor's Door: Union Organizing in the 1970s and the Roots of a New Economic Divide*. Chapel Hill: University of North Carolina Press, 2017.

Wyman, Mark. *Hard Rock Epic: Western Miners and the Industrial Revolution, 1860–1910*. Berkeley: University of California Press, 1979.

———. *The Wisconsin Frontier*. Bloomington: Indiana University Press, 1998.

Zieger, Robert H. *The CIO: 1935–1955*. Chapel Hill: University of North Carolina Press, 1995.

Zieger, Robert H., Timothy J. Minchin, and Gilbert J. Gall. *American Workers, American Unions: The 20th and Early 21st Centuries*. Baltimore: Johns Hopkins University Press, 2014.

INDEX

accidental deaths: Tri-State district rate of, 17, 85, 89, 134–35, 149, 174; of shovelers, 134, 146; and risk-and-reward ethic, 146; metal miners' power to sue employers for, 148–49; and health and safety regulations, 159

Admiralty Zinc Company, 172

Aetna Liability Insurance Company, 135

AFL Cement (United Cement, Lime, and Gypsum Workers International Union), 247–49, 250

African American miners: in Granby, 49–50, 63, 82; in Jasper and Newton Counties, 66, 80; as prospectors, 82; in Galena, 106; opportunities closing for, 106; Tri-State miners' exclusion of, 127, 169, 175, 176, 193, 231, 254; tuberculosis rates of, 157; hate strike of 1942, 255

African Americans: Tri-State miners attacking, 10; discrimination against, 12; Tri-State miners avoiding, 16; leaving Jasper and Newton Counties, 50; debates on rights of, 62; as strikebreakers, 94, 102, 110; epithets used against, 121; lynchings of, 126–27; violence against, 175. *See also* nonwhite people

Ageton, Richard, 181–84, 188–89, 194–95

American Federation of Labor (AFL): attempts to organize Tri-State miners, 1, 10, 11, 16, 99, 117, 144–45, 200–201, 251, 252, 254; and working-class conservatism, 3, 254; and skilled workers, 5, 113; capitalism accommodated by, 8, 16; WFM's challenge to, 9, 105, 122; as ally of white, native-born workingmen, 11; Mine Mill affiliated with, 13, 185, 201–2, 219, 223–24, 225, 234; WFM affiliated with, 99, 104, 114, 115, 116, 124, 152–54, 163, 164, 232; support for WFM strikes, 104, 124, 125; formation of local AFL unions, 112–

14, 116, 144–45; in Joplin, 112–18, 120, 122–23, 144–45, 146, 165; strike strategy of, 113, 231; and shovelers, 130; collective bargaining doctrines of, 161; employers' antiunion open-shop strategy, 185; and Farmer-Labor Reconstruction League, 185; Oklahoma State Federation of Labor affiliated with, 185; and nativism, 186; and cooperative relations with employers, 191; weakening of, 191; and CIO, 202, 220, 227, 229, 231–32, 233, 235, 247, 250, 251; Blue Card Union affiliated with, 202, 230–35, 251, 254; and NRA, 208; and mass production industries, 220; and NLRB, 234–35; challenges to Mine Mill, 247–49; and Americanism, 251

American Federation of Mineral Miners, 115

Americanism: and labor unions, 16, 186, 187, 247, 248, 250–51, 254; and World War I, 166; and white nationalist privilege, 175, 224, 253; of white working-class men, 253–54

American Labor Union, 122

American Metal Miners Union (AMMU), 161–63, 165, 166, 169

American Mining Congress, 139

American nationalism: conservative nationalism, 5; restrictive forms of, 5; and white masculinity, 10; racial and nativist advantages to Tri-State miners, 11–12, 13, 252–53; and federal economic policies, 12; and militarism, 12; and Theodore Roosevelt, 128; politics of, 154; and *Lusitania* sinking, 161; aggressiveness of, 165; of native-born white men, 165; and white working-class American men, 166; and labor unions, 251. *See also* white nationalism

American Plan, 185

American Smelting and Refining Company, 129, 172

American standard of living, 167, 187, 199, 245

American Zinc, Lead, and Smelting Company: mining operations on land holdings of, 107, 129, 168, 169; and piece rate for shovelers, 135–36, 141; exploration of steam shovels, 142; and wage cuts, 145–48, 149, 150, 169–70; and injury compensation law, 149, 153; and WFM, 151, 170; and foreign-born miners, 152; and concentration of mine ownership, 154, 170, 178; and health and safety regulations, 159–60; and AMMU strike, 162; liquidation of holdings, 171; and living conditions of Tri-State miners, 173; and American Zinc Institute, 179; Tri-State miners' opposition to, 251

American Zinc Institute (AZI), 179–82, 184, 194, 195, 212

Anna Beaver Mining Company, 174

antimonopoly reform, 8, 60–61, 70, 73, 85, 86, 88, 98, 252

anti-Semitism, 223, 254

Appeal to Reason, 15, 107, 145, 281n34

Arizona, 129, 164, 171, 245

Arkansas, 7, 42

Arkansas-Oklahoma Industrial Union Council, 235

Armer, T. L., 214–16

Askew, Robert, 114–15, 117, 122

Atlantic & Pacific Railroad, 45, 52, 53

Aurora, Missouri, 15, 80, 82, 117, 123

Austin, Moses, 31

authoritarianism, 201, 231

Avrutis, William, 234

back-to-work movements, 215–16, 247, 249, 253

banking reform, 71, 207

Barr, Lawrence, 192, 240, 243, 255

Barr, Theo Sisco, 240

Bartlesville, Oklahoma, 172

Battle of Newtonia, 44

Battle of Pea Ridge, Arkansas, 43

Battle of Wilson's Creek, 42

Baxter Springs, Kansas: mining memorial at, 18; investors from, 64; labor unions in, 188, 212, 217, 218, 220; and back-to-work movement, 215; munitions plant in, 243

Beck Mining Company, 218

Beggs, John J., 186–89

belligerence: political economy of, 12, 165, 253; of white nationalism, 12, 13; logic of, 171; and racial and nativist privilege, 253

Bendelari, A. E., 168, 179

Bisbee, Arizona, 142, 171

Black Hawk War (1832), 22

Black Hills, South Dakota, 148–49, 151, 156

Blow, Henry T., 37, 43–45, 49, 52, 59, 72

Blow, Peter E., 37, 43–45

Blow & Kennett, 37–45, 49, 54, 148

Blue, Rupert, 158

Blue Card Record, 226, 227, 228–29, 231, 232, 234

Blue Card Union: as company-controlled movement, 201, 216, 217–18, 222, 225, 227, 233–35, 248; affiliation with AFL, 202, 230–35, 251, 254; and Mine Mill strike, 216–17, 218, 223–24, 227, 254; and pick handles, 217, 219, 221, 223, 224, 225, 227–29, 238, 248; consolidation of authority, 219; destructive nature of, 222; picnic at Miami fairgrounds, 224–25, 232–33; status of, 226; and CIO, 227, 228–29, 231, 235, 254; disbanding of, 236, 237, 239; support for Franklin D. Roosevelt, 254

Blue Eagle Day, 209–10

Blue Eagle symbol, 208

Boyce, Ed, 104–5, 119–22, 123

Brady, Roy, 209–12

brass, 19, 74, 106, 128

Brock, Robert, 33, 36

Brophy, John, 223

Brown, Thomas, 209, 212–15, 218–19

Bruner, Sylvan, 17, 148, 234

Bryan, William Jennings, 7–8, 96, 104, 107, 147

Buffalo, New York, 117

Bunker Hill and Sullivan Mining and Concentrating Company, 108, 109, 111

Burkett, J. A., 114–15

Burns, William M., 123–24, 147

Butte, Montana, 60, 73, 100, 171, 185, 255

Butte Miners' Union, 104, 160

calamine (zinc silicate), 46, 59, 60

California, 19, 21, 24, 53

Campion, John, 99–101, 106
Cantwell v. Missouri (1905), 135
capitalism: American workers' challenges
to, 2; ideology of acquisitive individualism, 3; Tri-State miners' interests served
by, 4, 5, 6, 10, 11, 14, 16, 70, 85, 87, 95, 99,
144, 166, 252–53, 255; white men's interests identified with market functions of,
4, 20, 251; ideology of liberal capitalism,
5; support for policies favoring, 5; white
working-class faith in, 6, 13, 14; William
Jennings Bryan's popular challenge to,
7; and owner-operator miners, 7; wage
laborers' critical stance against, 8; labor
unions challenging, 8, 250; managerial
capitalism, 12; and metal mining, 20;
and mechanization, 21; and regimented
hierarchies, 21; and industrial consolidation, 21, 22, 47, 70, 73, 80, 89, 98, 99, 105;
and risk-and-reward ethic, 89, 107, 130;
Socialist criticism of, 118; and Theodore
Roosevelt, 127; anticapitalist politics, 147,
166, 251, 253; in Great Depression, 200,
253; and Blue Card Union, 223; labor
unions supporting, 251
Cardin, Oklahoma: as boom camp, 165;
population of, 166; living conditions in,
172–73; labor unions in, 186, 188, 215,
217, 245; U.S. Route 66 in, 192; Eagle-
Picher's Central Mill in, 206, 245, 246,
248, 249, 250
Carondelet, Missouri, smelters in, 36
Carter, T. Lane, 129, 140
Carterville, Missouri: as mining camp, 47,
63, 64, 69, 74, 129; population of, 65, 80,
106, 157, 166; zinc production in, 74–77,
79, 82–83, 129, 166; wage laborers of,
83–84, 117; labor unions in, 86, 87, 88,
151, 154; and leasehold system, 90; mine
closings in, 92; miners meetings in, 94;
and miner's consumption, 133, 158; and
election of 1908, 147; and AMMU, 162,
163, 164; and election of 1932, 207
Carthage, Missouri, 42, 51–52
Cartwright, M. E., 212, 215, 217
Cassell, Ed, 212, 233, 235
Cates, Clotis, 249
Center Creek, Jasper County, Missouri, 24,
26–27, 29–33, 42, 44, 56

Center Creek Mining and Smelting Company, 63, 74, 76, 77–79, 90, 91
Central City, Missouri, 107, 115
Central Labor Union (CLU), 114, 116, 118,
122–24, 126, 135, 144, 153
charitable poor relief, 12, 202–3, 206, 212
Chatham Mining Company, 91
Cherokee County, Kansas: metal mining in,
47; zinc smelter in, 58; African American
miners of, 66; coal mining in, 92; miner's
consumption rates in, 158; labor unions
in, 218
Cherry, Walter, 247–49
Chicago, Illinois, 28, 60
Chicago Zinc and Mining Company, 58
children: as scrappers, 56; odd jobs of, 83
Chinese workers, 20
Chitwood Hollow mining field, Missouri,
107, 123, 125, 162–63, 164
Chouteau, Oklahoma, 243
Cigar Makers' Union, 99
CIO News, 235
Civil Rights Act of 1875, 62
Civil War, 42–45, 50
Civil Works Administration, 208
Clark, Riley, 203
Clerc, F. L., 67–68, 74–75, 79, 80, 84–85
coal mining: and labor unions, 2, 3, 60,
73, 85, 87, 185; shallow coal mining, 7,
58; and inside contractor systems, 47;
and miner's freedom, 54; and depression of 1870s, 60; strikes in Missouri, 64,
85, 92–94, 98; and mine inspections, 85,
134; investors forming land companies,
89–90; strikes in Kansas, 92, 98; zinc
production tied to, 98, 108; and health
and safety regulations, 134; piece rates
in, 136; and foreign-born miners, 140;
mechanization of, 166; in Oklahoma, 174,
176, 185; strikes in Oklahoma, 176; and
open-shop campaigns, 191
Coeur d'Alene, Idaho: Joplin miners as
strikebreakers in, 9, 108–12, 118–22, 124,
125, 151; WFM in, 108, 114, 118–19, 120,
142; permit system in, 110–11, 118; nonunion Joplin miners hired in, 141–42
Cold War, 247, 249
collective bargaining: and New Deal era,
1, 10, 12–13, 201, 255; and solidarity

through collective action, 2, 7, 8, 15–16, 87, 104, 154–56, 161, 169; and Mine Mill, 202, 222, 244

Collier White Lead and Oil Company, 37

Colorado: silver-bearing lead mines of, 70, 72; Knights of Labor in, 73; coal mining in, 156, 160; and Tri-State miners, 252. *See also* Cripple Creek, Colorado; Leadville, Colorado

Colorado Fuel and Iron Company, 156

Colorado National Guard, 103, 124, 156

Commerce, Oklahoma: labor unions in, 156, 160, 164, 172, 186, 215, 217; as boom camp, 165; population of, 166

Commerce Mining and Royalty Company: Miami Royalty renamed as, 168; size of, 172, 199; and health and safety regulations, 174; and American Zinc Institute, 179, 183; and mergers, 193; and OPA, 194; and President's Re-Employment Agreement, 208; and Blue Eagle Day, 209; and Mine Mill's strike, 214; consolidation of power, 239

Committee for Industrial Organization, 201

Communism, 61, 235, 236, 246–48, 249, 250, 254

Communist Party, 204–5, 247

company towns, 47

Comstock Lode, 19, 21

Confederacy, 42–43, 44

Congress of Industrial Organizations (CIO): class-based solidarity practiced by, 1; as industrial union, 2, 201, 220, 249–50; truncated power of, 5; as social democratic movement, 5, 202, 253, 254; Mine Mill affiliated with, 13, 202, 220, 223, 225, 227, 228, 233, 235, 238, 250; radicalism of, 16, 227, 238; and power of white American masculinity, 16–17; and AFL, 202, 220, 227, 229, 231–32, 233, 235, 247, 250, 251; setbacks of, 233; and Taft-Hartley amendment, 247, 250; merger with AFL, 251

Connell, Lee, 173

Conover, J. D., 194

Cope, Marion, 155–56, 159–61, 163, 170, 172

Copley, D. C., 123–25

copper production, 19, 20, 128, 129, 142, 145

Corn and Wahl, 53, 55, 62

Cornish miners, 21–22, 31, 34, 49, 72, 100, 104

corporations: consolidation of power, 7, 8, 128, 252; and managerial capitalism, 12; workers battling with, 64, 71; and industrial capitalism, 73; in Leadville, Colorado mines, 73–74; political economy dominated by, 253

Cox, John, 24, 26–27, 42, 51, 53

craft unions, 2, 3, 220

Craig, Genevieve Stovall, 255–56

Cravens, Gobel, 236

Crawfish, Harry, 168

Cress, Solon, 116–17, 118

Cripple Creek, Colorado, 9, 121, 122, 123–26

Cullifer, Adam, 163

Daily Missouri Republican, 33, 36, 45

Daugherty, William, 56, 63, 77

Davey, Thomas, 63, 77, 84

Davey mine, Carterville, 63, 77, 147, 149

Davidson, L. S., 236

Deadwood, South Dakota, 9, 148

Debs, Eugene, 15, 145, 147, 155

Delaware people, 20, 24

Democratic Party: on capitalist prosperity of white workers, 7; in Tri-State district, 15; and election of 1860, 42; on bombings of mining companies, 62; and election of 1874, 62; racist campaign against Republican Party, 62; and election of 1880, 72; and election of 1896, 96; and election of 1900, 118; and eight-hour workday law, 122; and election of 1916, 185; and election of 1918, 185; and election of 1920, 185; and Farmer-Labor Reconstruction League, 185; Tri-State miners' support for, 185; and election of 1932, 206–7; and election of 1936, 226–27

Depression of 1870s, 7, 46–47, 60, 62, 63–64, 70, 71, 95

Depression of 1890s, 7, 9, 15, 92–96, 98, 99, 106

Dick, Lee, 237

Dick, Sheldon, 236–37

Dickey, Bill, 256

Disney, Wesley, 209

Doak, Clifford, 229

Dodson, S. G., 112, 122–23

Douthat, Oklahoma, 165, 172, 188
Drummond furnaces, for lead smelting, 28, 29
dry bone (zinc carbonate), 46
Duenweg, Missouri, 96, 115, 129, 155, 164
Dust Bowl, 222
Dutch Boy paint, 220
Dwyer, Dan, 97

Eads, James B., 45, 49
Eagle-Picher Lead Company: formation of, 168; size of, 172, 199; and living conditions of miners, 173; and health and safety regulations, 174, 179, 194, 239, 253; and American Zinc Institute, 179, 183; and studies of lung diseases, 181; as self-insured for injuries, 183; profits of, 184, 243; and mergers, 193, 206; and OPA, 194, 239; reaction to market collapse of Great Depression, 202–3; and rustling card proving work eligibility, 204; Central Mill of, 206, 245, 246, 248, 249, 250; and President's Re-Employment Agreement, 208; and Blue Eagle Day, 209; and Mine Mill strikes, 214, 222, 223, 225, 234, 239, 244, 246–47; and Blue Card Union, 217, 234–35; and AFL, 219; and NLRB ruling, 234, 235, 236, 239, 244; consolidation of power, 239, 253; and World War II, 243–44; smelters of, 244–45; and Mine Mill organizing, 244–46; closing of, 249, 250; Tri-State miners' opposition to, 251
Eagle White Lead Company, 168
Eastern Missouri district: history of, 21, 49; lead production in, 28, 29, 31, 36, 37, 70; leases granted by New Spain in, 31; miners leaving, 31, 32; mining camps of, 34; and mining companies, 40–41, 91; zinc production in, 57; and National Lead Company, 129; foreign-born miners in, 142; labor unions in, 145, 152; and WFM, 152, 160, 161; foreign-born miners ejected from, 171
Eisenhower, Dwight, 249
Eleventh Hour Mining Company, 91
Empire Zinc Company, 90
Engineering and Mining Journal, 107, 118, 149, 150, 188, 192–93
enslaved labor: in mining camps, 20; in

lead production, 21, 26–27, 28, 30, 43; in Jasper and Newton Counties, 24; skilled labor compared to, 41
Environmental Protection Agency, 258
Espionage Act, 171
ethnic divisions: in industrial unions, 2, 3; and privileges of white working-class American men, 4; mining companies' exploitation of, 100, 101, 111; and Tri-State miners, 125, 164. *See also* nativism
eugenics, 102, 152
Evans, Frederick W. "Mike," 172, 216–17, 218, 223, 225–26, 228–32, 234, 239, 248

Fall, Albert, 177
Farley, John F., 100–101, 105–6
Farmer-Labor Reconstruction League, 185
Farm Security Administration, 236, 242
fascism, 223
Fear, Charles, 146, 151, 153, 161
Federal Emergency Relief Administration (FERA), 208, 212, 231
federal government: labor unions promoted by, 1, 3, 201; economic policies of, 12; labor unions suppressed by, 12; and Native American lands, 12, 20, 22, 24, 253; support of railroad construction, 12, 38, 253; regulations of, 15, 18; land grants for internal improvements, 30, 38; selling of public mineral lands, 32; and military spending, 128, 249, 253; and consolidation of mining industry, 178
Federal Mining and Smelting Company, 172, 208, 239, 244, 248
Ferns, Jim, 236
Ferns, Rube, 156, 236
Fever River (now Galena River), 21
Finley, Ira, 185–89, 191, 215–16
Flinn, Frederick, 184, 194
Ford, F. N., 123
Fordney-McCumber Tariff, 179
foreign-born miners: and WFM, 9, 10, 105, 111, 152, 161, 170, 171; and Tri-State miners as strikebreakers, 10, 105, 111, 126, 144, 154, 156; Tri-State miners ejecting from district, 10, 144, 160, 163, 171; Tri-State miners' exclusion of, 11, 161, 163, 169, 175, 176, 193, 231, 254; from Great Britain, 34, 49, 65–66; as metal miners,

63, 142; Tri-State miners' opposition to, 111, 130, 144–45, 154, 160; radical unionism of, 130, 144; as coal miners, 140; shovelers' opposition to, 152, 160; AMMU on, 161, 162, 163

foreigners: suspicion of, 5, 165, 223; epithets used against, 121

Fort Smith, Arkansas, 29, 43, 172

Foster, William, 31, 33

Fowler, David, 233, 235, 242

Fox, Jack, 136–37, 155

Fox people, lead mines of, 21

Frazer, William, 40–41, 48–49

Fremont, John C., 45

French traders, lead mining of, 21, 31

Galena, Kansas: as mining camp, 47, 64–65, 69; population of, 65, 80, 106; and poor man's camp ethic, 69; zinc production in, 75, 76, 91, 95; lead production in, 76, 168; growth of, 79; labor unions in, 86, 117, 188, 209, 210, 211, 212, 217, 218, 219, 220, 229, 233; and wage labor, 117; and AMMU, 164; and railroads, 172; Tri-State miners commuting from, 173; and Mine Mill's strike, 214, 218

Galena & Chicago Union Railroad, 37

galena (lead sulfide) deposits: in Missouri, 20, 22, 24, 26, 30, 32–33, 36, 38–40; Native Americans' mining of, 21; preindustrial methods of extraction, 21, 26, 29, 34–35; prices paid for, 29, 36, 39, 40, 58; zinc found with, 46, 58, 60, 74; purity of, 48, 59; discoveries of, 51–52, 53; leasing terms for, 64

Galena Mining and Smelting Company, 64

Gazzam, Joseph, 100, 287n26

gender divisions: in industrial unions, 2, 3; and privileges of white working-class American men, 4, 5; and individualism, 13; and Tri-State miners, 13, 16, 111, 200, 224; and shovelers, 143

gender roles, 34, 187

General Agreement on Trade and Tariffs, 249

Girard, Kansas, 107, 145, 281n34

Golden Rod Mining and Smelting Corporation, 172, 193

Gold Hill, Nevada, 60

gold mining, 19, 20–21, 24, 53, 106

Gompers, Samuel: and organization of Tri-State miners, 99, 122, 163, 251; and Ed Boyce, 105; on immigration restriction, 113, 186; and organization of metal miners, 114–15, 117, 122; and WFM, 152–53, 163, 232; death of, 191

Granby, Missouri: stampede in, 33, 252, 256; lead production in, 33–35, 36, 37, 38–39, 42, 44–45, 49, 57, 64; centrality of family groups in, 34–35, 39, 49; and market for lead, 36, 37–38, 45, 47, 52; economic and social prosperity in, 39, 41–42, 50; census survey of, 41; and Civil War, 43–45; purity of galena at, 48; in post–Civil War period, 48–49; and Joplin Creek discovery shaft, 51; and Short Creek rush, 64; population of, 65; and railroads, 65; and election of 1880, 72; zinc production in, 75; labor unions in, 86; and wage labor, 117; Klondike mine of, 248

Granby Mining and Smelting Company: investors in, 45; and lease with Southwest Branch Railroad, 45; contract terms for skilled miners, 45, 47, 48–50, 52, 53, 54, 55, 57, 59, 62, 63, 71; and railroads, 45, 48, 71; in post–Civil War period, 46, 48; and market for lead, 47–48; connections in lead industry, 48; land purchased near Minersville, 48; smelters of, 48, 52, 74; companies competing with, 52, 53, 55, 65; and market for zinc, 58, 59; and zinc production, 60, 62, 74; and development of district, 65; holdings in Oronogo, 75; holdings in Jasper and Newton Counties, 90, 92; shallow holdings of, 91–92; and small mining companies, 91–92; American Zinc buying holdings of, 170, 171

The Grange, 8, 15, 61

Graves, S. E., 161

Great Depression, 1, 200, 201, 202–3, 253

Green, William, 191, 223, 230–33, 234, 251

Greenback-Labor Party, 8, 15, 70, 71–72, 86, 88, 96, 252

Greenback Party, 64

Hackett, Bess, 156–57, 158

Hain, Elwood, 244, 248

Hamilton, Alice, 157, 239
Harbaugh, M. D., 205, 206, 211–14, 224
Harklerode, William, 29, 36
Harrington, Daniel, 181–82
Haywood, Big Bill, 142, 145
health and safety regulations: Tri-State miners' opposition to, 1, 11, 12, 16, 88–89, 135, 158, 174, 175, 181–82, 238; and silicosis, 1, 158–59, 169, 182, 183, 202, 211, 236–37; class interests in demands for, 2; and Knights of Labor, 9, 85, 88–89; and lung ailments, 12, 133, 149–50, 154–58, 174, 182–84, 189–90; imposition of, 12, 134, 182; and insurance companies, 12, 135, 153, 183, 194, 202; defeat of, 16; Tri-State miners' support for, 16, 149, 150, 159, 161, 183–84; and mine inspections, 85, 88–89, 134, 148, 154–55, 156, 159, 189; and safety cages for hoists, 134, 174; and WFM, 134, 153, 154–55, 159–60, 169; and air quality, 148, 149–50, 157–58; and lead poisoning, 157; and reform of working conditions, 159, 169, 181–83; in Oklahoma, 174, 189, 194
Heckscher, August, 90
Heckscher, Richard, 90
Hell's Neck mining field, Missouri, 107, 113
Henryetta, Oklahoma, 168, 172, 245, 250
Herculaneum, Missouri, smelters in, 30, 36, 45
Hersey, B. K., 40
Hickman, Glenn, 216–17, 223, 231–32, 239
Higgins, Edwin, 158–59, 161, 174, 180, 181
Hilkene, W. H., 67, 85
Hills, Frank, 176, 193
Hockerville, Oklahoma, 165, 186, 193
Holibaugh, John, 90–92, 95
Homestake Mining Company, 148, 151
Hood, Harry, 172
Hoover, Herbert, 206, 207, 254
Hopkins, Hersey, and Company, 40
Hopkins, Joseph, 40, 48–49
Hopkins, Lyon, 110
Hunt, J. G. W., 86

Idaho, 245, 252. See also Coeur d'Alene, Idaho
Illinois, 21, 64, 73, 94, 168
Illinois Zinc Company, 58

immigrants: competition with Tri-State miners, 5; as nonwhite people, 5; Tri-State miners fighting in strike zones, 10; and xenophobic national culture, 10; and individualism, 13; and power of white American masculinity, 16–17; mining companies' hiring of, 20–21; population of, 106; as shovelers, 136; tuberculosis rates of, 157; language of anti-immigrant thought, 157. See also foreign-born miners
immigration restriction: and working-class conservatism, 5; and protection from foreign competition, 7, 12; and Republican Party, 15; and AFL, 113; and Tri-State miners, 166, 175, 253, 254; and Mine Mill, 186; and Charles Moyer, 187
Indiana, Knights of Labor in, 73
Indian Territory, 20, 24, 26, 29, 30, 107, 167
individualism, 3, 4, 5, 13
Industrial Workers of the World (IWW): labor historians on, 2; on capitalism, 9; radicalism of, 10, 122, 165, 170; federal government's suppression of, 12; founding of, 122; in Joplin, 145; and WFM, 147, 153, 160, 170; raids on, 171; and Mine Mill, 171, 185; in Oklahoma, 176; Metal Mine Workers, 185
injuries: Tri-State miners' suing over, 11, 147, 148–49, 153; Tri-State district rate of, 12, 17, 182, 184, 206, 222, 224; binding system of compensation for, 183; prevention methods, 194, 195
inside contractor systems, 47
insurance companies, 12, 135, 153, 183, 194, 202
Interchurch World Movement, 175
International Labor Defense, 236
International Union of Mine, Mill and Smelter Workers. See Mine Mill
Iowa, lead mining in, 21, 30, 34
Irish miners, 49, 72, 100, 104
iron production, 19, 20, 128
Italian miners, 105, 152

Jackson, Andrew, 4, 6, 54
Jackson, Claiborne, 42, 43
Jacksonian market revolution, 6, 13, 19–20, 21, 22

James River, Missouri, 22

Jasper County, Missouri: agricultural profile of, 14; mining output of, 14, 20, 22, 24, 27, 30, 33, 37, 42, 44, 47, 55, 60, 63, 171; white settlers of, 20, 24, 27; enslaved labor in, 24; geological environment in, 26; as public land, 27; G. C. Swallow's geological survey of, 30; and election of 1860, 42; Civil War in, 42, 44, 50; in post–Civil War period, 50; and leasehold system, 53, 90; aspiring miners arriving in, 56; and election of 1874, 62; and election of 1876, 64; African American miners of, 66, 80; and election of 1878, 71; and Greenback-Labor Party, 71; and election of 1880, 72; racial population breakdown of, 80; prospectors in, 82, 91; accidental deaths of miners in, 89, 134; and election of 1894, 96; and election of 1896, 107; and election of 1900, 118; and election of 1908, 147; miner's consumption rates in, 157, 158; and AMMU, 162; U.S. Public Health Service in, 181; and election of 1928, 254

Jasper County Anti-Tuberculosis Society, 156–58

Jasper Lead and Mining Company, 61, 62

Jefferson Barracks, 30, 36

Jenkins, Richard, 179, 180–81

Jinkerson, William, 147, 150–52

J. J. Luck Mining Company, 91

Jones, Mother, 163

Joplin, Missouri: and railroads, 14, 65, 172; as modern city, 14, 79–80, 157; Mine Mill's convention in, 17; mining memorial at, 18; streets named after successful miners, 18; as mining camp, 47, 57, 64, 65, 69, 71; reputation as poor man's camp, 55–56, 69; population of, 56, 65, 80, 106, 166, 173; and zinc market, 58; and Short Creek rush, 64; and election of 1880, 72; zinc smelter in, 74; zinc production in, 76, 106, 116, 167, 171; growth of, 79; as mining center, 79–80; lead production in, 80, 116; labor unions in, 86–87, 99, 101, 110, 112–18, 120, 122–25, 127, 141, 144–45, 146, 150–56, 169–70, 212, 214, 219, 220, 248; mine closings in, 92; and wage labor system, 98, 117; nonunion labor in, 108–11, 141–42; and eight-hour

workday law, 123; Labor Day parade in, 126; lynchings in, 126–27; tuberculosis mortality rate in, 155, 156; middle class of, 157; and AMMU, 162, 163, 164; Tri-State miners commuting from, 173; Ku Klux Klan in, 185; radio station in, 192; U.S. Route 66 in, 192, 255; U.S. Department of Labor conference on silicosis, 237–39, 241–42; Eagle-Picher's insulation works in, 244, 248; trucking companies of, 255; founding of, 294

Joplin Creek, Jasper County, Missouri, 24, 51–52, 53, 56, 57, 62

Joplin Daily News, 108

Joplin district. *See* Tri-State district

Joplin Globe/Joplin Daily Globe, 102, 109, 127, 138–40, 148, 168, 184, 224

Joplin Lead Mines, 53–55

Joplin Miners' Union, 60–61, 66–67, 113, 114

Joplin Mining Exchange, 89

Joplin Mining News, 102

Joplin Morning Herald, 93, 94, 102

Joplin News Herald, 131, 138, 156

Joplin Zinc Company, 74

Just, Evan, 238, 240

Kansas: as part of Tri-State district, 1; agriculture in, 42; border conflict with Missouri, 42; Short Creek mineral field in, 64; Knights of Labor in, 73; coal miners' strike in, 92, 175; smelters in, 172; workmen's compensation laws in, 204. *See also specific counties and towns*

Kansas City, Fort Scott & Gulf Railroad, 64

Kansas City, Missouri, 30, 64–65

Kansas City Mining Exchange, 89

Kansas National Guard, 218

Kansas State Federation of Labor, 220, 242

Kansas Territory, 20

Kennamer, Franklin E., 223–24, 234

Kennett, Ferdinand, 30–31, 37, 43

Kennett, Luther, 30–31, 37

Kimball, Harry S., 141–42, 143, 146, 150–51, 160, 164

King Jack, 175, 176, 192

Knights of Labor: and solidarity through collective action, 2, 8, 87, 104; dissolution of, 8; and antimonopoly reform and inflationary monetary policies, 8, 70, 73, 85, 252; strikes of, 8, 73, 87–88, 113; lack of success in Tri-State district, 8, 88, 94, 98, 101; mine safety laws advocated by, 9, 85, 88; and coal miners, 60, 73; and Leadville, Colorado, camp miners, 72–74, 86, 98, 100, 113; organizing efforts in Tri-State district, 85–88, 89, 94; Labor Day observances of, 86; cooperative common-wealth vision of, 86–87; and republican ideal of true manhood, 87; assistance fund supporting strikes, 87–88

Korean War, 249

Ku Klux Klan, 185

Labor Day, 86, 126, 189

labor movement, historians' narratives of, 2–4, 5, 14

labor unions: federal legislation supporting, 1, 3, 201; white working-class American men's opposition to, 1, 4, 5, 7, 10; radical unionism, 1, 10, 130, 202, 223, 224, 227, 252; craft unions, 2, 3, 220; solidarity through collective action, 2, 7, 8, 15–16, 87, 104, 154–56, 161, 169, 187, 188; mem-bership in, 3, 4, 61; and metal miners, 3, 93, 98, 99, 108–9, 114–15, 117, 122, 123, 130; Tri-State miners' opposition to, 4, 10, 12, 14, 18, 93, 101, 144–45, 166, 176, 185; challenges to capitalism, 8, 250; foreign-born union miners, 9; demands for safety and security, 11; federal legislation sup-pressing, 12; and labor organizers, 13–16; and Americanism, 16, 186, 187, 247, 248, 250–51, 254; conservative elements within, 16; and women, 17, 172, 188–89, 215; and precious metal camps, 60; asso-ciation with bombing attacks, 61–62; mixed assemblies, 86; antiunion ani-mosity, 93, 176, 185; and Woodrow Wil-son, 165; regulatory unionism, 188; and NRA, 208–9; and Taft-Hartley amend-ment, 247, 250; and American nation-alism, 251; and patriotism, 251. *See also* American Federation of Labor; Blue Card Union; Congress of Industrial Organi-zations; Knights of Labor; Mine Mill; Social democratic labor unions; United Mine Workers of America

LaGrave, Francis, 95–96

Land Act of 1820, 27

Landon, Alf, 218, 225

landowners: royalties of, 21, 32, 33, 36, 37, 38, 48, 54–55, 61, 63, 64, 66, 69, 91; steam-powered equipment rented to leaseholders, 77, 92; and investors in leasehold system, 89–90, 96, 107; opening land to prospectors, 95–96, 97

Landrum, Charles, 141–42, 146–49, 152–53

Langdon, Emma, 172, 191

Lanza, A. J., 158–59, 161, 174, 180, 181, 195, 204

Lawrence County, Missouri, 91, 134

Lead, South Dakota, 9, 148

Lead and Zinc News, 135

lead-mining industry: crisis in, 15, 70, 71, 80, 86, 94, 100; and ammunition, 19, 26, 30, 36, 43, 44; in Missouri, 20, 21, 22, 24, 29–30, 37; market price of lead, 20, 26, 28, 29, 31, 36, 38, 39, 40, 45, 46, 52, 54, 58, 60, 62, 64, 65, 70, 71, 72, 74, 75, 80, 98, 107, 165, 243, 249; history of, 21; destabilization of, 22; St. Louis as west-ern center of, 22; and yields on mineral lands, 32; and lead paint, 36, 46, 48; and zinc extraction, 59–60, 63; risks of, 68; silver-bearing lead mines, 70, 72, 98; strikes in, 94; market variables affect-ing, 98

lead poisoning, 157, 211

lead production: development of, 1; as dan-gerous work, 5, 22, 35, 40; and owner-operator miners, 6, 21, 63; and piece-work, 10; demand for, 10, 12, 19, 21, 22, 26, 27–28, 30, 36–37, 46, 47–48, 76, 190, 192, 243, 253; preindustrial methods used in, 21, 22, 26, 34–35; enslaved labor used in, 21, 26–27, 28, 30, 43; lack of govern-mental control of, 21, 32, 36; and free mining, 21–22; from surface mines, 26; and smelters, 26–32, 36, 37, 38, 40, 43, 44, 45, 46, 52, 53–55, 57, 60, 65, 66, 67, 70, 72, 157, 168; and preemption rights, 27, 39; investors in, 30, 45, 52–53, 64, 107; and finding's keeping principle, 32, 35–36, 38; and windlasses, 34–35, 45, 51, 75; and transportation costs, 36; and grubstake partners, 40, 49; and Civil War,

43–45; and miner's freedom, 54; profitability of, 57, 184; and health and safety regulations, 89; and prospecting methods, 91; and labor unions, 112; growth in, 128; and lack of ventilation, 133; and World War II, 202; decline in, 243

Leadville, Colorado: Joplin miners as strikebreakers in, 9, 100–104, 105, 110, 112, 116, 124, 131, 251; as mining camp, 72–73, 98; strike of May 1880, 73–74, 100–104; Knights of Labor in, 72–74, 86, 98, 100, 113; wage labor in, 79; WFM's negotiations in, 99–100, 104, 113, 120, 122

Leadville mines, Missouri, 27, 29, 42, 53, 57

Lewis, John, 115–16

Lewis, John L., 213, 219–20, 223, 228, 231, 233

Lincoln, Abraham, 43–44

Little, Frank, 170, 171

Little Steel, 233

Lochner v. New York (1905), 135

Loftus, John, 85–87

Lone Elm Mining Company, 61

Long, J. A., 213–14

Lord, James, 163

Los Angeles Times, 121

Lowney, James, 163

Ludlow Massacre (April 1914), 156

Lusitania sinking, 161

McAuliffe, Ed, 136–37

McCollum, Scott, 97, 99, 102–4, 109, 172

McIndoe, Hugh, 162

McKee, Andrew, 24, 26–27, 42

McKinley, William, 97, 106, 108

McTeer, Tony: in Picher, 174–75, 213; and Mine Mill strike, 215, 219, 220, 225, 235; and CIO, 227–28; and Galena Defense Committee, 233; and health and safety regulations, 236–38; and silicosis, 240, 244

Mahoney, Charles, 153

"The Man from Missouri" (song), 151

manhood: manly responsibility, 4, 5, 10–11, 34, 94–95, 187–88, 191, 222; risk-and-reward ethic associated with, 8, 9, 11, 80, 88, 136, 198–99; WFM's denunciation of strikebreaking as unmanly, 10, 103, 110,

119–20, 121; republican ideal of true manhood, 87; WFM's vision of respectable manhood, 126; white American manhood, 140; military's appeals to, 170; and Blue Card Union, 201

Mantle, Elven "Mutt," 226, 239, 255, 256–57

Mantle, Mickey, 18, 226, 255, 256–57

market prosperity: and Jacksonian market revolution, 6, 13, 19–20, 21, 22; democratic access to, 8; market price of metal, 11, 12, 20; Tri-State miners' access to, 11, 13

Marland, E. W., 217–18

masculinity: white masculinity, 4, 9–11, 13, 34, 127, 130, 140–44, 256; aggressive, heedless, 5, 9, 16, 124, 138, 174, 200, 224, 240; rough masculinity, 9–11, 13, 62, 130, 167, 240. *See also* manhood

Menace, 15

Men and Dust (documentary), 237

Meriwether, F. V., 194–98, 204–5

metal miners: and labor unions, 3, 93, 98, 99, 108–9, 114–15, 117, 122, 123, 130; dangerous labor of, 5, 22, 35, 40, 41; as prospectors, 6, 19, 48–52, 53, 54–55, 56, 58, 59, 60, 62, 63, 64, 65, 70, 73, 74, 76, 79, 80–84, 87, 89, 91, 95–96, 97, 115, 117, 132, 137–38, 167, 252, 253, 256; as wage laborers, 20–21, 73, 91; economic risks of, 22, 38, 40, 44, 67–68; in Missouri, 29–30, 31, 32, 33, 37, 43–45; geological knowledge of, 32; and leasehold terms, 32, 36, 38–40, 51, 55, 60, 62–63, 66–67, 69, 87, 89, 91, 108, 116, 137, 252; as squatting prospectors, 33, 37–38, 40, 148; native-born white men as, 33, 49, 63, 65, 69–70, 72, 107; families traveling with, 34, 67, 83; prosperity from skilled labor, 40–41, 131; and Civil War, 43, 44; recruitment of, 45; African American miners as, 49–50, 63, 66; relationship with smelting and land companies, 52, 53–55, 57, 59, 60, 61–62, 65, 66–67, 70, 71, 72, 107; self-government attempts of, 53–54; wildcat miners, 54; and market prices, 54, 65, 66, 69, 108; partisan divisions among, 62; and Knights of Labor, 72–73; and eight-hour workday law, 135; and piece rates,

136; oppositional strategies of, 147–50. *See also* African American miners; foreign-born miners; Tri-State miners

metal mining: growth of, 1, 18, 19–20, 52–53, 128, 159; and labor unions, 2, 3, 8, 16, 17; and tariffs, 7, 12; and market price, 11, 12, 20, 94, 96; of Native Americans, 19, 20, 21, 22, 24; in Great Britain, 34; and miner's freedom concept, 54; mechanization of, 165; downturn in, 199. *See also* lead production; Tri-State district; Tri-State miners; zinc production

Metals Reserve Company, 243

Metropolitan Life Insurance Company, 183, 195, 198, 204, 205

Mexico, copper mines in, 129

Miami, Oklahoma: labor unions in, 156, 188, 212, 215, 216–17, 224–25, 231; zinc production in, 167; Ku Klux Klan in, 185; national highway in, 192; U.S. Route 66 in, 192; and health and safety regulations, 194; and ground bosses, 203; Blue Eagle Day in, 209

Miami News-Record, 209, 210

Miami Royalty Company, 167–68

Michigan: copper mining in, 19, 21, 156; labor unions in, 98; WFM strike in, 160, 162

middle class: antituberculosis reformers, 156–58; and Ku Klux Klan, 185; political economy dominated by, 253

Miller, Guy, 153–55, 164

Miller, Lavoice, 229, 233

Mills, Charles Morris, 175, 176, 180

Mine Mill (International Union of Mine, Mill and Smelter Workers): and NLRB, 13, 17, 213, 220–23, 224, 229, 231, 233–34, 235, 236, 237, 244, 245, 248; strikes launched by, 13, 201, 213–16, 217, 218, 219–21, 223–26, 229, 238, 246–47; and CIO, 13, 202, 220, 223, 225, 227, 228, 233, 235, 238, 250; attempts to organize Tri-State miners, 16, 17, 172, 185–88, 190, 191–92, 199, 200–201, 209–12, 214, 227, 229, 233, 236–38, 240, 244, 251, 252; in Joplin, 17, 171; and adoption of new name, 170–71; and IWW, 171, 185; in Oklahoma, 185–92; and housing con-

ditions, 186, 189; women's support for, 188–89; and strikebreakers, 201, 214–17, 218, 219, 222–24, 225, 227; and collective bargaining, 202, 222, 244; and NRA, 208, 209–10, 211, 212, 213–14, 215, 219, 237; and mine inspections, 211; and health and safety regulations, 211, 222, 236, 242; and relief for unemployed members, 212; District Four, 212, 213, 214; legitimacy of, 233; and World War II, 244; membership of, 244–46, 247, 250; and Communist influence, 246–47, 248; AFL challenges to, 247–49; USW drive against, 251; and Americanism, 254; support for Franklin Roosevelt, 254

miner's consumption: prevalence of, 12, 149–50, 154, 155, 156–57, 190; deaths from, 133, 241; studies of, 158. *See also* silicosis

Miners Magazine, 116, 119, 121–22, 125, 155, 160, 185

Miners' Union of Granby, 60–61

Minersville, Missouri, 42, 44, 48, 50, 51, 53, 55, 56. *See also* Oronogo, Missouri

mine safety laws, Tri-State miners' resistance to, 9

mining companies: lack of control over Tri-State miners, 11; cost of workmen's compensation insurance, 12, 135, 153, 183, 194; risky workplaces of, 12, 177; and exploitation of Tri-State miners' racism and nativism, 13–14; deep mining organized by, 20, 76, 90–91, 129–30; and wage labor, 20–21, 55, 79, 81, 91, 96–97, 106, 128, 130, 145–46; and market price of lead, 37–38; and economic risks of metal miners, 38–41; and sublease mining, 49, 57, 61, 63, 76, 107; relationship with metal miners, 52, 53–55, 57, 59, 60, 61–62, 65, 66–67, 70, 71, 72, 107, 173, 180, 211; and subcontracting system, 55; metal miners' bombings of, 61–62; and mechanization of production, 77, 90, 91, 106–7, 116, 128, 129, 130–31, 136, 142, 166, 239; and Joplin mills, 77, 129; and ground bosses, 83, 84, 87, 194, 200, 203–4, 239; prospecting methods of, 90, 91, 96, 106, 129, 167, 168, 190–91; and intensification of production methods, 90–91, 128; production stoppages, 92, 96–97, 117, 119, 125, 145, 155, 164, 171, 184, 197, 200, 214, 245, 246–47, 249, 250; production restarts, 96, 106, 117, 249; antiunion campaigns of, 99, 104, 112, 190; consolidation of power, 128, 154, 178, 193, 206, 239; and steam shovels, 142, 146, 147; labor shortages of, 153; blacklists of union members, 159–60; and AMMU, 162; wartime government support of, 167; in Ottawa County, 172; and Tri-State miners' demands for exclusion of African Americans and foreign-born miners, 175, 176, 254; and industrial associations, 178–79, 184; and silicosis prevention measures, 182, 184; and flotation technology, 193; and Blue Card Union, 201, 216, 217; and charitable poor relief, 202; and Blue Eagle Day, 209; and NRA, 209; and Tri-State Metal Mine and Smelter Workers Union, 216; and NLRB, 244. *See also* small mining companies

Mississippi River, 24, 29, 45

Mississippi River valley, lead-mining districts of, 6, 21, 31, 34, 37, 49

Missouri: as part of Tri-State district, 1; metal mining in, 20, 21, 22, 24, 29–31; galena deposits in, 20, 22, 24; eastern Missouri district, 21, 28, 29; white settlers of, 22, 24; agriculture in, 24, 26, 27–28, 32, 42; federal land grant for internal improvements, 30; railroad development in, 30–31, 33; border conflict with Kansas, 42; Civil War in, 42–44, 50; and election of 1874, 62; and election of 1876, 64; strikes in, 92; eight-hour workday law, 122, 134, 135–36, 144; tuberculosis mortality rate in, 155; and WFM, 164; zinc production in, 166; federal and state relief for, 214. *See also* Eastern Missouri district; Tri-State district; *and specific counties and towns*

Missouri and Kansas Zinc Miners' Association, 108, 113, 117, 178

Missouri Board of Health, 155, 156–57

Missouri Bureau of Labor Statistics, 71, 81, 85, 154

Missouri National Guard, 170, 218

Missouri Ozarks, 7

Missouri Pacific Railroad, 120

Missouri River, shipping on, 24

Missouri State Federation of Labor, 144, 156

Missouri State Guard, 42–43

Missouri Supreme Court, on leasehold terms, 39

Missouri Trades Unionist, 171

Missouri Zinc Company, 57–58

Modern Woodmen of America, 135

Moffett, Elliott, 50–53, 56

monopolies: antimonopoly reform, 8, 60–61, 70, 73, 85, 86, 88, 98, 252; Tri-State miners' opposition to, 8, 60–61, 70, 252; and trusts, 128

Montana, 171, 185, 212, 245. *See also* Butte, Montana

Moon, Jasper, 49, 51

Moon Range, 51, 55

Moseley, George, 28–29, 36

Moseley, Oldham, and Company, 28–29, 31

Moseley, William, 28–29, 36

Moyer, Charles: and organization of metal miners, 123, 146–47, 150; imprisonment of, 125; and Joplin, 126, 156; and murder of Frank Steunenberg, 142; money raised for defense of, 145; on Tri-State miners as strikebreakers, 148, 152; and merger of WFM with UMW, 160; and AMMU, 163; and WFM, 164; and IWW faction, 170; as Mine Mill president, 185–86, 191, 192; and immigration restriction, 187

Mudd, Seeley W., 100, 287n26

Murphy, Patrick, 52–53, 64

Murray, Philip, 250

Muskogee, Oklahoma, 189

National Committee for People's Rights, 236

National Industrial Recovery Act, 12–13, 208

nationalism. *See* American nationalism; white nationalism

National Labor Relations Act (NLRA), 201, 218, 220–21, 223, 228–31, 235, 237, 247

National Labor Relations Board (NLRB): Mine Mill working with, 13, 17, 213, 220–23, 224, 229, 231, 233–34, 235, 236, 237, 244, 245, 248; and Blue Card Union, 202, 221, 233–35; reorganization of, 221

National Lead Company, 129, 136, 237
National Recovery Administration (NRA):
provisions of, 200, 208–9; implementation of, 201, 210; and OPA, 208; and
Mine Mill, 208, 209–10, 211, 212, 213–14,
215, 219, 237; and reemployment agreements, 208, 210, 211, 213; public support
for, 209; labor compliance board of, 213;
U.S. Supreme Court on, 218, 219; Tri-
State miners' use of, 251
National War Labor Board (NWLB), 244,
254
Native Americans: armed conflict with, 2,
19; federal government taking land from,
12, 20, 22, 24, 253; metal mining of, 19,
20, 21, 22, 24
native-born white men: AFL as ally of, 11,
113; racial advantages of, 16, 107, 110–11,
140, 152, 154, 190, 193, 252; as metal
miners, 33, 49, 63, 65, 69–70, 72, 107;
and WFM, 100, 104, 105; Joplin strikebreakers as, 105–6, 110, 126, 151; as percentage of population, 106; patriotism of
nonunionism, 111; as shovelers, 144, 160;
on foreign-born miners, 160; AMMU's
membership limited to, 161, 164; American nationalism of, 165; and Ottawa
County, Oklahoma, 166; Tri-State miners
as, 166, 211, 251; and Mine Mill, 186; and
Blue Card Union, 231
nativism: and Tri-State miners, 13–14, 16,
121, 152, 154, 160, 161, 164, 187, 199, 201,
253, 254; of shovelers, 144; and AMMU,
163; and OFL, 186; and Blue Card Union,
231
Neck City, Missouri, 123, 125, 147, 154–55,
162–63, 166
Neosho, Missouri, 42
Nevada, 21, 54, 70, 73
New Deal: collective bargaining rights promoted by, 1, 10, 12–13, 201, 255; and common class interests of American workers,
2; and working-class conservatism, 4, 13,
201, 222, 254; as short-term triumph for
labor unions, 5; Tri-State miners' atti-
tudes toward, 12, 13, 253, 254, 255; and
white working-class American men, 200,
255; labor regime of, 202, 212, 220, 221–
22, 223, 231, 254; and Tri-State district,

207–8, 226; and relief programs, 212; and
working conditions, 212; and economic
recovery, 226; and U.S. Supreme Court,
228; conservative opposition to, 236
New Jersey, 19, 33, 58
New Spain, mining leases granted by, 31
Newton County, Missouri: metal mining
in, 20, 22, 24, 28, 32, 33, 43, 47, 57, 60,
171; white settlers of, 20, 24, 27; enslaved
labor in, 24; geological environment in,
26; as public land, 27, 33; G. C. Swallow's
geological survey of, 30; Civil War in,
42–43, 50; in post–Civil War period, 50;
and railroads, 52; mining leases in, 53;
labor unions in, 60–61; and election of
1874, 62; and election of 1876, 64; African
American miners of, 66, 80; and election
of 1878, 71; and Greenback-Labor Party,
71; and election of 1880, 72; prospectors
in, 82, 91; and election of 1894, 96; and
election of 1896, 107; accidental deaths of
miners in, 134
Newtonia, Missouri, 44
Nieberding, Velma, 238, 257–58
Nolan, Joe: as Picher police chief, 203–4;
and Mine Mill strike, 215, 216, 217; and
Blue Card Union, 217, 223–24, 226,
228–29, 231–32, 234; and Eagle-Picher
Lead Company, 239; and AFL, 248; death
of, 256
nonwhite people: immigrants considered
nonwhite, 5; and racist politics, 10, 11, 12;
and white nationalism, 140; and labor
unions, 254. See also African Americans
Norman, Kelsey, 216, 231
Northern Mineral Mine Workers' Union,
98–99, 114, 123, 125
Northport, British Columbia, 120–22

Office of Price Administration, 243, 245
Office of Production Management, 243
Ogburn, Charlton, 234
O'Hare, Kate Richards, 145
Ohio, 73, 85, 117
Ojibwe people, 19
Oklahoma: as part of Tri-State district,
1, 165–66; boomtown in, 11, 167; zinc
production in, 156, 166, 174, 192; and
AMMU, 162; coal mining in, 174, 176,

185; health and safety regulations ignored in, 174, 189, 194; workmen's compensation law in, 183, 204, 211; labor unions in, 185–88; federal and state relief for, 214. *See also specific counties and towns*

Oklahoma City, Oklahoma, 173

Oklahoma Federationist, 186, 188, 191, 222

Oklahoma National Guard, 217

Oklahoma State Federation of Labor (OFL), 185, 186–88, 191–92, 199, 211, 220, 225, 229, 232, 235

Oldham, Simpson, 24, 26, 28

Oliver, Mildred, 236, 239, 241

Oronogo, Missouri: as mining camp, 47, 57, 63, 71, 129; naming of, 56, 270n21; and Short Creek rush, 64; and Granby Mining and Smelting Company, 75; labor unions in, 86, 117, 244; AFL in, 112, 113–14; wildcat strike in, 117; and election of 1908, 147; and AMMU, 164

Osage people, 20, 22, 24, 26, 27

Ottawa County, Oklahoma: and expansion of Tri-State district, 165; zinc production in, 166, 168, 192; Tri-State miners moving to, 171–72, 176–77; demographics of, 176; health conditions in, 180, 181, 189–90; living conditions in, 184; mortality rates in, 184; and election of 1916, 185; and election of 1918, 185; and election of 1922, 185; unemployment relief in, 206; and election of 1932, 206–7; and Blue Eagle Day, 209; FERA programs in, 212; labor unions in, 218; and election of 1936, 226; and election of 1928, 254; history of, 257

Ouray County, Colorado, 105, 112

owner-operator miners: in Tri-State district, 4, 6–7, 9, 15, 71, 80, 98, 252; in lead production, 6, 21, 63; diminished opportunities for, 9, 71, 90; historical accounts of, 18; and leasehold terms, 32, 47, 50–51, 95, 107; capital accumulated by, 47; hiring of wage laborers, 55, 67; success of, 69; decline in, 137

Pacific Railroad Company, 30

Panic of 1857, 36

Panic of 1873, 46, 60

Panic of 1907, 11, 141, 145–46

Parham, Charles, 135

Paris Commune, 61

Parran, Thomas, 12, 181, 196, 198

Parsons, Kansas, 243

patriotism: of Tri-State miners, 10, 11, 111, 121; WFM's denunciation of strikebreaking as unpatriotic, 121; and Missouri National Guard, 170; and Blue Eagle symbol, 208; and labor unions, 251

Pennsylvania: iron mining in, 20; zinc production in, 33; coal mining in, 54, 140; Knights of Labor in, 73; labor statistics in, 85; AFL in, 117

People's Party, 8, 15, 61, 62, 96

Peoria, Oklahoma, 107

Perkins, Frances, 235–39, 241–42

Pettibone, George, 142, 145

Picher, Oklahoma: as boom camp, 165, 198, 252, 256; population of, 166, 240, 249, 255, 257; living conditions of, 172–75, 177–78, 179, 181, 184, 192–93, 198–99; hospital of, 173, 179, 180; incorporation as town, 176–77; chat piles in, 177, 193; and health and safety regulations, 179; wildcat strike in, 180; working conditions of, 183; labor unions in, 186, 188–89, 209, 211, 212, 213, 214, 215–16, 217, 218, 219, 220, 224; cost of living in, 187; U.S. Route 66 in, 192; growth of, 192–93; mass meeting for unemployment grievances, 203; and Mine Mill's strike, 214; Mine Mill rallies in, 228–29, 245; collapse of, 249, 255–56; as Superfund site, 258

Picher, Oliver H., 51–53, 59, 61–63, 74, 90

Picher, Oliver S., 167–68

Picher, W. H., 61–63, 90

Picher Lead and Zinc Company, 62, 63, 72, 157, 167–68

Pickers, T. Z., 100–102, 131

Pittsburg, Kansas, 74, 86, 92, 176

Plummer, John, 38–39

poor man's camp: and small-scale producers, 6–7, 70, 252; as nonunion miners, 8; legacies of, 9, 252, 258; risk-and-reward ethic of, 10, 69, 70, 83–85, 89, 94, 103, 252; and federal government's forcing of Native Americans off land, 12; historical accounts of, 17–18, 71; ideals of, 47, 57, 64, 69, 70, 82–84, 87, 91, 92, 93, 94, 95, 102, 107, 115, 117, 118, 155,

190, 193, 216, 224; and wage labor, 47, 93; Joplin's reputation as, 55–56; and scrappers, 56; logic of, 56, 70, 83, 84, 89, 95–96, 134, 154; and entrepreneurial traditions, 57, 58, 63, 65, 70, 71, 72, 89, 130, 137, 140, 144, 162, 185; and Greenback-Labor Party, 71; Leadville, Colorado as, 72, 73, 74

poverty: of Tri-State miners, 4, 83, 189; poor men's fair economic opportunity, 8, 65, 222; and charitable poor relief, 12, 202–3, 206, 212. *See also* poor man's camp

Powderly, Terence V., 87

Pratt, George, 221–23, 234

Preemption Act of 1841, 27

President's Re-Employment Agreement (PRA), 208, 213, 217

Prosperity, Missouri, 94, 95, 147, 154, 155, 157, 164

Quapaw, Oklahoma, 188

Quapaw Nation, 167, 168, 172, 177

Quinby, George, 106–7

racial advantages: of white working-class American men, 4, 5, 6, 10, 16, 107, 110–11, 140, 253; of native-born white men, 16, 107, 110–11, 140, 152, 154, 190, 193, 252; of shovelers, 143, 144, 152

racial divisions: in industrial unions, 2, 3; and privileges of white working-class American men, 4, 5, 6, 10; in national culture, 10, 111; and individualism, 13; and Tri-State miners, 13–14, 16, 111, 126, 151–52, 161, 164, 186–87, 199, 200, 224, 251; and election of 1874, 62; and AMMU, 163

racist politics, 11, 12, 111

railroads: federal government's support of construction, 12, 38, 253; and Joplin, 14, 65, 172; and St. Louis markets, 30; public lands for, 30–31, 33, 35–36; and shipping, 37; and lease for Granby section, 37, 38; and Granby Mining and Smelting Company, 45, 48, 71; and growth in mining output, 53; and zinc production, 57–58, 65; and leases in Kansas, 64; strike of 1877, 64; strikes against, 88; bankruptcy of, 92; expansion of, 98

Randsburg, California, 123

Ray, Orville "Hoppy," 240

Reando, Francis, 29, 31

Recession of 1937–38, 239–40

Recession of 1957, 250

Red Scares, 250, 251, 253

reformers: struggles to change Tri-State miners, 16; antituberculosis reformers, 156–58

religion, 14, 135, 266n27

Republican Party: free labor ideology of, 7; in Tri-State district, 15; and investors in Granby Mining and Smelting Company, 45; Reconstruction policies of, 61; on bombings of mining companies, 62; and election of 1874, 62; and election of 1876, 64; and election of 1878, 71; and election of 1880, 72; and election of 1884, 86; and election of 1894, 96; and election of 1898, 107; Tri-State miners' support for, 107, 185; and election of 1900, 118; and election of 1916, 185; and election of 1918, 185; and election of 1920, 185; and election of 1922, 185; and election of 1932, 206; and election of 1936, 226; and nativist exclusion, 254

Rex Mining and Smelting Company, 90–91, 97

Rice, George, 182

Richardson, Albert, 39–40

Rich Hill, Missouri, 74

right to work, 15

Ringer, William, 234–36

risk-and-reward ethic: and democratic spirit of fairness and opportunity for white men, 6, 80, 94; manhood associated with, 8, 9, 11, 80, 88, 136, 198–99; and Tri-State miners' performance of rough masculinity, 10, 11, 13; of poor man's camp, 10, 69, 70, 83–85, 89, 94, 103, 252; and market-based incentives, 11, 47, 54, 65, 66, 67–68, 69, 70, 75, 144, 145; and shovelers, 11, 130, 136, 137–38, 139, 144, 145, 149, 166, 174, 190; and accidental deaths, 146; and compensation lawsuits, 149

Robinson, Reid, 227–29, 231, 233, 245–47

Rockefeller, John D., Jr., 156

Roosevelt, Eleanor, 235–37

Roosevelt, Franklin D.: Tri-State miners'
expectations of, 12, 227; New Deal of,
200, 202, 207–10, 212, 220, 221–22; on
tariffs, 206; and election of 1932, 206–7,
231, 254; and election of 1936, 226–27,
254; proposal for reform of federal judiciary, 228
Roosevelt, Theodore, 10, 15, 127, 128, 130,
140, 155
Rossland, Washington, 120
Rossland Great Western Company, 120
rough masculinity, 9, 10, 11, 13, 62, 130, 167,
240
Ruhl, Otto, 134, 152–53, 154
Russian miners, 152
Ruth, Babe, 256

safety: belittling of, 11; lack of regard for, 85.
See also health and safety regulations
Sapp, William F., 93, 95
Sarcoxie, Missouri, 52, 164
Sauk people, lead mines of, 21
Sayers, Royd, 181, 195, 198, 204–5
Scammon, Kansas, 86, 92, 156
Schasteen, Ted, 212–13, 218, 219, 225,
227–28, 240
Schoolcraft, Henry Rowe, 22
Scott, Dred, 37
Sergeant, John, 50–53
sexually transmitted diseases, 180–81, 195,
196, 198, 205
Shepherd, Thomas, 24, 26, 28
Sheridan, Thomas, 124, 126, 144
Sherwin-Williams paint, 220
Shoal Creek, Newton County, Missouri, 24,
26, 31, 33
Short Creek, Jasper County, Missouri, 64
shovelers: physical labor as capital, 10, 137,
139, 143, 175; piece rates of, 10, 11, 130,
135–38, 139, 141, 142–44, 145–46, 147,
149, 154, 158, 169, 171, 190, 191, 192, 194,
208, 226, 252; as unskilled, 10, 131–33,
137, 143; as embodiment of white masculinity, 10–11, 130, 140–41, 182; risk-and-
reward ethic of, 11, 130, 136, 137–38, 139,
144, 145, 166, 174, 190; historical accounts
of, 18; and zinc production, 130; oppositional power of, 130, 141, 142, 143, 153; as
percentage of miners in ground, 131; as

strikebreakers, 131, 141, 142, 148, 152; demand for, 131–32, 133, 138, 141, 147, 239;
accidental death rate of, 134, 146; gang-
labor model of, 137; contests for, 138–39;
and compensation lawsuits, 149; and
health and safety regulations, 150, 155,
174; opposition to foreign-born miners,
152, 160; mobility of, 153; sustainability
of culture of, 154; wildcat strikes of, 155,
252; and silicosis, 158–59, 182, 184, 196;
AMMU's daily wage scale for, 161; in
Ottawa County, Oklahoma, 166, 184; mechanical loaders replacing, 255
silicosis: and health and safety regulations, 1, 158–59, 169, 182, 183, 202, 211,
236–37; Tri-State district rate of, 12, 17,
180–82, 184, 187, 196, 206; causes of, 158,
184, 240; public hospitals for, 173–74;
and U.S. Public Health Service, 181, 194;
effects of, 186, 189, 190, 222, 225, 237,
252, 256; U.S. Bureau of Mines, 194–98;
and workmen's compensation law, 204;
and women, 222, 241
silver mining, 19, 20, 54, 60, 70, 98
Small Hopes Mining Company, 100
small mining companies: Tri-State miners'
opportunities as owner-operators of, 4,
6–7, 10, 74, 115; threats of corporate consolidation, 8, 20, 117, 252; outside investors foreclosing possibilities for, 9; and
right-wing populism, 15; and lead production, 22, 40, 75; and leasehold terms,
32, 95; employment of miners, 47, 63,
81; and African Americans, 63; success
of, 69, 74, 129; wage laborers hired by,
70, 78–79, 80; increase in scale, 70–71,
75–76; and Greenback-Labor Party, 71;
and zinc production, 75, 77–79; machine
installations of, 77, 80, 129; landowners'
renting equipment to, 77, 92; and production stoppages, 79, 92; growth of,
81; earnings of, 82, 95; and health and
safety regulations, 89, 134; associations
of, 183; and compensation insurance, 183;
closing of, 200, 246, 248, 250; and NLRB,
244; and Mine Mill, 245. *See also* owner-
operator miners
smallpox, 180
Smith, J. R., 109, 118–19

Taylor, John H., 52–53, 56
Taylorism, 139
Telluride, Colorado, 123
Thew Automatic Steam Shovel Company, 142
Thiel Detective Agency, 100–101
Tingle, William, 24, 26–27, 29, 36
Tonopah, Nevada, 123
Trappe, Edwin, 117–18
Travis, Maurice, 247
Treece, Oklahoma: as boom camp, 165; population of, 166; living conditions in, 172–73; labor unions in, 186, 188, 212–13, 218, 220, 229
Tri-State Council, 188–90, 192
Tri-State district: owner-operator miners in, 4, 6–7, 9, 15, 71, 80, 98, 252; and speculative mining leaseholds, 6–7; competition in, 7, 10; smelting and land companies buying interests in, 8, 52, 53–55, 57, 59, 60, 61–62, 65, 66–67, 70, 129; outside investors foreclosing owner-operators, 9; profits of, 11; working-class culture of, 11, 192–93; physical toll of life in, 12, 235–36; prosperity of, 14; politics in, 15; progressive success in, 16; open shop conditions in, 17, 93, 190, 191; historical accounts of, 17–18; development of, 31, 65; skilled miners of, 40–41, 131; and Civil War, 42, 43, 44–45; as national leader, 47; and risk-and-reward ethic, 47; and wildcat miners, 54; survey of 1873, 54–55; railroads in, 57–58; geological survey of, 58, 60; leasing terms in, 60, 64, 66; and jug handle contracts, 67, 71, 137; and mine inspections, 85, 88, 134; native-born whites as percentage of population, 106; producers' association of, 129; and World War I, 165; balance of power in, 204; and New Deal, 207–8, 226; health crisis in, 235–36, 240–42, 252; industrial structure of, 239; decline in, 250, 255; environmental fallout from mines, 257–58. See also lead production; poor man's camp; zinc production
Tri-State Industrial Examining Bureau, 205, 210–11, 222, 225, 226, 239
Tri-State Metal Mine and Smelter Workers Union, 216. See also Blue Card Union

Tri-State Mine and Mill Workers' Association, 247
Tri-State miners: class interests of, 1; as strikebreakers, 1, 5, 9, 10, 11, 15, 99, 100–104, 105, 108–12, 116, 118–21, 124, 126, 131, 141, 142, 147–49, 150, 151–53, 154, 201, 220, 222, 225, 228, 251, 252, 253; opposition to health and safety regulations, 1, 11, 12, 16, 88–89, 135, 158, 174, 175, 181–82, 238; persistent patterns of actions, 3, 4; explanations of actions of, 4; logic of self-interest of, 4, 5, 13, 14, 16, 70, 111, 231, 251, 252; conservative ideas of, 4, 5–6, 13, 14–15, 17, 238, 254; opposition to labor unions, 4, 10, 12, 14, 18, 93, 101, 144–45, 166, 251; wage labor embraced by, 5, 9, 11, 15, 72, 91, 92; entrepreneurial ambitions of, 6, 7, 8, 57, 58, 63, 65, 70, 71, 89, 101, 130, 137, 144, 145, 161, 165, 167, 251; as broader class of business men, 8; on monopolies, 8, 60–61, 70, 252; performance of rough masculinity, 10, 11; demands for share of district's profits, 11; and power to exclude, 11, 13, 127, 161, 163, 169, 175, 176; wildcat strikes of, 11, 117, 147, 149–50, 153, 155, 164, 180, 251; legal claims for workplace injuries, 11, 147, 148–49, 153; switching employers without notice, 11, 153, 251; white nationalism of, 11–12, 13, 16, 163, 167, 175; rates of infectious diseases among, 12; company union supported by, 13; and return to earlier, greater era, 13; and racial divisions, 13–14, 16, 111, 126, 151–52, 161, 164, 186–87, 199, 200, 224, 251; and nativism, 13–14, 16, 121, 152, 154, 160, 161, 164, 187, 201, 253, 254; relationship with employers, 14, 153; political support of candidates, 15; support for government measures preserving prerogatives of white men, 15; support for health and safety regulations, 16, 149, 150, 159, 161, 183–84; communities of, 17–18; opposition to government regulation, 18; housing of, 66, 67, 167, 173; daily/weekly wages of, 67, 79, 83–84; and western metal camp opportunities, 72–73, 79; stories of poor men who hit pay dirt, 84–85, 137; public debates of, 94–95; and sense of exceptionalism, 98;

and eight-hour workday law, 135–36; and white supremacy, 153, 254; and World War I, 166; as native-born white men, 166, 211, 251; living conditions of, 166–67, 172–75, 177–79, 181, 184, 192–93, 198–99; as commuters, 173, 203; and grade cards of health clinic, 194–98, 199, 200, 201, 202, 204–6, 210–11, 225–26, 241, 246; and mass meetings for airing grievances on unemployment, 203–4; and AFL, 232, 251; and World War II, 243–44; opposition to mining companies, 251. *See also* hard rock miners; metal miners; shovelers

Tri-State Mines Safety and Sanitation Association, 179

Tri-State Sanitary District, 181

Tri-State Survey Committee, 236, 237

Tri-State Zinc and Lead Ore Producers Association (OPA): and compensation charges, 183, 204, 205; and silicosis research study, 184; on weekly ore sales, 191; and American Zinc Institute, 194; and control of insurance costs, 194; mining companies' cooperation with, 194, 201, 202; and health and safety regulations, 194, 236; and Eagle-Picher Lead Company, 194, 239; and silicosis health clinic, 194–98, 200, 202, 204–5, 210; and charitable poor relief, 202–3, 206, 212; and mass meetings for unemployment grievances, 203–4; and NRA, 208; and labor unions, 209, 210, 211–12, 213, 240; reemployment agreement, 211; and back-to-work movement, 215–16; and Blue Card Union, 217–18, 224, 225, 226, 227; on employment levels, 226

Truman, Harry, 245–46, 249

trusts: as monopolistic cartels, 128; and lead industry, 129

tuberculosis: Tri-State district rate of, 12, 133, 156, 157, 158, 180, 191, 196, 206; prevention efforts, 156–57, 179, 189–90, 195, 196, 198. *See also* miner's consumption; silicosis

Tulsa, Oklahoma, massacre of 1921, 175

Turkey Creek, Jasper County, Missouri, 24, 26–27, 29–33, 51, 53, 80

Turney, John, 186, 188–89

unemployment: and grade cards of health clinic, 194–98, 199, 200, 201, 202, 204–6, 210–11, 225–26, 241, 246; and Great Depression, 202–3; and ground bosses, 203; and mass meetings for airing grievances, 203–4; and rustling cards, 204–5; and injuries, 206; Franklin Roosevelt's interventions in, 207–8; federal and state relief for, 214, 215; and closing of mining companies, 248, 249

Union Labor Party, 15, 88, 252

Union of America, 247

United Auto Workers, 227

United Auto Workers–AFL, 234

United Cement, Lime, and Gypsum Workers International Union (AFL Cement), 247–49, 250

United Copper Company, 145

United Mine Workers of America (UMW): as social democratic vanguard, 2; and class-based confrontation of company domination, 7; emergence of, 8; in Kansas, 92; strikes of, 92–93, 94, 101, 102, 108, 112, 156, 160; in Joplin, 156; WFM's merger with, 160; and Farmer-Labor Reconstruction League, 185; in Oklahoma, 185, 191, 235; and Mine Mill, 213, 219, 233; and mass production industries, 220

United Steelworkers of America (USW), 250–51

Upper Mississippi field, 28, 29, 31, 32, 36, 37, 50

Upper South, mining districts of, 6

U.S. Army, and demand for lead, 36

U.S. Army of the Southwest, 43–44

U.S. Bureau of Mines: paternalistic strategy of, 12; studies of silicosis, 181; on accident prevention, 182, 184; and insurance costs, 194; and silicosis health clinic, 194–98, 205, 235

U.S. Congress: public land grants for internal improvements, 30; public land grants for mine exploration, 31–32; and mining of Native American allotted lands, 168; and selling of Native American allotted lands, 177; and Wagner labor disputes bill, 222; and price controls, 245, 246

U.S. Department of Defense, 249

U.S. Department of Interior, 167, 168, 177

113–14, 123, 125–26, 144–45, 146, 147,
151, 152, 154, 155–56, 212; and leasehold
system, 90; mine closings in, 92; miners'
meetings in, 94; and wage labor, 117; and
miner's consumption, 133, 158; strikes in,
146; and election of 1908, 147; tuberculosis mortality rate in, 155, 156; miners'
wildcat strikes in, 160; and AMMU, 162,
163, 164; Tri-State miners commuting
from, 173; sanitarium in, 173–74, 179; and
silicosis crisis, 173–74, 181; and election
of 1932, 207
Webb City Lead and Zinc Company, 77
*Webb City Register/Webb City Daily Regis-
ter*, 125, 133, 140, 142–45, 158
Webb diggings, 56, 63
Weber, Frank, 114–17, 122
Weir City, Kansas, 58, 92
Welsh miners, 34, 49
Western Federation of Miners (WFM):
class-based solidarity practiced by, 1, 251;
as industrial union, 2; as social democratic vanguard, 2; percentage of metal
miners in, 3; AFL as alternative to, 8, 16,
113, 114, 117; emergence of, 8, 98; foreign-
born miners of, 9, 10, 105, 111, 152, 161,
170, 171; Tri-State miners as strikebreakers against, 9, 99, 100–104, 108–12,
122, 142, 151–52, 251, 252; challenge to
AFL, 9, 105, 122; attempts to organize
Tri-State miners, 9–10, 16, 117, 150, 251–
52; radicalism of, 10, 16, 104, 115, 165,
170, 252; denunciation of strikebreaking
as immoral act, 10, 111–12; as threat to
Tri-State miners, 10, 252; Mine Mill as
successor to, 13, 248; native-born members following strikebreakers, 16; affiliation with AFL, 99, 104, 114, 115, 116, 124,
152–54, 163, 164, 232; strikes of, 99–106,
108–9, 110, 112, 113, 120, 121, 123–24, 125,
156, 160, 162, 170; ethnic divisions within,
100; AFL's support for strikes, 104, 124,
125; in Coeur d'Alene, 108, 114, 118–19,
120, 142; and closed shop, 111; regional
strategy of, 114; in Joplin, 115–16, 117, 118,
120, 123–26, 133, 141, 147, 150–56, 160,
164, 165, 169, 170; and shovelers, 130; and
health and safety regulations, 134, 153,
154–55, 159–60, 169; on piece rates, 136;

in eastern Missouri field, 145; as diverse
union, 152, 251; implosion of, 160–61; and
AMMU, 161, 163, 164; sectarianism in,
170; in Oklahoma, 172, 176
Western Journal, 28, 29
Western Labor Union (WLU), 9, 105, 114,
122
white masculinity: social and political fraternities of, 4; loss of self-determination
as threat to, 9; shovelers as embodiment
of, 10–11, 130, 140–41, 182; strikebreakers
on WFM's denunciation of strikebreaking as unmanly, 10; privileges of, 13;
family groups reflecting respectability of,
34; Theodore Roosevelt championing,
127, 130; and Mickey Mantle exemplify-
ing, 256
white men: interests identified with market functions of capitalism, 4, 20, 251;
democratic spirit of fairness between, 6,
80, 101–2, 163; and freedom from control, 130. *See also* native-born white men;
white working-class American men
white nationalism: of Tri-State miners,
11–12, 13, 16, 163, 164, 167, 175; belligerence of, 12, 13; and AFL, 16; and shovelers, 130; on white American manhood,
140; privileges, 199; and Blue Card
Union, 201, 224
whiteness: white working-class faith in, 14;
and assertions of dominance, 20, 54
white supremacy, 5, 64, 153, 175, 251, 254
white working-class American men: opposition to social democratic labor unions, 1,
4; as prospective unionists, 3; and percentage of union membership, 3, 4; and
privileges of, 4, 5, 6, 10, 13, 14, 50, 54, 70,
252; racial advantages of, 4, 5, 6, 10, 16,
107, 110–11, 113, 140, 253; motivations of,
5, 6–7, 13, 17; subordination of, 9; economic independence and social status
of, 22, 40–41, 70, 83, 164, 165, 166, 201,
251, 254; factions formed by, 38; social
equality of, 47; and sublease mining, 49,
57; and election of 1874, 62; shovelers
as exemplars of, 139–40; and New Deal,
200, 255; commitments of, 202; ideals of,
226; Americanism of, 253–54
Wilson, Woodrow, 15, 165, 171, 176

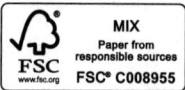